T0186448

LOWLANDS
DEVELOPMENT AND MANAGEMENT

INSTITUTE OF LOWLAND TECHNOLOGY
SAGA UNIVERSITY

LOWLANDS
Development and Management

Edited by
NORIHIKO MIURA
Saga University, Saga, Japan

MADHIRA R. MADHAV
IIT, Kanpur, India & Saga University, Saga, Japan

KENICHI KOGA
Saga University, Saga, Japan

CRC Press
Taylor & Francis Group
Boca Raton London New York

CRC Press is an imprint of the
Taylor & Francis Group, an **informa** business

A BALKEMA BOOK

Published by:
CRC Press/Balkema
P.O. Box 447, 2300 AK Leiden, The Netherlands
e-mail: Pub.NL@taylorandfrancis.com
www.crcpress.com – www.taylorandfrancis.com

© 1994 by Taylor & Francis Group, LLC
CRC Press/Balkema is an imprint of the Taylor & Francis Group, an informa business

No claim to original U.S. Government works

ISBN 13: 978-90-5410-603-6 (hbk)

DOI: 10.1201/9780203748596

Visit the Taylor & Francis Web site at
http://www.taylorandfrancis.com

and the CRC Press Web site at
http://www.crcpress.com

CONTENTS

FOREWORD

The population of the world is estimated to reach 6 billion by the end of the century and is likely to stabilize around 8.5 to 10 billion people; pessimistic scenarios would, however, indicate a saturation figure as high as 12 billion. This growth in world population is essentially taking place in the developing countries. An important feature of this growth is the continuing migration from rural to urban areas. Along with this urbanization, there is a significant increase in the number of megacities with a population of over 10 million inhabitants. The total number of megacities in the world has increased from three in 1972 to thirteen in 1992; of these only one used to be in the developing world in 1972; in 1992 there are nine. This trend will continue.

Another significant aspect is the great attraction of coastal areas for human activities. It is expected that in the next quarter of a century, about 70% of the world's population would be living within 50 km of the coastlines. A striking example of the development of a super–megacity can be seen in Japan, with significant coastal interaction, of the Tokyo–Yokohama–Nagoya–Osaka–Kobe–Hiroshima belt which accounts for the most significant proportion of the country's population and of its activities.

Coastal areas have an uniqueness in that three main components of the earth system, the land, the sea and the atmosphere come together. A large percentage of the coastal areas are lowlands as they are seriously affected by fluctuating water levels. These coastal lowlands are spread out in the developing countries, e.g. Bangladesh, China, India, Indonesia, Korea, Philippines, Thailand, Taiwan and small island states. The coastal lowlands are subject to many natural hazards such as tides, floods, cyclones, typhoons or storms, storm surges, tidal waves (tsunamis), etc. These events are not isolated but involve strong coupling between the oceans and the atmosphere and are related to topography.

Engineers and scientists have been able to minimize damages resulting from many natural disasters to a small fraction of their former impacts. Notable amongst them is the mitigation of damages from earthquakes, cyclones and typhoons. In the case of earthquakes, this has come about through greatly improved seismic monitoring, (which enables a delineation of hazardous zones and forewarns about stress build–up), as also through development of relevant seismic codes and regulations. Of course, much of this is not yet in place in the developing world. In the case of cyclones, typhoons and hurricanes, one can track the events using satellite and radar systems. One knows well in advance about the belt through which these disturbances will pass and, therefore, can put all precautionary systems in place. There are also well–designed shelters to take refuge in, when such catastrophes strike.

A proper study of the characteristics of coastal lowlands is urgently needed for the benefit of societies living in them to enable proper planning for the management of the regions including disaster mitigation. The present State–of–the–Art Volume on Lowlands is thus most welcome and should greatly assist planners, engineers, scientists and administrators involved or concerned with lowlands, in their efforts for sustainable development of such areas.

In addition to tackling existing hazards faced by lowland areas, we will now need to consider new types of hazards arising as a result of the impact of human activities on the environment. First, as mentioned earlier, the world's population has been growing and continuing to do so. There is the aspect of meeting the basic human needs of this population, but additionally, of catering to their rising expectations. In the world as a whole, these expectations are defined by the affluent standards of life in the developed nations. Since the Industrial Revolution, industrial production has grown more than 100 times. Energy generation, which is principally from fossil fuels, and the needs for transportation, have resulted in an enormous increase in the quantity of carbon dioxide put into the atmosphere. Just in the last twenty years, the annual increase of carbon dioxide put into the atmosphere has grown from 16 to 23 billion tons. The increasing amounts of carbon dioxide and other greenhouse gases in the atmosphere is expected to result in global warming. The exact magnitude of this warming is subject to uncertainties arising from present incomplete understanding of the various parameters of the earth system, and the manner in which they interact. An important consequence of global warming is a rise in sea level. These issues are being discussed at various fora, notably in the Second World Climate Conference held in Geneva, toward the end of 1990, and the U.N. Conference on Environment and Development held in June 1992 in Rio de Janeiro.

A substantial rise in sea level can lead to disastrous situations, such as large scale inundation of low lying areas in many parts of the world; but even a small rise, of a few tens of centimeters can lead to serious consequences, e.g. weakening of seafront structures and coastal defenses, ingress of saline water into ground water systems and much greater impact of storm surges and cyclone–like conditions. Large megacities in coastal areas will become more vulnerable because the pressure from the sea will impact particularly on their weak foundational characteristics, arising from the pressure of aboveground structures, and the need for large underground development for electricity, water, sewerage, communications, transportation, storage and the like.

It is, therefore, important to keep these various facts relating to the lowlands under continuing review. As already pointed out in this foreword, these areas will be a focus of increased activity as also of stresses of various types. The Editors and the Contributors of this volume have made an effort to draw our attention to the problems that need to be tackled, and have suggested methodologies for the development and management of lowland areas. They need to be complimented for a pioneering effort relating to the important region on the earth's surface, which will become of increasing interest to more and more people as we move into the future.

M.G.K. MENON
F.R.S. & Member of Parliament, India

PREFACE

Lowlands defined as lands affected by fluctuating water levels and where human activities are already existing or are being proposed, have always been of concern to human societies. By constructing dikes, reclaiming the land below shallow sea and cultivating it, man, communities, villages and even countries have battled and tamed the seas and the rivers in the last few centuries. Initially, lowlands were developed mostly for agriculture. In the subsequent phase, human settlements grew into full fledged villages, towns, cities and regions with the land being utilized for industrial, cultural and recreational purposes and economic activities. In recent years, the sea has become a construction site wherein many offshore platforms and artificial islands have been constructed for oil related projects. This idea is now being extended to cover the area near the coast termed as either lowland or shallow sea, out of which many manmade coastal islands are raising out of the sea. Land that is affected by tides and floods, can be reclaimed to meet the needs for new areas to ease demographic pressures, for urbanization, construction of industrial facilities, airports, etc.

Developmental objectives for lowlands encompass reclamation of the land, provision of access and facilities and promotion of human activities – agriculture, industry, and housing. Reclamation of large areas or separating them from tidal action, restricting the flow regime of rivers, could have serious effect on the ecosystem in the inter–tidal areas. Therefore, the developmental activities in lowlands have to be integrated at each stage to maintain balance with the ecological and environmental aspects or concerns of the region. Land and water management is of paramount importance for lowlands. In particular, the level of the ground surface in relation to adjacent water level, salinity as a function of distance from the sea, seasonal fluctuations in ground or surface water levels and salinity, acidity of the soil, subsidence due to withdrawal of groundwater and/or due to reclamation and sea level rise due to global warming, are some of the major concerns. The soil and water (both quality and quantity) conditions, get altered in lowlands systems. Management of lowlands requires planning, design and maintenance of the system for sustainable utilization.

The Institute of Lowland Technology, Saga University, has been set up recently to foster research into problems of lowlands, to evolve rational and efficient methods for their development and management and to focus attention on their environmental aspects, through a systematic study and for dissemination of knowledge. An International Symposium on Problems of Lowland Development had been organized at the Institute in November 1992, which drew participants from countries all over the world but especially from many of the developing countries which are, of late, facing these concerns. This State–of–the–Art Volume is a logical follow up with the express purpose of collating and presenting relevant information concerning lowlands into a single volume. The contributors to this volume hailing from across the globe have readily agreed to share their vast experiences and promptly sent in their contributions in spite of their various pressing commitments. The editors have greatly benefitted from the advice of the members of the advisory group especially from Dr. H. Araki and Dr. W. Liengcharernsit, feedback from whom was enlightening and always readily forthcoming. The tedious and enormous task of checking for editorial correctness and ensuring uniformity of presentation has been shouldered by Dr. R. Shivashankar who was assisted by Mr. M. Alamgir and Mr. M.C. Alfaro. The editors place on record their deep appreciation to all the contributors to this volume, the members of the advisory group and many others who have given their best to make this venture possible and successful.

NORIHIKO MIURA
MADHIRA R. MADHAV
KENICHI KOGA

LIST OF CONTRIBUTORS

ir. P. ANKUM*, *Sanitary and Water Management Division, Faculty of Civil Engineering, Delft University of Technology, Delft, The Netherlands.*

DR. K. AOKI, *Institute of Environmental Systems, Kyushu University, Fukuoka, Japan.*

DR. H. ARAKI*, *Institute of Lowland Technology, Saga University, Saga, Japan.*

DR. D.T. BERGADO*, *School of Civil Engineering, Division of Geotechnical and Transportation Engineering, Asian Institute of Technology, Bangkok, Thailand.*

PROF. T. ESAKI*, *Institute of Environmental Systems, Kyushu University, Fukuoka, Japan.*

DR. N. FUKUDA, *Fukken Co. Ltd., Hiroshima, Japan.*

MR. Y. HOSOKAWA, *Purification Hydraulic Laboratory, Port and Harbour Research Institute, Ministry of Transport, Yokosuka, Japan.*

PROF. K. JINNO, *Department of Civil Engineering, Kyushu University, Fukuoka, Japan.*

PROF. K. KOGA*, *Department of Civil Engineering, Saga University, Saga, Japan.*

DR. W. LIENGCHARERNSIT*, *Department of Civil Engineering, Kasetsart University, Bangkok, Thailand & Department of Civil Engineering, Saga University, Saga, Japan.*

PROF. M.R. MADHAV*, *Department of Civil Engineering, Indian Institute of Technology, Kanpur, India & Institute of Lowland Technology, Saga University, Saga, Japan.*

PROF. N. MIMURA, *Department of Urban and Civil Engineering, Ibaraki University, Ibaraki, Japan.*

PROF. N. MIURA*, *Department of Civil Engineering & Institute of Lowland Technology, Saga University, Saga, Japan.*

PROF. V.V.N. MURTY, *Irrigation Engineering and Management Program, Asian Institute of Technology, Bangkok, Thailand.*

DR. A.J.M. NELEN, *Center for Operational Water Management (COW), Delft University of Technology, Delft, The Netherlands.*

PROF. H.B. POOROOSHASB*, *Department of Civil Engineering, Concordia University, Montreal, Quebec, Canada.*

DR. K. SHIKATA, *Department of Civil Engineering, Kyushu University, Fukuoka, Japan.*

DR. T. TANAKA, *Fukken Co. Ltd., Hiroshima, Japan.*

PROF. T. TINGSANCHALI, *Division of Water Resources Engineering, Asian Institute of Technology, Bangkok, Thailand.*

DR. I. TOHNO, *Water and Soil Environment Division, The National Institute for Environmental Studies, Tsukuba, Japan.*

DR. T. TSUCHIDA, *Port & Harbour Research Institute, Ministry of Transport, Yokosuka, Japan.*

PROF. J.C. van DAM, *Faculty of Civil Engineering, Delft University of Technology, Delft, The Netherlands.*

PROF. S. VALLIAPPAN*, *School of Civil Engineering, University of New South Wales, Kensington, N.S.W., Australia.*

PROF. K. WATANABE, *Emeritus Professor, Faculty of Agriculture, Saga University, Saga, Japan.*

DR. T.P. XU, *School of Civil Engineering, University of New South Wales, Kensington, N.S.W., Australia.*

DR. C.B. ZHAO, *School of Civil Engineering, University of New South Wales, Kensington, N.S.W., Australia.*

*Members of Advisory Group

1 INTRODUCTION

N. MIURA* and M. R. MADHAV**

*Saga University, Saga, Japan

**I.I.T., Kanpur, India & Saga University, Saga, Japan

LOWLANDS

The origins of many civilizations can be traced to deltas and coastal regions of the world. Presently about 80% of the cities or regions with the largest populations are located in these areas. The term 'Lowland' covers a broad spectrum of lands affected by fluctuating water levels, in which human activities such as agriculture, industry and living are pursued or being proposed. They encompass a wide range of terms such as 'Swamp', 'Wetland', 'Tideland', 'Mangrove Forest', etc. which have already been developed or are being developed by the concept of 'wise use' with concern for the protection of and in harmony with nature. These tracts of lowlying ground are lands (i) on which water collects and makes the ground too moist for cultivation, (ii) the ground is wet and spongy, consisting chiefly of decayed organic matter and too soft to bear any load, or (iii) which are flooded during some part of each year and more or less wet throughout the year, as in coastal lands, deltas, tropical lands, etc. Figure 1.1 and Table 1.1 depict the locations and extents of some of the lowlands.

Lowlands can be classified into the following physiographic units (Table 1.1) based on drainage characteristics of the ground.

The concern in this book is mainly with alluvial and marine plains. The formation of coastal lowlands, particularly in the tropics, is influenced by many

Table 1.1. Classification of lowlands based on drainage characteristics.

No.	Unit	Drainage Characteristics of Ground
1	Low foothills	Well drained; never inundated.
2	Pleistocene Terraces	Poorly drained, occasionally inundated for short periods.
3	Alluvial Plains ¦	Very poorly drained, regularly
4	Marine Plains ¦	inundated for prolonged periods.
5	Peat Formations	Permanently inundated, except for some dry periods.

factors related to natural processes but lately to human activities. The important factors amongst them are:

i) Climate – the humid tropical climate leads to rapid mostly chemical weathering, erosion and subsequent deposition of clayey materials downstream of river channels;

ii) Neotectonic movement – in tectogenic areas, the seaward growth of the lowland zone is affected by the subsidence caused by neotectonic movements (Verstappen 1986);

iii) Tidal range – the coastal configuration and the river courses in the lower reaches are affected by the tidal range. If the tidal range is small, delta tends to build out into the sea, in the form of birdfoot delta. If the tidal range is one meter or more, estuaries and relatively deep river channels with well developed natural levees may be found;

iv) Wind and currents – in coastal development, winds, currents and longshore drifting patterns are important. In the wet season, the sediments reach the sea while in the dry season, part of this material is redistributed along the coast, depending on the orientation of the coast and its exposure to the onshore winds;

v) Changes in the position and/or discharge of the rivers – in lowland rivers, a new delta may be formed resulting from breaches in the natural levees when the river is in spate or through human interference; and

vi) Human activity – deforestation in the last century, has accelerated the erosion, transportation and deposition of sediments by rivers, thus extending the floodprone zones and the deltas. Subsidence locally or extensively results from extraction of groundwater or natural gas/oil.

Figure 1.1 Some typical lowlands of the world

Lowlands along the near shore waters include the coastal zones, estuaries, inlets, etc. They are subjected frequently to natural phenomena such as tides, waves, and currents, and exposed storms, cyclones or hurricanes and tidal waves, earthquakes and tsunamis, endangering people, livestock, agriculture and engineering works. The geomorphic, pedogenic and hydrologic elements are interwoven and difficult to distinguish in the lowland environment. Some typical lowlands of the world are depicted in Figure 1.1 and listed in Table 1.2.

FLOODING AND DRAINAGE

Frequent flooding by storm water is one of the most serious problems in lowland. The flooding in lowland generally consists of the wild flood through destruction of river banks and inundation. The former occurs when the dike or bank is poorly designed or constructed or not improved appropriately from time to time. This case is more serious than the latter because of the flash floods. In the latter case, the flood water level rises slowly because of difficulty of drainage in lowland. This "mild inundation" is observed in the case the interior land which is protected by banks and dikes but the drainage systems are not established. Not only should the prevention facility such as dikes and banks be provided, but the functional drainage systems should also be established including channels, pipes, sluiceways, pumping stations, gates, regulating reservoirs, etc. However, the damage resulting from the lowland–type flood finally depends on the drainage system for storm water and its management. The drainage systems should be operated appropriately taking account of tidal effect, rainfall intensity and run–off coefficient/reaching time of the storm water. Such integrated operating systems, unfortunately, have not been established yet in many developing countries.

ENVIRONMENTAL ASPECTS

The quality of water available in lowland areas is a major concern. Pollution of surface and ground waters and salt water intrusion into aquifers due to human activities are the root causes for the deterioration in the water quality. Water pollution problems in lowland result mainly from stagnated water in interior land and inadequate water management without considering water quality. High BOD concentration and anaerobic conditions might appear easily in water bodies through long retention time of wastewater in the interior land. On the other hand, even if wastewater is treated in a secondary wastewater treatment plant, the treated water will be the cause of eutrophication in water courses which tend to have the characteristics as closed water bodies. In lowland, advanced wastewater treatment systems might have to be introduced for preserving the water environment.

Table 1.2. Lowlands around the world (Extracted from Stoutjesdijk 1982).

Country	Estimated area (x1000 ha)	Country	Estimated area (x1000 ha)
AFRICA		**CENTRAL AMERICA & THE CARRIBEAN**	
Egypt	1800		
Chad	15	Cuba	150
Ghana	25	Jamaica	10
Guinea–Bissau	400	Mexico	*
Kenya	118	Trinidad & Tobago	13
Madagascar	30		
Morocco	43	**EUROPE**	
Mozambique	28		
Nigeria	100	Belgium	*
Rwanda	10	Bulgaria	75
Senegal	500	C.I.S.	1400
Somalia	10	Denmark	50
Sudan	200	Germany	450
Zambia	520	France	400
		Great Britain	100
ASIA & OCEANIA		Greece	45
		Hungary	300
Australia	10	Irish Republic	10
Bangladesh	8500	Italy	1250
Burma	650	Netherlands	2000
Cambodia	40	Poland	171
China	5600	Portugal	28
India	4400	Romania	30
Indonesia	43000	Slovakia	500
Iran	*	Spain	150
Iraq	*	Yugoslavia	65
Israel	*		
Japan	2600	**NORTH AMERICA**	
Korea	1600		
Malaysia	420	Canada	120
New Zealand	*	United States	88000
Pakistan	30		
Philippines	*	**SOUTH AMERICA**	
Singapore	14[+]		
Sri Lanka	70	Argentina	3500
Taiwan	48	Brazil	5500
Thailand	85	French Guyana	100
Turkey	50	Guyana	14
Vietnam	1100	Paraguay	*
		Surinam	45

* Estimates not available.
[+] Reclaimed ground.

Venezuela   120

Another issue on the environment is the preservation of ecosystems in tidal flats and marshy lands. Recently, the Environment Agency of Japan reported that about 4000 ha of tidal flats, or about 8% of the total, have disappeared in Japan during the past 10 years because of rapid reclamation and urban development.

SUBSIDENCE AND FLOODING

Subsidence (lowering of ground surface) and sea level rise are two major problems developers and managers of lowlands have to contend with. The hazards, costs, ecological, environmental and social impacts of these two phenomena can be enormous. The land subsidence caused by over withdrawal of groundwater is observed in many lowland areas. Geotechnical peculiarity, i.e. soft clay, and shortage of water resources are the causes of such land subsidence. In general, there are tidal rivers in lowland and hence utilization of fresh water is difficult. In case of the Saga plain, Japan, for example, the catchment area is too small to irrigate the vast paddy fields which have been developed through long term reclamation works. As a result, serious land subsidence has been occurring. The sea water intrusion into groundwater through the over withdrawal is a serious problem. Subsidence due to withdrawal of natural gas and/or oil is also occuring in few places, e.g. Japan, Italy, U.S.A. and Venezuela

In coastal areas, flooding can be directly correlated with subsidence. The sea level and the storm surge levels remain constant while the land or ground surface level is lowered. Each one meter of subsidence results in a corresponding increase in the depth of flooding. In riverine flooding the relationship between subsidence and increased flooding is not clear nor apparent. In this case, the channel capacity and the rate of flow are the controlling factors, rather than the ground elevation. Channel capacity is a function of the geometry of channel cross–section and the gradient of flow. Land subsidence affects only the latter, i.e. the slope of the energy gradient. Potok (1991) reports that in riverine flooding, the maximum increase in flooding depth is less than one–third of the related subsidence of the ground.

Kinds of damage resulting from subsidence, flooding, poor drainage, reclamation of land in coastal regions can be:
1) Loss in production in farming and agriculture due to increased wetness of soil;
2) Decrease in the safety of dikes;
3) Increased erosion of the dunes;
4) Decrease in the quality of subsurface water;
5) Ecological damage;
6) Disruption of water courses and drainage channels; and
7) The enormous quantities of earthfills involved in reclamation works, may in

some rare cases, induce earthquakes, just as Reservoir Induced Seismicity (RIS) is now a reality.

MANMADE COASTAL ISLANDS

The need for reclamations and construction of manmade coastal islands (M.C.I.) has escalated in the last two decades due to urban congestion, pollution and high land, especially waterfront land, prices. The advantages of near shore islands (Frankel 1992) are
(i) removal of some undesirable activities from the inner city waterfront;
(ii) provide access to large ships;
(iii) free the waterfront land for recreational, residential and commercial uses;
(iv) development facilities for treatment and deposition of waste;
(v) improve logistics of urban communications; etc.
   Since 1972, more than 100 large manmade coastal islands, with a total area in excess of 20,000 ha have been reclaimed from the sea near urban areas. Out of these, 67 M.C.I. covering an area of 10,000 ha or more, have been developed in Japan. Kobe and Osaka have moved industrial and transportation activities to near shore M.C.I. and also developed integrated urban island communities which incorporate new residential, commercial, industrial and port activities (e.g. Port Island, Kobe). The Kansai International Airport, one of the largest airports in the world, is on an M.C.I. 775 ha in area.

SEA LEVEL RISE

   The development of lowlands in coastal regions under present predictions of global climatic change, is affected by (i) the potential for sea level rise, (ii) possible modifications of the frequency and intensity of storms and tides, and (iii) planning time frame and intensity of the proposed developments.
   For example, intensive developments such as for agriculture, industry and housing, require more allowance for potential sea level rise than for open spaces (parklands, recreational areas, etc.). Gerstle (1992) gives high, low and median potential sea level rise in the years 2050 and 2100, respectively as 0.5 m and 1.6 m, 0.1 m and 0.4 m, and 0.3 m and 0.8 m. Of greater concern than the sea level rise, is the potential for increased intensity and frequency of flooding due to storms and tides which cause extensive damage.

ECOLOGICAL ASPECTS

The preservation of ecosystems in tidal flats and marshy lands, is one of the major concerns. Some of the lowlands are designated as protected wetlands which encompass water bodies such as seagrasses, coral reefs, estuaries, rivers, lakes and reservoirs, etc., apart from the lands effected by fluctuating water levels (Isozaki et al. 1993). These wetlands are productive ecosystems and generate many important benefits, e.g. groundwater recharge, flood control, energy production (fuel wood), removal of pollutants, habitat for many plant and animal species including some rare species of the eco–system, recreation, etc. Wetlands provide extremely diverse benefits locally as well as downstream and sometimes even offshore (e.g. sediment retention). These benefits are essential for conservation of nature and human societies in particular. This problem of the diminishing tidal flats which are habitats for wildlife such as birds and fish, is enthusiastically discussed as a global environmental problem in various organizations of the world, e.g. the "Convention on Wetlands of International Importance Especially as Waterfowl Habitat" (Ramsar Convention which was adopted in 1971). In this book, the aspects related to the ecology of lowlands which are important by themselves are implicitly considered in that the development and management of lowlands is carried out by including the above concerns. It is proposed that with further accumulation of knowledge and expertise, a similar publication on the environmental and ecological aspects of lowlands will appear in the near future.

DEVELOPMENT AND MANAGEMENT

The need to reclaim lowlands for agriculture, communications, infrastructure and industrial development, is receiving increasing attention in many countries over the past few years as part of their overall development.

The general aim of development and management of lowlands, is to mitigate damage to and control the risks relating to:
(i) human safety;
(ii) structural reliability;
(iii) construction methods;
(iv) functioning of operating systems;
(v) environmental impacts; and
(vi) economic feasibility.

The purpose of this State of Art Volume on Lowlands, is to explore the various facets of development and management of lowlands, and to compile them into a volume that would serve as a reference text on a subject that was not widely studied in the past. This volume is divided into five sections, viz., I Description of lowlands: geotechnical, hydrological and environmental aspects; II Development: methods, coastal protection, ground improvement, wave action, flood control and construction of manmade coastal islands; III Groundwater: hydrology, modelling and numerical solutions; IV Water management: for agriculture and water quality; and V Subsidence and Monitoring: groundwater, oil and natural gas withdrawal, monitoring with digital processing and sea level rise and consequences.

A broad spectrum of topics have been addressed by the various contributors to this volume which could serve as a useful sourcebook or reference on lowlands. The need for a broad interdisciplinary approach for the study of development and management of lowlands, is aptly demonstrated by these contributions. Difficulties encountered in their development are highlighted particularly with reference to the environment, water management and the geotechnical aspects. Development of lowlands involves consideration of their ecology, engineering aspects of design, construction and management, and assessment of environmental impact. Since most lowlands border the coastline between sea and land, i.e. coastal lowlands, they are often underlain by alluvial or marine sediments which are soft, highly compressible, and very sensitive. The determination of the in situ properties of the soils, design and construction of structures on them, and if required, to improve or modify the original ground, is a challenging task. Geosynthetics, the newly developed materials, have an important role to play in the over all development of lowlands.

With the vast experience gained over a period of time, drainage works and water management techniques have been developed to a highly refined state. The environmental concerns are also a major part of the overall development of the lowlands. Subsidence is now a well recognized hazard. Determination of aquifer systems with soil strata capable of creating subsidence, and prediction of its ultimate values under different rates of withdrawal of the fluids, are essential for the development and management of newly created lowlands and remediation of existing ones. If subsidence is inevitable and has taken place, one of the ways to accommodate the same in the regional development is to use part of the area as a reservoir. The low lying area can function either as a temporary storage reservoir for flood waters if it is close to an urban area or as a permanent one to store liquids – water, oil, etc.

REFERENCES

Frankel, E.G. 1992. Artificial island city developments. *Aquapolis:*5/92:20–25.
Gerstle, B. 1992. Sea level rise and the effects on Australia. *Aquapolis:*4/92:18–20.
Isozaki, H., M.Ando, and Y.Natori 1993. Towards wise use of Asian Wetlands. *Proc. of the Asian wetland symposium, Otsu and Kushiro, Japan.*
Potok, A.J. 1991. A study of relationship between subsidence and flooding. *Proc. 4th Int. conf. land subsidence, Houston, U.S.A.*:389–396.
Stoutjesdijk, J.A. 1982. Polders of the world. *Compendium of Polder Projects. Delft University of Technology, the Netherlands.*
Verstappen, H.Th. 1986. Rural land use, urban stress and environment in coastal lowlands. *Proc. symp. on lowland development in Indonesia, Jakarta:*supporting papers:491–502.

2 GENERAL DESCRIPTION AND GEOTECHNICAL CHARACTERISTICS

M. R. MADHAV[*] and N. MIURA[**]
[*]I.I.T., Kanpur, India & Saga University, Saga, Japan
[**]Saga University, Saga, Japan

INTRODUCTION

Lowlands are defined as lands affected by fluctuating surface water level, e.g. by tides, floods, etc. Subsiding ground by itself, however vast it may be, as in Mexico and some parts of the United States, would not qualify as lowland if it is not affected by a relatively higher elevation of water level at least for some time periodically. Large tracts of coastal lands at or below the mean sea level, existing all over the world, e.g., in the United States, the Netherlands, Japan, Bangladesh, India, Indonesia, Thailand, Venezuela, etc., qualify as lowlands. So also the flood plains of major rivers which are inundated periodically. Lowlands could be created by earthquake induced subsidence and upheaval of the sea bottom.

The term 'wetlands' commonly used in the United States of America is synonymous with the term 'lowlands'. The following subdivisions distinguish different types of lowlands such as bogs, prairie potholes, Cypress swamps, riverine bottomlands, and coastal marshes.

(i)Bogs – wet, spongy, soft, waterlogged ground consisting chiefly of decayed organic matter too soft to bear any load;

(ii)Prairie Potholes – shallow seasonally flooded lowlands left in the northern plains by the retreating glaciers;

(iii)Cypress Domes – tracts of lowlying ground in which tall Cypress trees grow;

(iv)Riverine Bottomlands – floodplains, levees, sloughs, and oxbows which get inundated with flood waters; and

(v)Coastal Marshes – marshy coastal lands submerged by the tides.

Lowlands are also created by human activities such as withdrawal or removal of natural resources, reclamation of ground and construction of manmade coastal islands. Human settlements are encroaching onto the flood plains in many of the developing countries due to heavy population pressure compounding the problems in lowlands. Pumping of groundwater, oil or natural gas from aquifers below soft clay deposits, and extraction of coal are the most common causes of subsidence of ground creating lowlands by human activities. Japan, Mexico, United States, Italy and Thailand are countries with prime examples of such phenomenon.

In some of the developed countries, due to paucity of land, the same is being reclaimed from the sea, as in Japan and Singapore. These manmade coastal islands or reclaimed ground are founded usually on soft marine deposits and subjected to tides and storm surges. Global warming is presently a major threat. Even one meter rise in sea level could submerge large land areas. In this chapter, the geographical, geomorphological and geotechnical characteristics of lowlands are discussed and some of the lowlands of the world described in particular.

FORMATION OF LOWLANDS

Oele (1986) presents a detailed description of the formation of lowlands. Coastal lowlands are those lands whose surface is at or close to the sea level and are therefore found between the sea and the more elevated inland terrain. Mountain ranges, their foothills and old terraces constitute the latter. Two types of lowlands, erosive and depositional, can be distinguished. Erosive lowlands are associated with marine and glacial activities and are less common than the depositional ones. They are normally found in regions of tectonic uplift.

The depositional type of lowlands are composed of materials produced by weathering and erosion of soils and rocks from the higher terrains in the uplands. Temperature, rainfall, seasonal distribution of precipitation and evaporation influence the weathering processes. In the tropics, the weathering is predominantly chemical since both temperature and rainfall are very high and vegetation develops easily. In spite of the cover of vegetation, the sediment quantities in the rivers are large because of the high rates of precipitation. The rivers transport these products of erosion into the depositional lowlands.

Cotecchia (1986) presents a review of lowlands created by earthquake induced subsidence. If instead of the ground subsiding, the sea bottom rises, tsunamis may result and submerge large areas.

Riverine lowlands

Rivers tend to form as either braided or meandering systems upon entering the flatter alluvial or coastal plains. Braided systems are formed in areas with seasonal peak discharges and originate with non–cohesive sediments. If the sediments are cohesive and the discharge more regular, the river forms a meandering course. In the tropics, conditions are favorable for the development of meanders where the river banks are stabilized by dense vegetation. Rivers in mature valleys frequently have extensive marshes and swamps along their banks. Along a meandering river, erosion on the outer edge and deposition on

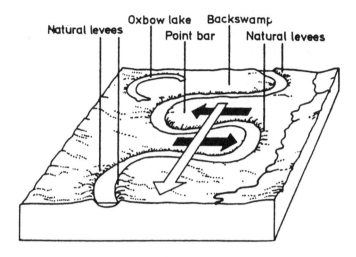

Figure 2.1. Lowlands in the meandering river system.

the inner side of each loop take place, to form point bars (Figure 2.1). The radius of the meander increases continuously until a sharp loop is cut-off to form an oxbow lake. The system of meanders travels downslope creating point bars, levees, backswamps and oxbow lakes. In the humid tropics, the meander patterns are fairly stable due to the steady high runoff and sediment load. Floodplains elevated only one or two meters above the river level, abandoned river channels, and oxbows having standing or sluggishly flowing water for appreciable part of the year support swamps and marshes. Topography and supply of water are the two most important features that determine the distribution of fresh water swamps. Flood waters overflow the normal river bed as a sheet over the surrounding areas during peak flows. The consequent

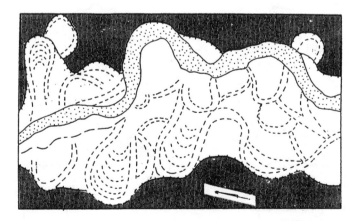

Figure 2.2. Lowlands from transition of a river from meander to braided pattern (after Verstappen 1986).

reduction in the velocity of flow results in the coarsest fraction of the sediment to be deposited as levees close to the river bed. The finer particles are deposited farther away. Dense vegetation develops in these backswamps which may eventually form peat deposits. In a meandering river system, point bar deposits are composed of coarse particles, natural levees of sand and silt, backswamps of fine grained soils and sometimes peat. However, the system is dynamic, with an intricate sedimentation pattern and subsequent fragmentation. The subsoil profile can be highly variable in the areas adjacent to the meandering river. The increased bedload in the rivers can in some instances, transform the river from a meandering to a braided pattern, e.g. lower Palu river, Sulawesi, Indonesia, and create riverine lowlands (Figure 2.2).

Tidal lowlands

As a meandering river reaches the sea, it forms a delta extending into the latter by splitting into several branches and distributing the sediment over the delta. The mechanisms responsible for the distribution of the sediments at or along the coast and the formation of the delta, are strongly dependent on the waves, the tides and the currents. The coastal morphology is the net result of these factors and the sediment load. The sediments are swept ashore or carried along the coast to form beach ridges, dunes, estuaries and lagoons. High wave energy conditions favor the transport of the material to the deep ocean floor while the low energy environment is particularly favorable for the formation of deltas. The morphology of deltas is determined by the influx of sediments, tides and action of waves on the delta front. Deltas are composed of distributaries with natural levees, marshes and lagoons between the distributaries and on both sides of the delta, and tidal flats, salt marshes and mangrove forests in the delta front

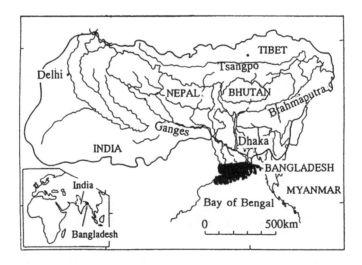

Figure 2.3. Deltaic lowland: Ganges river delta.

e.g. the delta of the river Ganges. Beach ridges develop between high tide and spring tide levels along sandy beaches close to tide and wave affected deltas. These ridges running parallel to the coast extend over long stretches. Sandy ridges of limited dimensions, called Cheniers, are formed where muddy prograding coasts are partially eroded and reformed. Salt water swamps are formed by tidal flooding and drainage on the exposed intertidal land. Regularly flooded but protected areas develop into mangrove swamps in the tropics and the subtropical regions. Extensive swamps develop mainly where the land runoff is very large and brings sediments that accumulate and extend the swamp. The deltas of Mekong, Amazon, Congo and Ganges (Figure 2.3), the north coast of Australia and of Sumatra have notably extensive mangrove swamps.

The coastal lowland also is a dynamic system where sedimentation or deposition and erosion alternate and are influenced by fluviatile, coastal and marine systems. A coastline that protrudes implies extensive fluviatile influence of sedimentation while a retracted one, marine influence on erosion.

INFLUENCE OF SEA LEVEL CHANGE

The history of the earth in the last 20,000 years, has greatly influenced the formation of lowlands. In the last two and half million years, i.e. the Quaternary period, the development of ice caps led to the expansion of the land – ice mass in the higher latitudes. The sea level dropped by more than a hundred meters but it rose with the melting of the ice during the interglacial periods. The sea

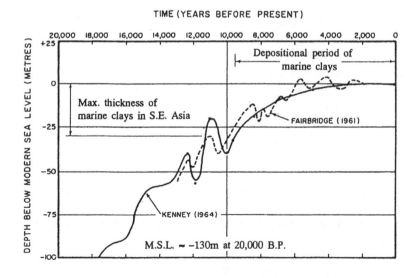

Figure 2.4. Sea level changes in Recent times (after Cox 1970).

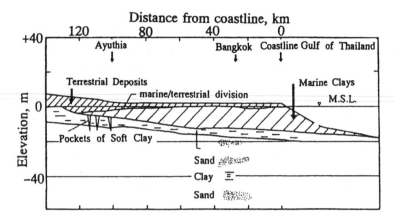

Figure 2.5. Occurrence of soft clay pockets, Chao Phraya delta (after Cox 1970).

level dropped by 120 m below the present level in the last glacial period. The rivers continued their course across the shelf, e.g. Sundra shelf in Indonesia that emerged the during the glacial period, to reach the sea. The glacial period ended about 18,000 years ago with the climate becoming warmer. The rise of sea level (Figure 2.4) was initially at a faster rate than that at later times. The lowlands presently known are formed as a result of the geologically the most recent sea level rise. However, the changes in sea level were different at different places. According to Clark et al. (1978), the Indonesian archipelago itself lies in two zones of sea level rise. The shoreline emerged in some places, e.g. Java, while they were at submergence level in other areas. Local tectonic movements together with the sea level rise should be considered for each lowland formation.

While Kenney (1964) predicts a continuous rise of sea level (Figure 2.4), the prediction by Fairbridge (1961) shows periods of submergence and emergence with +/- 4 m sea level fluctuations. Geyh et al. (1979) estimate a rise of about 5 m above the present sea level for the Strait of Malacca. The deltaic areas which are normally within 3 m to 4 m elevation from the present sea level, would thus have been submerged with possible marine sedimentation. The perturbations of the sea level changes during 13,000 to 11,000 B.P., are correlated with the occurrence of the soft clay pockets (Figure 2.5) in the underlying stiff marine clay.

The rate of sea level change with time is depicted in Figure 2.6. The rate decreased from about 25 mm/yr 10,000 years B.P. to about 1 mm/yr at present. The advance or retreat of the shoreline or the delta, is governed by the net difference between the rates of sea level rise and sedimentation, and other tectonic movements.

The most important factor for the formation of the top soil layers or crust in case of lowlands, is the hydrological regime. Where the water level fluctuates strongly (alternating between dry and wet), soils are firm; they are oxidized and

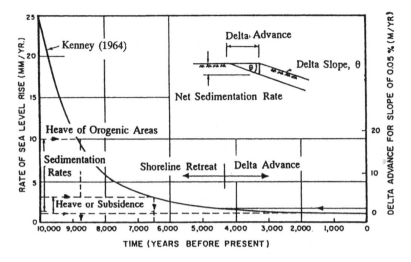

Figure 2.6. Rate of sea level rise in Recent times (after Cox 1970).

reduced during the dry and wet seasons. If the water level is more stable, the oxidation/reduction process does not take place. Stable water conditions tend to promote peat formation.

TIDAL FLATS

Tidal flats are depositional environment along flat coasts which are subject to significant tides, e.g. Saga plain (Figure 2.7). Much of the area gets exposed during low tide and flooded during the high tide. This typical rhythm of in and

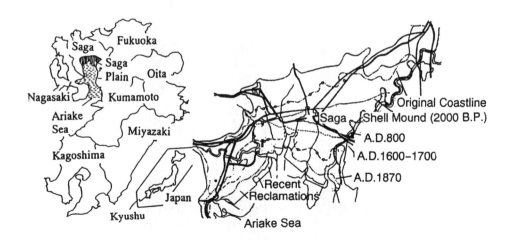

Figure 2.7. Tidal lowland – Saga plain, Japan.

out flow of tidal sea, characterizes the tidal flats. These are subdivided into supratidal, intertidal and subtidal areas, depending on the tide level in relation to the ground. Supratidal zone gets covered by water only few times in a year and thus, can be vegetated. This zone, therefore, is sometimes termed as a marsh and in the tropics as mangrove forest. The subtidal zone is perennially below water level. The intertidal zone is an intermediate area between supratidal and subtidal zones and is further classified into mud flats, sand flats and mixed flats, based on the sediment characteristics. The deposition and accumulation of fine grained soils in tidal flats, is governed by what are termed as lag effects. Settling lag which occurs when the tidal current slows down, causes the particles to be carried some distance away from the point of deposition. A faster current velocity is required to move the same particles during scour lag that results from flow reversal. A net landward transport of sediments into the tidal flat results with each tidal cycle. Most marine clays are formed in tidal flats which form parts of large deltas particularly in the tropical and subtropical regions. In the marshes, the sediments are trapped by vegetation and, hence, are organic.

BAYS AND LAGOONS

A bay is a recess in a coastline that is well protected from the near–shore currents, while a lagoon is a shallow water body that is connected by one or more inlets. They have similar characteristics from sedimentological and geomorphological considerations, and may contain brackish, normal to

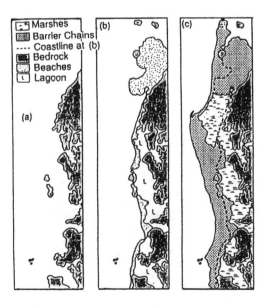

Figure 2.8. Sequential development of coastal lowlands (after Weerakkody 1992).

hypersaline waters. The sequential development of coastal landforms (Figure 2.8) is described by Weerakkody (1992). In phase I, about 6000 to 5000 B.P., bay beaches were formed during mid Holocene sea level transgression. These were about 1 m to 5 m above the mean sea level. Regression in phase II (5000 to 4000 B.P.) of coastal development, led to the formation of fluviomarine plains. Further regression in the third phase (4000 B.P. to present times) resulted in the formation of barrier chains and lagoons. Mangrove forests and marshes may surround a lagoon, if the latter dries up partially. Fine grained soils predominate in deep bays, while silty clays and clays are found in extensive lagoons. Where the water velocity is high, sands get deposited.

DELTAIC DEPOSITS

The sediments built up at the mouth of the river where the velocity of water reduces practically to zero, are termed as deltaic deposits. They extend deep into the sea or the ocean, and grade themselves from coarse to fine grained soils towards the continental slope.

The depositional processes in deltaic deposits are governed by (i) river regime (ii) coastal processes (iii) currents and (iv) climatic factors. The complex sedimentary structure of large deltas consists of topset, foreset, and bottomset deposits (Figure 2.9). Deltas can also be subdivided as subaerial and subaqueous zones. The former consists of a lower and an upper deltaic plain, while the latter is that portion of the delta which is below the low tide level. Prodelta or the bottomset is the most seaward stratum of the subaqueous delta wherein the fine grained material is deposited. In some large deltas, the topset and foreset deposits are difficult to distinguish and hence, the terms prodelta, intradelta, and interdistributory environment, have been suggested (Brenner et al. 1981). The intertidal flats and shallow bays constitute the subaqueous zone. Swamps and marshes consisting of highly organic deposits, constitute the subaerial zone of the delta.

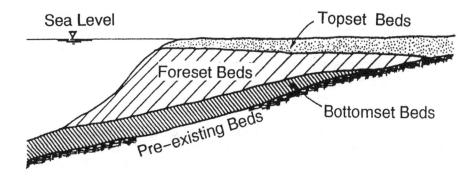

Figure 2.9. Components of delta (after Brenner et al. 1981).

Table 2.1. Rate of growth of river deltas and coastal plains.
         (after Cox 1970).

River/Region	Growth rate (m/yr)	Deltaic slope (%)	Net sedimentation rate (mm/yr)
(a)Rivers			
Chao Phraya, Thailand	12–25	0.2	24–50
Irrawaddy, Myanmar	50	0.2	100
(b)Coastal plains			
Chao Phraya	4–5	0.05	2–3
West Malayan	1	0.1	1
North Borneo	9	0.1	9
Indonesia	12–30	0.1	12–30

Prodelta clays usually appear to be homogeneous and are highly plastic. Radiographs, however, reveal thin laminations, fractures and complex displacements. The topset and interdistributory zones of a delta are heterogenous since they are formed partly in subaerial and partly in subaqueous zones. Deposition in deltas takes place with water salinity ranging from that of fresh to sea water. Major deposits of soft marine clays (Brenner et al. 1981), encompass flat deltaic lowlands of North America, South and Southeast Asia, Japan, etc., at the mouths of rivers such as Mississippi, Ganges, Irrawaddy, Chao Phraya, Mekong, Kelantan, Pahang, etc.

Deltaic areas can be subdivided (Cox 1970) geomorphologically into: (i) alluvial fans, composed of terrestrial deposits; (ii) natural levees, usually 1 m to 4 m above the surrounding plains with terrestrial deposits on the surface; (iii) swamps between levees, the underlying sediments being finer and organic; and (iv) lower deltaic areas which are nearly at sea level, flat and with marine deposits. The salinity of the soils is highest at the coastline and decreases with distance inland. The groundwater level is close to the surface and often rises above the ground due to floods and tides.

Transgression of the sea onto the land results if the rate of sea level rise and other tectonic movements exceed the rate of sedimentation. Conversely, regression takes place if the rate of deposition of sediments exceeds the rate of sea level rise. Table 2.1 presents the rate of growth of river deltas and coastal plains in Southeast Asia.

The rate of growth of individual deltas has been greater during the initial periods due to shallower waters and narrow deltaic front (Cox 1970). For Chao Phraya coastal plains, the initial growth rate was reported to be 70 m/yr while presently it is only 4 to 5 m/yr. The rates of growths and net sedimentation in deltas are much higher at river mouths than along the adjoining coastal plains.

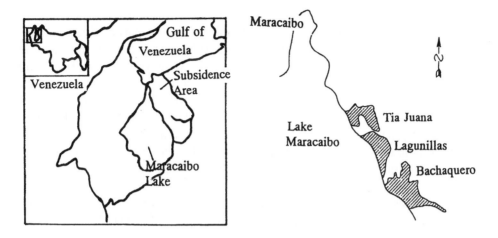

Figure 2.10. Lacustrine lowland – Lake Maracaibo, Venezuela.

LACUSTRINE LOWLANDS

Lands adjacent to or surrounding a lake, may form into 'Lacustrine Lowlands'. Lacustrine deposits are controlled largely by the hydrodynamic conditions of the lake, the topography and the depth of the lake, relief and supply of materials from the surroundings, etc. Clays get deposited towards the center of the lake. The composition of lacustrine deposits can be varied as they comprise of both clastic and nonclastic materials. Lake Maracaibo, Venezuela (Figure 2.10), is a typical example of lacustrine lowland.

FLOODPLAINS

The river as it leaves the mountains and flows towards the sea, forms what are termed as flood plains which can attain great depth and extent. The floodplains have been morphologically divided into (Brammer 1990): (a) Piedmont plains, very gently sloping outwash fans at the foot of the hills; (b) Active river plains,

Table 2.2. Classification of floodplains of Bangladesh (after Brammer 1990).

Land type	Highland	Medium Highland	Medium Lowland	Lowland	Very Lowland	Water Settlements
Depth of flooding, cm.	Above flood level	<90	90–180	180–300	>300	
Area, %	17	40	15	9	2	17

ephemeral alluvial land within and immediately alongside major rivers; (c) River meander floodplains, older alluvial land formed by earlier river courses, seasonally flooded upto 2 m deep on the ridges and upto 5 m deep in depressions; (d) Major flood plain basins, extensive old back–swamp depressions which stay wet even in dry season and are deeply flooded (5 m$^+$) in the rainy season; (e) Estuarine floodplains, almost flat thick silty alluvium with 2 m to 5 m deep seasonal flooding. Part of the coastal areas become saline during the dry season; and (f) Tidal flood plains, almost at sea level (<1 m high) predominantly clayey alluvium. These are flooded in the rainy season and at high tide in the dry season. The deposits are formed during periods of high floods.

The FAO classification of floodplains (of Bangladesh) based on water depth (Table 2.2) can be utilized for quantifying the degree of lowland flooding.

LOWLANDS FROM HUMAN ACTIVITIES

Lowlands can also be classified as natural and created lowlands. 'Natural lowlands' are those formed by the natural/geological processes near the surface of the earth, while the 'created lowlands' result from various human activities. Created lowlands can be further subdivided, based on type of activity: a) groundwater withdrawal; b) extraction of oil and gas; c) extraction of coal and d) reclamation of ground from the sea (coastal islands). Vast amounts of water, natural gas and oil, are being extracted from aquifers underlying many soft deposits that are located in lowlands. Resulting subsidence of the softer and compressible deposits is creating many coastal and lacustrine lowlands. With increased utilization of geothermal energy, it is possible some more lowlands may emerge (Finnemore & Gillam 1976). Japan and Singapore are reclaiming land from the sea in the form of extension of mainland or as manmade coastal islands which have characteristics similar to those of lowlands.

GEOTECHNICAL ASPECTS

The formation of soil deposits in the coastal lowlands and their engineering characteristics are closely related to the sedimentation history of the site. The sea level fluctuations that have occurred in the Pleistocene, Holocene and Recent epochs probably have had the most important influence on the sedimentation history of these deposits. The groundwater table rises as the sea transgresses over the land, producing swampy brackish water environment ahead of the advancing coastline. An organic layer may be found at the base of the marine clay layer. The rise of the groundwater level may produce fresh water lakes in the lowlying basins inland, which may get filled up with organic matter. Transgression of the sea brings about a change from fresh or brackish water conditions to marine environment. Marine sediments continue to be deposited as the sea transgresses to the maximum extent.

As the ground level rises to the limits of tidal range due to sedimentation, the environment will revert to brackish water condition. The soil genesis is affected by tidal movements. Alternate wetting and drying cycles cause oxidation and desiccation of the soil. The structure of the clay is less flocculated because of deposition in brackish water conditions. Further, sedimentation from annual or periodic flooding builds up terrestrial sediments until the surface of the ground is above the high tide level. Desiccation, low water content, dispersed structure and low salt content of porewater characterize these terrestrial sediments.

Coastal/Marine soils

The soils of the lower deltaic plains can be categorized as: (a) Coastal saline clays, these soils often flooded at spring tide, are soft and slightly acidic; (b) Fresh clays, slightly alkaline and soft. The strength and profile development of these clays is controlled by the leaching of salt water by flood waters; (c) Organic soils, found in swamps, marshes and localized basins where the water table is high and drainage poor. If the organic content exceeds 30%, they are termed as peat soils; and (d) Acidic clays, with pH values less than 4.5 found in backswamps.

The maximum thickness of coastal marine clays, especially in Asia, is in the range of 5 m to 35 m. From the apparent stable soil conditions beneath, it appears that these marine clays would be younger than 10,000 years as the sea level could not have reached the base of these deposits until then.

Tidal clays

As mentioned in the preceding paragraphs, soft clays are deposited in tidal flats, e.g. the Ariake bay. The thickness of the clay deposits can vary between 10 m to 40 m while the water content ranges from 50% to about 200%. The natural water content of the soil is often much higher than its liquid limit and the liquidity index of these soils can be as high as 2.5. The soil samples lose strength on remolding and cannot be tested without a confining stress. The sensitivity of these soils is very high and is strongly influenced by the salt content (Figure 2.11). While the undisturbed strength of these clays depends mainly on the water content but not on the salt content, the strength of the remolded soils decreases with decrease of salt content. Consequently, the tidal flat clays are, in general, quick clays.

Floodplain soils

The floodplain soils grade from coarse grained to fine grained deposits as the river traverses to the sea. Silt, silty clay and clay layers are found in the lower reaches of the river. However, flood plain deposits are intercalated with fine sand layers. These deposits are subject to alternate cycles of wetting and drying, desiccation, weathering, shrinkage, etc. Therefore, flood plain deposits are much stronger and stiffer than those in the deltas and the tidal flats.

Peats

With the rise in sea level and its subsequent stabilization, large coastal plains were formed by the rapid deposition of sediments. As a result of prograding coastline, fresh water swamps and lakes replaced saline or brackish mangroves and lagoons. The shallow fresh water lakes got filled up gradually with organic matter produced by lake vegetation, which subsequently transformed into peat deposits. Peats and peaty deposits of varying thickness are extensively distributed in Indonesia, parts of Bangladesh, etc.

Acid sulphate soils

Sediments from swamps which are organic and contain brackish water, develop into acid sulphate soils. The bacteria which obtain their energy from oxidation of organic matter, reduce the sulphides in the sea water into sulphates and sulfuric acid as a result of aeration or by leaching. The pH values are lowered to less than 4.5. The clay minerals are broken down in the acidic conditions. The liberated aluminum ions prevent the intake of phosphorus and become toxic to plants. If the sulphate content is more than 0.5%, concrete structures built in these soils would be subject to sulphate attack and need special protective measures. Mud clay which is neutral to slightly acidic and non-oxidized, exists at depth. Sulphate soils or 'cat clays' are widely prevalent in humid tropics, and particularly in the Plains of Jonnes, Vietnam, Sarawak, the swampy coastal plains of Borneo and east Sumatra, and in the Chao Phraya delta.

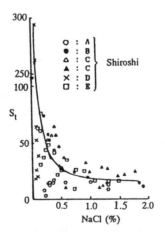

Figure 2.11. Relationship between sensitivity and salt content (after Onitsuka 1988).

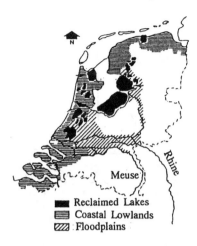

Figure 2.12. The Netherlands.

EXAMPLES OF NATURAL LOWLANDS

The Netherlands (Figure 2.12) is a country with the largest percentage of land area (about 30%) below mean sea level and 65% of the land would be flooded at high tide and with rivers but for the dikes provided (Ankum et al. 1988). The country is located in the deltas and flood plains of the rivers Rhine, Meuse and Scheldt. The usual range of tides from the North Sea is about 2 m. The sea dikes are presently designed for a sea level rise of 5 m and an additional 10 m for wave run up. The history of the development of lowlands spans nearly 2000 years though the real scientific management of these lowlands, is of recent origin. Not only the coastal areas, but the lands adjacent to lakes and rivers have been empoldered. The stratigraphy of the land consists of about 20 m of soft and highly compressible clay and/or peat underlain by a sand layer.

Ariake sea (Figure 2.7) is a large gulf in the northwestern part of Kyushu island of Japan, covering an area of 1700 sq.km. The sea bottom is characterized by an undersea cauldron at the bay's entrance, an undersea plain spreading from the center, sandbars and channels in the bay's interior and tidelands along the coast. The tidal fluctuations in the sea are about 3 m at the entrance and 6 m in the interior. The average velocity of the tidal current is 6.6 knots at spring tide at the sea's entrance and 1.0 to 1.5 knots at the interior and coastal areas. Vast tidelands (230 sq.km. at spring tide and 110 sq.km. at neap tide) exposed at low tide of Ariake sea form part of the Saga plain with an area of about 400 sq.km. Many rivers, Chikugo, Kase, Rokkaku, etc. flow into the Ariake sea bringing in substantial clay and sand deposits. Due to successive sedimentation and the wide range of tidal fluctuations, a beach emerges from the water at ebb tide some 5 to 7 km offshore. The original coastline of the Ariake sea is understood to be approximately 20 km inland from the present one (Watanabe 1988).

Figure 2.13. Natural and created lowlands of Japan.

Reclamation in this part of the world parallels that in the Netherlands and dates back to the 7th century. Dikes with stones were built from the 17th century onwards to reclaim land. The plain is underlain by the soft sensitive Ariake clay, 5–30 m thick. The natural water content ranges between 50% and 200%, the liquidity index between 0.5 and 2.5 and sensitivity from 8 to 300. Sensitivity increases (Figure 2.11) with a reduction in salt content of the pore fluid (Onitsuka 1988). For a site reclaimed about 1500 years ago, the salt content is less than 0.1 % and the sensitivity is as high as 200 to 300. Below the clay layer water bearing sand strata exist, excessive pumping from which resulted in subsidences in the range of 30 to 80 cm and intrusion of salt water into the aquifers (Miura et al. 1988).

Hachirogata, Japan (Figure 2.13), is a tidal lowland formed by the ground movements and the development of sand bars which led to the transformation of Oga island into the present Oga peninsula. A semi–saline lake with a total area of 221 sq.km. is connected to the Japan sea through a narrow channel, the Funakoshi channel. The depth of water is around 1.0 m upto about 400 m from the shore and 4.5 m at the deepest point. This lowland was reclaimed (Ohtsuki et al. 1981) by developing 173 sq.km. as central and surrounding polders. The clay stratum on which the front dike was constructed is 20 m thick, had water contents in the range of 100–200%.

Bangladesh (Figure 2.14) spreads over an area of 14.4 Mha. Out of this, about 6.0 Mha is subject to submergence under 0.3 m to 2.0 m of water annually (Altaf Ali 1991). The frequent submergence results from flooding from three major rivers – Ganges, Brahmaputra and Meghna, and flash floods from several tributaries, heavy rainfall, poor drainage, high tides, and tropical cyclones. In the active flood plains, the major rivers meander or shift constantly threatening the adjoining towns and fields. Vast areas of the country including Dhaka were flooded with water depths in the range 0.3 m to 4.5 m during 1987 and 1988 (Shafi 1991). The coastal regions subjected to storm surge flooding and tidal waves associated with tropical cyclones, are depicted in Figure 2.14.

The Indian subcontinent (Figure 2.15) has more than 5000 km long coastline with the Bay of Bengal on the east and the Arabian sea on the west. Coastal lowlands less than 2 m above the mean sea level exist almost all along the coast, particularly close to the major ports – Bombay, Calcutta, Visakhapatnam, Cochin, etc. The east coast is affected by cyclones (typhoons) every year. The resulting floods and tidal waves upto 4 m high, inundate 10–15 km of coastal lowlands. The Rann of Kutch in western India gets submerged by the sea for more than 3–4 months in a year. In addition, the alluvial rivers, Ganges, Brahmaputra, Godavari, etc., flood vast tracts of land almost every year during the monsoon months. The total area affected by floods is estimated as 8.2 Mha. At Visakhapatnam, a lowland on the east coast, extensive weathering marks the basement rocks which are overlain by transported soils and recent marine clays. The liquid limit of the soils ranges from 55 to 110%, the plasticity index 25 to 70%, clay content 30 to 70%, initial void ratio 2.0 to 2.7, virgin compression index, C_c, 0.8 to 0.9, and undrained strength of 5 to 40 kPa with liquidity index close to 1.0. They can be described as inorganic, soft, highly plastic, compressible and impermeable clays of very low to low strength.

Ranging from moist prairie depressions of about one hectare to flooded grasslands stretching to hundreds of square kilometers in Florida, lowlands are located in every region of U.S.A. (Figure 2.16). Riverine and tidal swamps are abundant along the coastal plains in the southeastern United States. Although not associated with a large river, the Dismal swamp of North Carolina and

Figure 2.14. Lowlands of Bangladesh.

Figure 2.15. Lowlands of India.

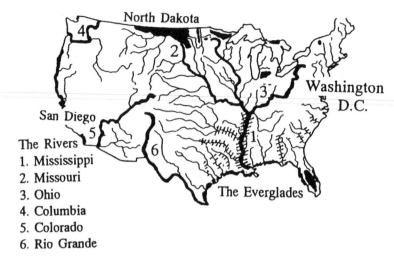

The Rivers
1. Mississippi
2. Missouri
3. Ohio
4. Columbia
5. Colorado
6. Rio Grande

Figure 2.16. Extensive wetlands of U.S.A.

Virginia, is a combination of marshes, swamps and waterways. The Mississippi and its tributaries in the low reaches, have extensive swamps in the coastal plain. The Everglades, Florida, constitute a unique combination of marshes, swamps and Cypress domes, as the region is at sea level and the water from abundant rains does not drain but stagnates. Out of an estimated 88 Mha in the lower 48 states, more than 50% of wetlands have been developed (Mitchell 1992) mostly for agriculture, but also for dry land, residential and commercial purposes, reservoirs, highways and airports. About 68 Mha of wetlands in Alaska and 52,000 ha in Hawaii add upto a staggering 160 Mha of lowlands in the United States. Significant amount of development of lowlands has taken place along the coasts of California, Florida, Louisiana and Texas, and along the Mississippi river and delta. Presently 120,000 ha of lowlands are being reclaimed every year to meet the ever increasing demand for land.

Very soft underconsolidated clays (Pamukcu et al. 1983) have accumulated in the delta of Mississippi river, a typical site, as a consequence of large quantities and fast rate of deposition of the sediments. Large scale delta front movements in the form of mudflows and turbidity flows are associated with these deposits which overlie stiff overconsolidated marine clays. A crustal zone of silty and sandy sediments is formed on the soft clay when the water depth in front of the distributory bar of the river decreases after the mudflows. The top 7 m of the clay is very soft, olive colored with a spongy structure indicating presence of gases. Traces of organic matter in the form of fibrous plant growth is also encountered. Thin layers of silty and sandy formations occur from 7 m to 14 m. Below this depth, soft clay extends down upto 60 m depth beneath which dense sand and stiff clay are encountered. The natural water contents of the soil in the top 60 m of the deposit, are close to or higher than the liquid limits and indicate underconsolidated state rather than high sensitivity.

Figure 2.17. Lowlands of Southeast Asia.

The lowlands of Southeast Asia (Figure 2.17) are spread over the coastal plains of Malaysia, the Chao Phraya and the Mekong deltas, and the Indonesian coastal plains. Mangrove covered foreshore of Malacca straits, Selangor, Malaysia, started forming around 6000 B.P. when the sea level was about 5 m above the present level. Marine clays accumulated at the site and the water near the shore became very shallow due to continuous sedimentation. Very soft clay 18 m thick overlies soft to stiff clay of 16 m thickness (Risseeuw et al. 1986).

It is estimated that about 20% of Indonesia (Figure 2.18) i.e. about 40 Mha is lowland, spread over the most populated islands of Sumatra, Kalimantan and Irian Jaya. They comprise of seasonally or periodically inundated or submerged under fresh or brackish water or high tides inland, coastal and near coastal swamps. From 25 Mha of lowland that have been identified, 8.8 Mha are earmarked for development, out of which 3.3 Mha are already reclaimed along the east coast of Sumatra, and the south, central and west coast of Kalimantan (Supriyanto 1990). Jakarta, Indonesia, is located on a coastal plain where 13 estuaries drain into the Java sea. Heavy rains during December to February combined with high tides and inadequate drainage system, cause inundation of lowlying areas of the city (Tingsanchali 1988). A polder system consisting of embankments, bypass flood channels, and flood control gates, is in operation since 1920. Recently another scheme for the Sarinah – Thamrin area (inner core of Jakarta) covering an area of 7.5 sq.km. is proposed and taken up as an additional measure to prevent flooding.

A typical and geotechnically interesting lowland in Indonesia, is the huge swamp area near Banjarmasin city, along the Barito river in south central Kalimantan island. The ground is being reclaimed by 2.5 m to 4.0 m thick fill.

Figure 2.18. Coastal swamps of Indonesia.

The site is composed of 35 m thick marine clay, underlain by diluvial clay and sand with N – values of 12 and 30 respectively. The top 18 m of marine clay (Figure 2.19) is soft while the lower 17 m is stiff indicated by a sharp increase in the static cone resistance. It is surmised that the lower clay layer was deposited more than 10,000 years ago (Subagio 1991), and subjected to desiccation due to lowering of the sea level. The lowlands of Nagara basin, Kalimantan, cover an area of 485,600 ha and are bordered in the north and east by low hills and alluvial terraces. They are generally of Pleistocene age and have developed podsolic soils (Klepper & Asfihani 1990). The entire lowlands are underlain by marine clay deposits. In the eastern part of the area, the marine

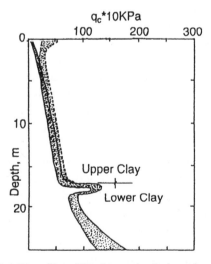

Figure 2.19. Typical soil profile of Banjarmasin, Indonesia.

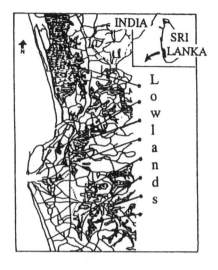

Figure 2.20. Lowlands in Sri Lanka.

Figure 2.21. Lowlands in South Korea.

sediments are covered by alluvial deposits. West of the alluvial zone, the terrain is very flat and poorly drained. A shallow organic layer (thickness < 0.5 m) covers large areas of the marine plain. Peat deposits reach depths of approximately 2 m over extended areas, and in the north–west a peat–dome of 5 m thickness is found.

Manila delta plain is a lowland (Benjamin 1990) formed near the Manila bay with large quantities of fluvial deposits of Pasig and Marakina rivers and water from Laguna lake. The soils, in general, are typical fluvial deposits consisting of (i) loose and poorly compacted sands, (ii) soft silty clay and (iii) stiff to hard clay or silt and dense sand. The thickness, lateral extent and the engineering properties of the individual layers are highly variable. The liquidity index of the 12 m thick soft clay layer is greater than 1.0. Below this layer, is a 10 m to 18 m thick layer of clayey or sandy silt of low to medium plasticity.

Colombo, Sri Lanka, is surrounded by the Indian ocean on one side and rivers, estuaries, and large areas of lowlying marshy lands on the other (Perera 1988). Most of these lowlands (Figure 2.20) are upto 2.5 m lower than the sea level and get submerged with monsoon rains. Poor drainage causes the flood waters to stagnate and convert the land into a marsh. Hence, the subsoils are highly organic (peat and muck), compressible and extend upto depths of 20 m. In addition, the south–west coastal region stretching over 100 km is intercepted by four major rivers and is subject to flooding and coastal erosion. Tidal waves upto 12 m high, during May to July are not uncommon (Perera 1991). About 350 m wide strip of land has been lost in the last 60 years. Erosion has damaged and caused parts of Hikkaduwa and Beonwela coasts to retreat inland.

Comparatively extensive lowlands (Figure 2.21) exist along the lower parts of the rivers Han, Kum and Naktong, on the southern and western coasts of

Republic of Korea. The western coastline is an extremely complicated ria with many islands. The Yellow sea and the complex coastline cause one of the highest tidal range, 9 m, in the world at Inch'on near Seoul. The Kyunggi plain in the estuary of the river Han, about 30 km from Seoul (Miura et al. 1990), is a lowland built up by reclamation upto an elevation of 1.5 m to 6.0 m. More than 50% of this area is lower than the mean tide level of 4.7 m of Kyunggi bay, requiring the construction of polders. The delta is formed by the three rivers Han, Imjin, and Yaesung, joining the Yellow sea. A very high (8 m) macrotidal range and annual precipitation in excess of 1150 mm, contribute to the complexity of this lowland development. The geotechnical profile of Kyunggi plain consists of 8–14 m thick layer of soft silt and clay underlain by a 4 m thick gravelly sand containing diluvium and weathered rock. The liquidity index of the silty clays ranges from 0.7 to 3.0 with an average value of 2.0.

The Amazon, Parana and the Paraguay rivers in South America have extensive lowlands (swamps and marshes) along their course. The region called Patanal covering an area of 100,000 sq.km, is an area of swamps and marshes in north–western Mato Grasso do Sul and southern Mato Grasso states of Brazil. In northern Iran, marshes 27 m below the sea level border the Caspian sea. The Tigris and the Euphrates rivers in Iraq, tend to build up their beds to a level considerably above the surrounding plains. They merge in a shallow lake Hawr al–Hammer before reaching the sea. Marshes occupy nearly 10% of Soviet territory. They amount to 40 – 50% of the land surface in the central and northern Karelia and 80% in some parts of western Siberian Taiga. Wetlands located in the floodplains of the major rivers, Rhine, Elbe, Moselle, Main and Danube in Germany, have all been turned into agricultural and forest land. Only a few square kilometers of the original floodplain wetlands remain in the upper Rhine valley (Kern 1992). There are many areas such as Taipei basin, Taiwan; along the Mekong river, Laos; Wairakie, New Zealand; Far West Land, South Africa; Latrobe valley, Australia; Tianjin, China; Klang valley, Malaysia; etc., which can be considered to be lowlands.

LOWLANDS CREATED BY GROUNDWATER WITHDRAWAL

Even though the deltaic clays are of low permeability, large quantities of groundwater are pumped out from the underlying sand strata. Excessive withdrawal of groundwater results in subsidence, consequent flooding and salt water intrusion into the aquifers as in Bangkok, Saga plain, Venice, etc.

Bangkok, Thailand, is located about 48 km from the mouth of the river Chao Phraya where it meets the Gulf of Thailand. The entire city and the surrounding areas are flat and at an elevation slightly above (1.0–1.5 m) mean sea level. The area affected by subsidence, floods and tidal changes is shown in Figure 2.22(a) (Bergado et al. 1988). Regular flooding of the area is because of (i) high

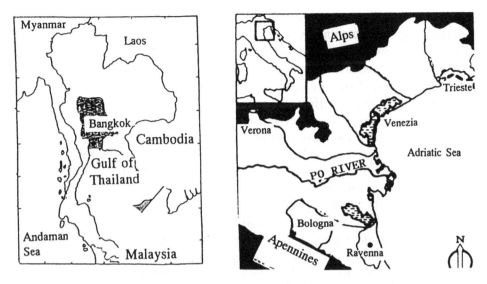

Figure 2.22. Lowlands created by groundwater pumping (a) Bangkok, Thailand and
          (b) Adriatic coast, Italy.

discharge of Chao Phraya river; (ii) high tides from the river mouth; (iii) heavy
rainfall in the city; and (iv) subsidence due to groundwater withdrawal
(Tingsanchali 1988). Subsidence rates of 10 cm/year have been reported
(Bergado et al. 1988) in some areas. Thus, widespread subsidence coupled with
heavy rain and high tides from the sea aggravate the problems of Bangkok. The
typical soil profile consists of alternate layers of clay and sand with gravel upto
a depth of 1000 m. The top most layers comprise of 2.0–2.5 m thick weathered
crust, 5.0–6.0 m thick very soft to soft clay and about 5.0 m thick stiff clay.
The city of Bangkok and the adjoining areas are protected by (a) 60 km long
flood bypass channel, (b) diversion dam, (c) tidal barrier near the river mouth
with a large pumping station and (d) embankment.

Thessaloniki, Greece, is an almost flat area at a mean elevation of 4.0 to 5.0
m above the sea level, and is underlain by marine and lacustrine deposits. Three
main rivers, Aliakmonas, Axios, and Gallikos, drain into the basin. Due to
intense groundwater extraction since the 1950s, extensive subsidence
phenomenon manifested in the area (Andronopoulos et al. 1991). The
groundwater table originally at 3.0 to 6.0 m from the surface has been lowered
to a maximum depth of 40.0 m. The resulting subsidence is reported to have
been caused by the rearrangement of mica platelets from a flocculated to an
oriented structure. As a result, the sea has intruded inland by about 500 m. A
dike was constructed during the 1970s as a protective measure, but its height
had to be raised continually. Presently, the ground is below the sea level and
a pumping system is used for the drainage of the area.

The famous and historic city of Venice, Italy (Figure 2.22b), is also at a lower elevation relative to the sea level. The problem of Venice arises essentially from the subsidence due to withdrawal of groundwater and periodic tides. Subsidence in Venice has not only tectonic origins, but is also due to natural consolidation of sedimentary deposits, lower piezometric levels and rise in sea level (Marchini & Tomiola 1976). Tides are called locally as 'high waters'. Even 90 cm rise in water level significantly affects large parts of the city. Settlements of the order of 10 to 14 cm have been recorded during the period 1952 to 1969. The stratigraphy of the ground consists of layers of sand bounded by silt and clay layers. The lowland surrounding the city of Bologna, Italy (Figure 2.22b), has been subject to subsidence due to exploitation of underground water resources (Balestri & Villani 1991). The land at the foot of Apennines is well compacted and bonded, while the land in front is a lowland formed by the water courses. The composition of the alluvial fan system alternates between sand and gravelly lenticular sediments juxtaposed with finer clayey deposits. The existence of highly compressible material between two sectors with large granular accumulations, leads to anomalous development of subsidence. The magnitude of subsidence in the city center has been more than 20 cm while in some parts it exceeds 1.4 m. The rate of subsidence was as high as 11 cm/year.

Shanghai, China, is situated in the Changjiang (Yangtze) river delta. The alluvial sediments 150–400 m thick were formed during the Quaternary period under conditions of alternating warm and cold climates, with the sea advancing and regressing, and through interactions of river, lake and sea (Bao 1988). A highly compressible and soft clay layer 7–10 m thick exists at a depth of 10–12 m. The natural water content of this soil is greater than its liquid limit implying liquidity index greater than one and low shear strengths. Subsidence in

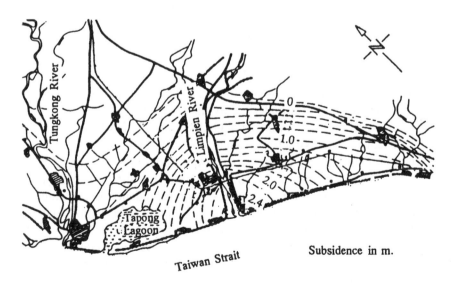

Figure 2.23. Lowland from subsidence in Taipei basin, Taiwan (after Liao et al. 1991).

Shanghai started in 1921, and reached about 2.7 m in 1966 with a maximum annual rate of 98 mm (Gu et al. 1991). Minor subsidence is continuing in spite of reduction in groundwater pumping and artificial recharging of the aquifers.

In Taiwan, significantly large amounts of subsidence (Figure 2.23) are occurring in the Taipei basin, and Chosui river alluvial and Pintung coastal areas since 1950, due to withdrawal of groundwater. The Pintung coastal area (Liao et al. 1991), geographically part of the Lili river alluvial fan, is circled by Linpien, Lili, and Peishui rivers and occupies an area of 73 sq.km. The average groundwater withdrawal rate of 165 million cu.m. per year amounts to 6.19 mm/day. The maximum ground subsidence measured at Wenfong, is 2.43 m for the period 1970 to 1988. The rate of subsidence in the coastal plain of southwest Taiwan is 0 to 3 cm/year. The overall effect of subsidence can be noted from the increase in the area of fishery farms from 6 ha. in 1972 to 980 ha. in 1982. The alluvium is 165 to 200 m thick and consists of poorly graded sand and clayey sands.

LOWLANDS CREATED BY WITHDRAWAL OF OIL AND GROUNDWATER

The eastern edge of Lake Maracaibo, Venezuela, known as Bolivar coast (Figure 2.10), extends over an area of low ground and swamp (Nunez & Escojido 1976). Before the start of withdrawal of oil, the swamps were separated from the lake by a comparatively narrow strip of land which was slightly higher than the mean water level in the lake. These strips of land got inundated during high tides and storms. A number of drainage canals connect the swamps to the lake. The extraction of oil from the Lagunillas, Bachaquero, and Tia Juana fields, resulted in the foreshore to be permanently submerged. Subsidence extended over an area of 452 sq. km. Ground subsidence of more than 4.0 m is attributed to consolidation of soft clay layers and the adjoining unconsolidated oil producing sand layers (Murria 1991). To protect the area from flooding, coastal dikes and a drainage system have been constructed. The dikes have been built in stages and raised according to the projected subsidence. The dike system consists of 44 km of coastal dikes and 58 km of inland dikes. The drainage system has 345 km of ditches and 24 drainage stations.

The Po river, Italy (Figure 2.22b), in its lower reaches, has very flat slopes. The debris from the upper reaches has filled up the river bed and the water level is higher than the surrounding country side (Gambardella et al. 1991). The northern Adriatic coast has a natural subsidence of 15–30 cm per century due to glacial eustatism. Since 1950, this has been aggravated by water and natural gas extraction from layers found upto 600 m in depth. Presently, the maximum lowering of ground has been noted to be 3.5 m and the area covered is about 800 sq.km.

Ravenna is located 60 km south of the river Po and 120 km from Venice

(Figure 2.22b), and is also on the Adriatic sea. The underlying deposits 1500–3000 m thick, consist mostly of sandy and silty clays, deposited under continental, lagoonal, deltaic, and marine environments. The upper multi – aquifer system overlies a transition zone of brackish and salt water below which several gas reservoirs exist. With increased withdrawal of natural gas and groundwater, subsidence of the order of 1.1 m is threatening the industrial zone, the urban districts, historical monuments and the reclaimed marshlands (Gambolati et al. 1991).

Potok (1991) reports nearly 3 m subsidence near Houston, Texas, U.S.A. resulting in permanent flooding of large areas of valuable land close to the coast and substantially increased flooding in areas subject to tidal surges associated with tropical storms. The subsidence is caused initially by withdrawal of oil and gas, and later groundwater. Presently, the rate of subsidence is about 4 cm/year. A region within 32 km of Houston has subsided atleast 30 cm during 1963 to 1978. The maximum subsidence for this period has been 1.5 m. The effect of subsidence on riverine flooding and drainage is reported.

LANDS CREATED FROM THE SEA

Increasing demand for the urban waterfronts is forcing traditional urban activities such as port terminals, water access dependent industries, transportation systems, etc., on to the near shore. As a result, major coastal cities all over the world, especially in land starved countries like Japan and Singapore, are reclaiming large coastal islands to accommodate the above activities, and consequently improve the social, the economic and the environmental conditions of urban development.

Manmade coastal islands

Manmade coastal islands are usually constructed by reclamation which implies build up of the island by deposition of soils or other construction materials until the water surface is penetrated and an island surface is created. Reclaimed coastal islands require long construction time and become very costly if constructed in water depths greater than 20 m. They could suffer from large total and differential settlements, to minimize which, some form or a combination of several ground improvement techniques need to be adopted. To improve the island foundation response, reinforcement of the bottom material by sand, structural or cement hardening piles, sand or gravel mats, etc., are the preferred alternatives.

For developments along the coast, many manmade coastal islands have been created in Japan and Singapore. Some more including one for the new Hongkong airport, are in the offing. In Japan (Figure 2.13), these islands are situated off Tokyo, Osaka, Kobe, etc., and measure more than sixty thousand

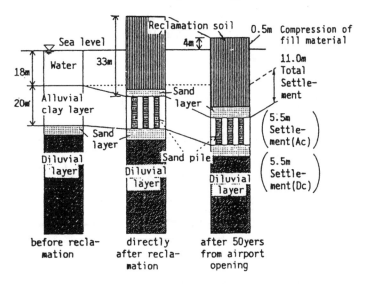

Figure 2.24. Settlement profile at manmade coastal island, Kansai, Japan(after Arai 1991).

hectares in area (Osaka South Port, Port Island, Rokko Island, Nanko Island, Koshien, etc.). The Kansai International Airport would be the first airport in the world built entirely on land reclaimed from the sea, about 5 km from the coast. The manmade island 5.11 sq. km. is being constructed in 18 m of water depth, and involving 180 million cu.m of fill for reclamation (Arai 1991). The sea bed consists of 20 m thick soft alluvial clay and about 400 m thick diluvial stratum consisting of alternating layers of sand and clay. As a result of the construction of the Kansai airport, the vertical stresses increase by 450 kPa. The highest high water level is +3.2 m and the airport is being constructed to maintain a minimum elevation of 4.0 m above this level after accounting for almost 11 m of settlement (Figure 2.24) 50 years from the airport opening. Endo et al. (1991) report that even the underlying diluvium settled by 2.7 m in 21 months from the start of the reclamation work. A similar scale of operation is involved in the expansion of the Tokyo International Airport at Haneda (Katayama 1991). Many geotechnical problems are associated with manmade coastal islands due to the nature of the sea bed. Large deformations both vertical and lateral, differential settlement, etc., not only govern the design and construction of the structures but affect the adjacent waterfront structures.

In Singapore (Figure 2.25), more than 5000 ha of land has been reclaimed from the coast from 1964 onwards. For the Changi International Airport, about 700 ha of land was reclaimed from the sea. The second runway and the terminal buildings were constructed on the reclaimed land. A shallow stratum of loose sand and gravel overlies cemented sandstone and shale in the center of the reclamation site, while marine clay 2 m to 40 m thick fills the channels cut into the sandstone in the north and south of the site. In the early stages hill cut

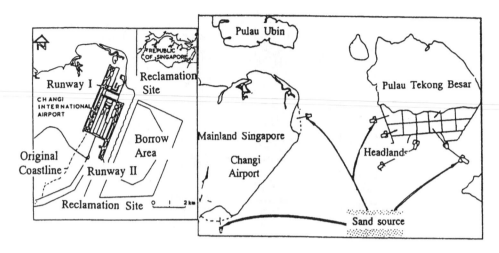

Figure 2.25. Reclamation works in Singapore.

hydraulic fill available locally were utilized. Subsequently, a new technique of layered clay – sand reclamation scheme, wherein a thin layer of sand is spread on the surface of soft clay slurry, has been developed (Lee et al. 1987), to reduce the length of drainage path and to increase the strength of the clay layers. The process of alternate placement of clay and sand is repeated until the desired ground elevation is achieved.

SEISMICITY

A large number of lowlands are located in seismically active zones in the world. Many important structures such as buildings, bridges, tank farms, quay walls, piers, breakwaters etc., are built as port facilities and coastal or flood protection works. Counter measures against seismic damage to these structures are provided by designing them to withstand an anticipated earthquake, minimization of earthquake damage, e.g. liquefaction, by improving the ground conditions, and/or by a disaster preparedness program (Tsuchida 1991). The latter is a program to deal with some expected damage to the structures which is not of catastrophic nature.

Seismic damage to many of the port and coastal structures due to liquefaction of sandy deposits, is one of the most common occurrences. Therefore, the potential for liquefaction of the soils should be assessed and appropriate measures taken to prevent or minimize the damage arising from it. For earthquake resistant design, the effects of seismic inertial forces and displacements on the stability and performance of the structures and lifeline support facilities are to be considered. The importance of preparing people living in areas subjected to natural disasters such as floods, cyclones (typhoons), earthquakes, etc., is being realized by many agencies. These programs are relevant especially for lowlands where natural disasters have a compounding effect.

CONCLUDING REMARKS

Defining lowlands as lands affected by fluctuating water level, a brief summary is presented of the different types of lowlands of the world such as riverine, deltaic, tidal, lacustrine, etc. The characteristics, geomorphology and the geotechnical aspects of lowlands are discussed. They are also categorized as natural lowlands and those created by human activities such as withdrawal or extraction of natural resources such as groundwater, oil, gas, coal, etc. Manmade coastal islands created out of sea for various purposes are also subject to the same kind of problems as other lowlands. Some typical examples of lowlands are presented.

REFERENCES

Altaf Ali, A.H.M. 1991.Lack of land use planning and natural disaster. *Proc. int. meeting on disaster and human settlements in urban areas in developing countries, Toyohashi, Japan:*5–30.

Andronopoulos, B., D.Rozos & I.Hadzinakos 1991. Subsidence phenomenon in the industrial area of Thessaloniki, Greece. *Proc. 4th int. symp. land subsidence, Houston:*IAHS publ.200:59–69.

Ankum, P., K.Koga, W.A.Segeren, & J.Luijendijk 1988. Lessons from 1200 years impoldering in the Netherlands. *Proc. ISSL:*101–108.

Arai, Y. 1991. Construction of an artificial offshore island for the Kansai International Airport. *GEO–COAST '91:*2:927–943.

Balestri, M. & B.Villani 1991. Study of subsidence in the Bolognese area. *Proc. 4th int. symp. land subsidence, Houston:*IAHS publ.200:71–80.

Bao, C.M. 1988. Geotechnical properties of Shanghai soils and several engineering practices. *Proc. ISSL:*185–194.

Benjamin Jr, B.R. 1990. Properties of soils in Metro Manila. *Proc. ISGWPL:*33–42.

Bergado, D.T., C.L.Sampaco, B.N.Lekhak & P.Nutalaya 1988. Lowland disaster prevention technology–study of AIT campus ground subsidence, Chao Phraya Plain, Thailand. *Proc.ISSL:*125–134.

Brammer, H. 1990. Floods in Bangladesh: I.Geographical background to the 1987 and 1988 floods. *The Geographic J.:*156(1):12–22.

Brenner, R.P., P.Nutalaya, G.V.Chilingarian, & J.O.Robertson 1981. Engineering geology of soft clay. In E.W.Brand & R.P.Brenner (Eds), *Soft clay engineering:*159–240.Elsevier Publ. Co., Amsterdam.

Clark, J.A., W.E.Farrell & W.R.Peltier 1978. Global changes in postglacial sea level: a numerical calculation. *Quaternary Research:*9:265–287.

Cotecchia, V. 1986. Subsidence phenomena due to earthquakes: Italian cases. *Proc. Venice symp. on land subsidence:*IAHS publ.151:829–840.

Cox, J.B. 1970. The distribution and formation of recent sediments in Southeast Asia. *Proc. 2nd Southeast Asian conf. soil engrg., Singapore:*29–47.

Endo, H., K.Oikawa, A.Komatsu & M.Kobayashi 1991. Settlement of diluvial clay layers caused by a large scale man–made island. *GEO–COAST '91:*1:177–182.

Fairbridge, R.W. 1961. Eustatic changes in sea level. *Physics and Chemistry of the Earth:* 4:99–186.Pergamon Press, London.

Finnemore, E.J. & M.L.Gillam 1976. Compaction processes and mathematical models of land subsidence in geothermal areas. *Proc. second int. symp. on land subsidence, Anaheim:* IAHS publ.121:157–166.

Gambardella, F., S.Bortolotto & M.Zambon 1991. The positioning system GPS for subsidence control of the terminal reach of the Po river. *Proc. of 4th int. symp. land subsidence, Houston:* IAHS publ.200:433–441.

Gambolati, G., G.Ricceri, W.Bertoni, G.Brighenti & E.Vuillermin 1991. Numerical analysis of land subsidence at Ravenna due to water withdrawal and gas removal. *Proc. 4th int. symp. land subsidence, Houston:* IAHS publ.200:119–128.

Geyh, M.A., H.R.Kudrass & H.Streif 1979. Sea level changes during the late Pleistocene and Holocene in the Strait of Malacca. *Nature:*278:441–443.

Gu, X.Y., S.I.Tsien, H.C.Huang & Y.Liu 1991. Analysis of Shanghai land subsidence. *Proc. 4th int. conf. on land subsidence, Houston:*IAHS Publ.200:603–612.

Katayama, T. 1991. Meeting the challenge to the very soft ground – the Tokyo International Airport offshore expansion project. *GEO–COAST '91:*2:954–967.

Kenney, T.C. 1964. Sea level movements and the geological histories of the post glacial marine soils at Boston, Nicolet, Ottawa and Oslo. *Geotechnique* 14:203.

Kern, K. 1992. Reclamation of lowland rivers: the German experience. In P.A.Carling & G.E.Petts (Eds),*Lowland floodplain rivers: geomorphological perspectives:*279–297. John Wiley & Sons, New York.

Klepper, O. & Asfihani 1990. Legend to the reconnaissance soil map of the S.Negara Basin, South Kalimantan. *Proc. workshop on the conservation of the Sungai Negara wetlands, Asian wetlands bureau, Bogor, Indonesia:*95–125.

Lee, S.L., G.P.Karunaratne, K.Y.Yong & V.Ganeshan 1987. A field test for layered clay–sand reclamation scheme. *Proc. 8th A.R.C., Kyoto:*1:57–60.

Liao, J.S., K.S.Pan & B.C.Haimson 1991. The monitoring and investigation of ground subsidence in Southwest Taiwan. *Proc. 4th int. symp. land subsidence, Houston:*IAHS publ.200:81–96.

Marchini, S. & A.Tomiola 1976. The use of mudjacking for the upheaval of urban zones, computer control of the works, experimental application to the problem of Venice. *Proc. 2nd int. symp. land subsidence, Anaheim:*IAHS publ.121:83–94.

Mitchell, J.G. 1992. Our disappearing wetlands. *National Geographic:*182(4):3–45.

Miura, N., Y.M.Park & H.I.Kim 1990. Geotechnical properties of soft soils in Kyunggi, Korea. *Proc. ISGWPL:*25–32.

Miura, N., Y.Taesiri, A.Sakai & M.Tanaka 1988. Land subsidence and its influence to geotechnical aspects in Saga plain. *Proc. ISSL:*151–158.

Murria, J. 1991. Subsidence due to oil production in western Venezuela: engineering problems and solutions. *Proc. 4th int. symp. land subsidence, Houston:*IAHS publ. 200:129–139.

Nunez, O. & D.Escojido 1976. Subsidence in the Bolivar coast. *Proc. 2nd int. symp.*

land subsidence, Anaheim:IAHS publ.121:257–264.

Oele, E. 1986. Coastal lowland development on a firm basis. *Proc. symp. on lowland development in Indonesia, Jakarta:*supporting papers:87–101.

Ohtsuki, Y., K.Tanaka & M.Sato 1981. Enclosing embankment on soft ground – Hachirogata tidal reclamation project. *Proc. 10th ICSMFE, Stockholm,* case history vol:555–598.

Onitsuka, K. 1988. Mechanical properties of the very sensitive Ariake clay. *Proc. ISSL:*159–168.

Pamukcu, S., J.K.Poplin, J.N.Suhayda & M.T.Tumay 1983. Dynamic sediment properties, Mississippi delta. *Proc. spec. conf. on geotechnical practices in offshore engineering., ASCE:*111–132.

Perera, A.K.S.A. 1988. Reclamation of low–lying marshy land near the city of Colombo. *Proc. ISSL:*135–138.

Perera, R. 1991. Impact of coastal erosion on the urbanization trend of Colombo – Galle growth corridor in Sri Lanka. *Int. meeting on disaster and human settlements in urban areas in developing countries, Toyohashi, Japan:*95–109.

Potok, A.J. 1991. A study of the relationship between subsidence and flooding. *Proc. 4th int. symp. land subsidence, Houston:*IAHS publ.200:389–396.

Risseeuw, P., R.Volders & H.Teunissen 1986. Soil Improvement by preloading and vertical drains, Port Kelong Power Station, Malaysia. *Proc. 3rd int. conf. on geotextiles, Vienna:*3:599–604.

Shafi, S.A. 1991. Disaster & human settlements in urban areas: Flood in Dhaka city–its causes and effects. *Int. meeting on disaster and human settlements in urban areas in developing countries, Toyohashi, Japan:*122–141.

Subagio, H. 1991. Evaluation of engineering properties of Banjarmasin clay, Indonesia. *GEO–COAST '91:*1:93–98.

Supriyanto, H. 1990. Swamp reclamation projects in Indonesia: An integrated approach to rural development–a case study of the Rawa Sragi Project, Lampung province. *Proc. 2nd seminar on ICATDC, the development of reclamation technology and the utilization of reclaimed Land, Kwangju, Korea:*129–137.

Tingsanchali, T. 1988. Flood control investigations in lowland areas in Southeast Asia. *Proc. ISSL:*55–66.

Tsuchida, H. 1991. Disasters caused by earthquakes in coastal areas and countermeasures. GEO–COAST '91:2:918–926.

Verstappen, H.Th. 1986. Rural land use, urban stress and environment in coastal lowlands. *Proc. symp. on lowland development in Indonesia, Jakarta:*supporting papers:491–502.

Watanabe, K. 1988. Over view of the Ariake Sea and its surrounding areas. *Proc. ISSL:*1–8.

Weerakkody, U. 1992. The Holocene coasts of Sri Lanka. *The Geographic J.:*158(3): 300–306.

GEO–COAST '91: Proc. of the International Conference on Geotechnical Engineering for Coastal Development, Yokohama, Japan.

ICSMFE: International Conference on Soil Mechanics and Foundation Engineering.

ISGWPL: International Symposium on Geotechnical and Water Problems of Lowland, Saga, Japan.

ISSL: International Symposium on Shallow Sea and Low Land, Saga, Japan.

3 HYDROLOGICAL AND ENVIRONMENTAL CHARACTERISTICS

K. KOGA[*], H. ARAKI[*] and W. LIENGCHARERNSIT[**]
* Saga University, Saga, Japan
** Kasetsart University, Bangkok, Thailand & Saga University, Saga, Japan

INTRODUCTION

Hydrological and environmental characteristics in lowland areas are more or less different from other regions. Most lowlands are located in the downstream parts of river basins which receive runoffs from the whole watersheds prior to discharging into the seas or lakes. Hydrologic regime in a lowland river is influenced by watershed characteristics as well as tidal phenomena in the sea. Since topography in the lowland is rather flat, deposition of suspended sediment usually occurs and periodic channel dredging is needed to enable navigation. Tidal fluctuation in the sea also has an influence on pollutant dispersion in the river. Saltwater intrusion into upstream rivers and coastal groundwater aquifers is also one important issue in coastal lowland areas, since it affects the available sources of water supply. Usually, the drainage capacity of the lowland is poor due to its flatness. As a result, flooding always occurs when there is heavy rainfall in the catchment area of the basin. Flood protection and drainage works are very important for lowland development. Water pollution in lowland rivers and coastal waters is also a significant issue since most lowland areas have dense population. In addition, some other phenomena such as land subsidence due to excessive extraction of groundwater, sea level rise resulting from global warming, etc., will cause greater impacts on the lowland areas than other regions. Therefore, it is necessary that the problems in lowland development and management should receive high attention from the decision makers, especially those who are concerned with water resources and environmental planning and management.

In this chapter, general features of hydrological and environmental characteristics in lowlands are described with an aim at providing background information on some specific water resource and environmental issues normally encountered in the lowland areas. In addition, water quantity and water quality management systems usually employed to handle these problems are introduced. Some related topics including self−purification of lowland water, restoration of rivers and floodplains, lake eutrophication, etc. are also mentioned.

HYDROLOGY

Precipitation

Hydrologic regime of a region is closely related to climatic condition, particularly precipitation which is the original source of surface water and groundwater. Precipitation may occur in a variety of forms including rainfall, snow, hail, freezing rain, etc. Precipitation characteristics in various parts of the world are different, either in terms of its form, intensity, or distribution pattern. The complex pattern of precipitation depends upon several interacting influences such as the latitude of the region, geography of surrounding areas, distance from a water source, as well as regional environmental conditions. In general, precipitation decreases with increasing latitude because decreasing temperatures reduce atmospheric moisture (Linsley & Franzini 1979). Among various forms of precipitation, rainfall is of great concern for sanitary and hydraulic engineers, since it is directly related to runoff characteristics of the areas.

Rainfall measurement can be made either by using the U.S. Weather Bureau standard gage, the weighing–bucket gage, or the tipping–bucket gage. The standard gage is used to measure manually the depth of rainfall over 24–hour period. The weighing–bucket gage measures the accumulated weight of rainwater in a bucket. It is normally connected with a recorder which provides a continuous strip chart record. The tipping–bucket gage records the amount of rainfall by counting the times the bucket tips. It consists of a rainfall collector which is funneled into a two–compartment bucket. Each bucket is designed to collect the equivalent of 0.01 in. of water over either an 8–in. or 10–in. diameter collector. Once one of the buckets is filled, the filled bucket will tip and empty its contents. The bucket on the other side is now in the position to collect water from the funnel. When the bucket tips, an electrical signal is sent to a recording unit (Wanielista 1990).

Hydrologists and engineers are concerned mainly with two kinds of rainfall information, i.e. records of total amount of rainfall in some fixed periods, e.g. days, months, and years; and records of intensities and durations of individual storms. In water resources planning and management, time variations in rainfall are more important than regional variations. Data on monthly variations of rainfall are useful for irrigation planning and selecting suitable cropping patterns, whereas variations in rainfall from year to year make it important to have a reservoir to store water for the year of low rainfall. Records on rainfall intensities and durations will be useful only when they are related to frequency of occurrence. With records from many individual storms, the generalized intensity–duration–frequency relationship can be determined. The obtained relationship can be expressed in graphical form or in the form of empirical equation. The rainfall intensity–duration–frequency relationship is useful for the design of storm sewers and other drainage facilities.

As the lowlands are normally located in the downstream parts of the river basins, rainfall patterns in the whole basin will affect the hydrologic regime of the areas and must be taken into consideration in water resources planning.

Evapotranspiration

More than half of the precipitation which reaches land areas of the earth will return to the atmosphere by the combined evaporation and transpiration processes, the so–called evapotranspiration. Evapotranspiration rates vary from place to place and from time to time depending upon air temperature, relative humidity, wind, surface water, soil moisture, as well as type of vegetation cover. Normally, the rate of evaporation from a water surface is proportional to the difference between the vapor pressure at the surface and the vapor pressure at the overlying air. Turbulence caused by wind and thermal convection will greatly increase the evaporation rate.

Measurement of evaporation can be made by using evaporation pans. The U.S. Weather Service Class A pan is the most common. This pan is 4 ft (1.22 m) in diameter and 10 in. (25.4 cm) deep (Linsley & Franzini 1979). Evaporation from the pan is measured daily with a hook gage. Determination of evaporation loss from a storage reservoir or lake can be made by using data from the evaporation pan. However, the rate of evaporation from a pan is considerably different than that from a lake surface due to differences in actual conditions including surface turbulence, heat transfer, etc. Therefore, in estimating evaporation loss from a lake using the evaporation pan data, some correction factor known as a pan coefficient must be considered.

Measurement of transpiration for small plants can be made in a closed container or greenhouse in which humidity changes are measured. The soil surface should be sealed to prevent evaporation from the soil, otherwise some corrections are needed. Precise determinations of transpiration are difficult, and extrapolations to other areas can be misleading (Wanielista 1990). The water balance analysis is one method of estimating monthly or yearly evapotranspiration in a basin provided that other variables can be estimated.

Seasonal variations in evapotranspiration are normally much less than variations in precipitation and streamflow. In tropical regions where evapotranspiration rates are rather high, extreme maximum and minimum monthly evapotranspirations are about 150 and 70 percent, respectively, of the average monthly value.

Evaporation may consume a large portion of water stored in reservoirs. Loss of water through evapotranspiration is an important factor to be considered in water resources development planning.

Infiltration

When rainwater reaches ground surface, some portion will infiltrate into the ground. The infiltration rate depends upon soil texture, soil moisture, ground cover and drainage condition, depth of water above the ground or rainfall intensity. The infiltration rate in sandy soil can be as high as 10,000 times or more than that in clayey soil. At the beginning of rainfall, when soil moisture is low, the infiltration rate is usually high. The infiltrated water will gradually fill the pore space of the soil until it becomes saturated. Then, the infiltration rate will be proportional to the hydraulic gradient of the soil following the Darcy's law. The excessive rainfall will accumulate on the ground or drains as surface runoff to lower areas.

Measurement of the infiltration rate can be made by conducting laboratory tests or field experiments. The results obtained from the field experiments are usually more reliable than those from the laboratory, since in the laboratory the soil sample is rather small and disturbed, and the testing conditions are often different from those in the field.

Surface runoff

The portion of rainwater that flows overland to the nearest stream is called surface runoff. Normally, this water reaches the stream and drains out from the basin in a short period. Surface runoff characteristics depend upon many factors including rainfall intensity, duration, and distribution pattern; topography, size, shape, and ground cover of the catchment area; and soil texture and permeability. In some regions, surface runoff is derived from snowmelt and the runoff characteristics also depend upon the heat available for snow melting. Correlation between rainfall and runoff in a given area can be determined if adequate records of rainfall data and runoff measurements are available. The simplest correlation is a plot of average rainfall versus runoff. The proportional factor between runoff and rainfall is called runoff coefficient which is dependent on area characteristics.

Determinations of streamflow can be made either from stage records which can be converted into flow rates, from velocity measurements using a current meter or a float, from an outflow control device such as a broad–crested weir or a flume, or by using some empirical equations.

In estimating storm runoff or flood flows for engineering designs, one of the following methods can be applied: 1) statistical analyses based upon observed records of adequate length; 2) statistical augmentation of available information; 3) rational estimates of runoff from rainfall, of which two are of general interest, i.e., the rational method and the unit hydrograph method; and 4) calculations based on empirical formulations not devised specifically from observations in the design area but reasonably applicable to existing watershed conditions (Fair et al. 1966).

Flow characteristics of lowland waters

A lowland river or estuary normally receives runoff from the whole watershed of the river basin. Therefore, distribution patterns of rainfall, land use, water use, vegetation cover, soil characteristics, topography, hydrogeology, and all water resources development projects in the main river and its tributaries will have an influence on the flow characteristics of the lowland river. In the estuary which drains runoff from the river basin into the sea, flow characteristics are also influenced by tidal fluctuation in the sea. Not only does the flow pattern vary from time to time, but also water quality characteristics in the estuary always change. During the spring tide, sea water will intrude the upstream reach of the estuary, while during the ebb tide, the saline water will be drained back into the sea. With the tidal effect, dispersion phenomenon in the estuary is much different from a free flowing river.

The lowland river usually has a very flat slope, so flow velocity is rather low while the discharge is high, since it receives water from most parts of the basin. Flow pattern in the downstream reach of the lowland river is mainly governed by tidal fluctuation which characterizes the back–and–forth flow condition. Usually, the flow across any cross–section of the downstream reach of the lowland river is much higher than the net freshwater discharge from the upstream river.

Open channel network

In some lowland areas there exist open channel networks which have been constructed for some purposes including irrigation water supply, navigation, drainage, etc. These open channel networks are found to be an effective drainage facility in the areas which are often subject to heavy rainfall. They can serve as retarding reservoirs for flood mitigation. In some areas, these open channel networks are used as an intermediate water body for receiving wastewater from the communities prior to discharging into the main river. In order to serve their main functions properly, these open channel networks must be carefully designed, constructed, and operated. The factors that must be taken into consideration include water sources, topography, land use, water demand, rainfall and runoff characteristic, as well as water level fluctuation in the main river or the sea which receives runoff from these open channel networks.

Groundwater

Groundwater is one main source of water supply either for domestic consumption, industrial or agricultural use. In the lowland areas, there usually exist good groundwater aquifers. Rainfall in the catchment area may penetrate the surface soil directly to the groundwater aquifers or may enter the streams and percolate from these channels into the aquifers. Groundwater will slowly flow to the area with lower piezometric head which usually occurs in the lowland area where the elevation is lower and there is intensive extraction of groundwater for various purposes.

In the deltaic plain where the subsurface soil is formed by deposition of river sediment, the level of groundwater table in the upper aquifer is normally high compared with ground surface. This condition together with low permeability of clayey soil results in low rate of water seepage to the underground zone. This property affects drainage characteristics of the area and also affects the function of seepage pits used in the on-site wastewater treatment and disposal.

In deltaic and coastal areas where population density is great and there exist many human activities, the problem of saltwater intrusion into groundwater aquifers has frequently occurred. This is caused by excessive withdrawal of groundwater which results in lowering of piezometric head in the groundwater extraction zone, then, seepage of saltwater into the groundwater aquifers occurs. The problem of saltwater intrusion is not always immediately recognized, but once it is obviously observed it is often evident that intensive and costly measures are required to solve the problem. In some regions, many groundwater wells have been abandoned due to their saline water. In order to prevent this phenomenon, the rate of groundwater extraction must be reduced, or in some cases, the location of groundwater extraction zone must be changed.

Groundwater plays an important role in water resources development and management. In many regions it is the only source of water for various uses. The large volume of water stored in the aquifers can serve as a buffer which supplies water during the drought periods. Seasonal fluctuations in groundwater levels and storage volumes are fairly small compared with surface water resources. While regulations of surface water resource requires costly hydraulic structures such as dams, weirs and storage ponds, regulations of groundwater resources can be made by some appropriate management schemes such as a proper distribution of groundwater extraction wells and recharge wells. As regards water quality, groundwater quality is normally better than surface water quality since it is less susceptible to man-made pollution and due to the natural purification process – subsurface filtration. Many groundwater wells provide water of good quality which can be directly consumed without any treatment.

Actually, surface water and groundwater are closely interrelated. During flood periods, some portion of the stream water will infiltrate into the underlying groundwater aquifers through the stream banks. In the dry periods, water released from groundwater aquifers serves as a base flow in the stream. Spring discharge is one example of groundwater emerging at the ground surface and becoming surface runoff. Thus, the management of regional water resources should consider both surface water and groundwater simultaneously.

WATER QUALITY CHARACTERISTICS

Characteristics of water quality in lowlands

Most lowlands are the downstream part of a river basin which collects surface runoff from the whole catchment area prior to discharging into the sea or other receiving water bodies. Most of the pollutants from various sources in the basin will find their ways to reach the downstream rivers or estuaries. In addition, most lowlands are densely populated areas with several kinds of human activities being undertaken and large amounts of wastes are generated. As a result, water quality in the rivers or estuaries in lowlands is usually poorer than in other parts of the regions. Nowadays, water pollution problem has become one of the major environmental issues in most lowland areas.

In the lowland coastal area, not only does the flow pattern in the estuary be influenced by tidal fluctuation, but also estuarine water quality always changes from time to time. Fluctuation in flow velocity in the estuary due to tidal phenomena affects dispersion of the discharged pollutants. During the spring-tide period, these pollutants can propagate to the upstream river for a considerable distance, while during the ebb-tide period the pollutants are transported back to the downstream river and are finally drained out into the sea. Fluctuation in estuarine water quality has great influence on the ecosystem and also affects water uses in the area.

Saltwater intrusion into estuaries and groundwater aquifers in the coastal area is one important issue in water resources development. A lot of groundwater wells have been abandoned due to their saline water. Saltwater intrusion into the coastal aquifers is caused by overpumping of groundwater which results in a decrease in piezometric head and the interface between freshwater and saltwater moves upward from the seaside. Farther intrusion of saltwater into the upstream river usually occurs during the period of low freshwater discharge which may cause damage to agricultural products and restrict water uses.

In the lowlands, it can be also pointed out, as a general characteristic, that water quality tends to become eutrophic in open channel networks, lakes, storage reservoirs, ponds, etc. in which water is stagnant due to low flow velocity. Eutrophication will become more severe if remarkable amount of fertilizer or nutrients is discharged from farm lands or households.

Water pollution

Nowadays, conservation and enhancement of environmental quality are of major concern of worldwide public. Among various environmental issues, water pollution problem has received great attention. It is evident that water quality in most natural streams, rivers, lakes, estuarine and coastal waters has been deteriorated to some extent due to disposal of wastewater resulting from human activities. Pollution of surface water and groundwater is a threat to public health, aquatic ecology, aesthetic value, and various types of water uses. To conserve and improve the water environment, it is necessary to protect the available water resources from any acts, such as the disposal of organic and inorganic pollutants, and any form of the harmful substances.

A variety of polluting substances may enter a receiving water body or public water supply. These include organic materials from households, commercial establishments and industries, toxic substances from mineral productions, heavy metals in industrial wastewater and sludge, pesticides from farm lands, poisonous leachate from solid waste landfill sites, etc. Some of these substances may cause acute poisoning in human beings and animals, but some may accumulate in cell tissues and be responsible for chronic poisoning.

Water–borne diseases can be caused by several types of microorganisms including bacteria, protozoa, fungi, viruses, and worms. Some of them require some intermediate hosts in water to complete their life cycle. Others are transmitted by water from man to man.

Besides toxic contaminants and pathogenic microorganisms, some minerals in water may possess laxative properties. Highly mineralized or hard water consumes much soap before lather is formed and also lays down scales in a boiler when it is heated. Water containing high magnesium and sulfate ions can cause intestinal effect. Though fluoride in water can help reduce dental caries, too high fluoride concentrations may cause dental enamel. Though there is no evidence that large amounts of natural organic matter in water are harmful, bad tastes, odors and appearance are accepted as warnings of danger.

Seepage of wastewater from various sources into groundwater aquifers will deteriorate groundwater quality. However, even under undisturbed conditions, and without human's intervention, groundwater already contains a certain amount of dissolved matter, sometimes reaching levels which render the water unsuitable for certain usages (Bear 1979). For example, significant amounts of carbonate can be dissolved in groundwater that flows through carbonate rocks. Intrusion of saltwater into coastal aquifers is also one type of groundwater pollution.

Self–purification of lowland waters

Self–purification of natural water plays an important role in water pollution control. In water quality management, the engineers should understand the significant forces of self–purification and recognize the limitations of this natural mechanism, in order to decide how much pollutants can be discharged into the receiving water without causing objectionable conditions.

The natural forces of purification are closely related and mutually dependent. When organic pollutants are discharged into a stream, the initial effect is to degrade the physical quality of the water. As decomposition becomes active, a shift to chemical degradation is biologically induced. At the same time biological degradation also takes place. When the energy values of the discharged pollutants are used up, the biochemical oxygen demand is decreased and the rate of oxygen utilization falls below the rate of absorption from the atmosphere again, which results in an increase in dissolved oxygen concentration and the stream water is gradually returned to its normal purity (Fair et al. 1968).

The self–purification process of polluted groundwater differs from that of surface water. The variety of microorganisms which utilize organic matter for food is restricted in the confinement and the darkness of the pore space of the soil (Fair et al. 1968). The filtration mechanism plays an important role in groundwater self–purification. In general, the rate of self–purification of groundwater is much faster than that of surface water.

In addition to the physical, chemical and biological degradation of pollutants in water, turbulent dispersion also has significant effect on self–purification. In the turbulent stream, the discharged pollutants can disperse away from the discharging point in a short period resulting in significant dilution, while the rate of absorption of oxygen from the atmosphere is rather high due to renewal of water surface. On the other hand, the natural dispersion in lakes and coastal waters is often poor. In the absence of wind and tide there is little turbulent mixing. In tidal estuaries, dispersion is mainly influenced by tidal current. However, difference in the densities of saltwater and freshwater may cause saltwater creep, i.e. the upstream progress of a tongue of saltwater moving inland while overriding freshwater still pours towards the sea. This phenomenon will affect dispersion and is important as a concern for the disposal of wastewater.

WATER MANAGEMENT

Water use

Many large cities in the world are located in lowlands which are formed by deposition of river sediments as it drains into the sea. These areas are normally flat plains with fertile soils. So most ancient civilizations settled and developed them as agricultural lands. The existing water resources in the areas were mainly utilized for farming. Later, as population increased, these agricultural communities gradually developed to urban areas with many human activities. As a result, the demand for water for various purposes including domestic consumption, commercial and industrial uses has increased. At present, in many lowland areas, shortage of water supply is a major problem, and allocation of the available water resources for various types of water use is an essential task.

Agricultural water demand depends upon size of cultivated lands, climatic conditions, existing irrigation facilities, types of crops and cropping patterns, farming practices, as well as water cost and metering system. In some regions where water control is not efficient, farmers may withdraw water directly from nearby irrigation canals to their farm lands, which is a cause of imbalance in water use, i.e. farmers in the upstream part of the irrigation system can be supplied with excessive water while those in the downstream part lack water. Collection of water charge is not possible in many countries due to socio-economic and political problems.

Domestic water demand depends mainly upon size, characteristics and economic status of population, as well as climatic condition, water cost and metering system. The demand can vary from less than 50 liters per capita per day in poor areas to as high as 2000 liters per capita per day in wealthy areas.

Commercial and industrial water demands depend upon types, sizes and characteristics of commercial establishments and industries, and also depend upon water cost and metering system. Manufacturing plants often require large amounts of water, and the water demands are usually proportional to their production. The location of industry is often greatly influenced by the availability of water supply, particularly those which require large amounts of water. Industry is an important factor which affects the per capita water consumption.

Water cost and metering system are major factors that affect water use. If water cost is high, people will be more conservative in water use. Some commercial establishments and industries may develop their own water supply systems to reduce costs. With good metering system, people will be more aware of water loss due to leakage and unintentional use. This can help reduce the amounts of water use to a remarkable extent.

Another aspect which affects water consumption is system management. In those regions where the system management is poor, water losses through leakage and unaccounted water use may be as high as 50 percent of the total water supply. Management of the water supply system includes detection and repair of leaks in the water distribution system, prevention and detection of unauthorized connections, efficient collection of water charge, etc.

Control of water use is necessary especially in those areas with limited water supply. Too high pressure in water distribution network will increase water use. Therefore, the distribution system which includes pumping stations must be properly designed and managed. Effective metering and water charging systems can help reduce water use to some extent. Attempts should be provided to minimize water losses through leakage from the system and unauthorized connections. In some countries, some water laws are issued for the purposes of water use control.

Storage and runoff control

Naturally, runoff from a river basin is stored, in varying degrees, in lakes, ponds, stream channels, subsurface soils, etc. During wet periods these natural storages will usually be filled, while during the dry periods some water will be released and can help reduce severity of low flow in the downstream river. Usually, the capacities of these natural storages are not high enough, and so great seasonal variation in runoff volume in the river exists. For those areas with high population density, the water demand may exceed the available river flow during the dry periods. Therefore, it is necessary to create some storage facilities for storing excessive runoff occurring in the wet season and gradually releasing to cope with the high water demand in the dry season. There are many types of storage systems. If topographic and geologic conditions are suitable, storage dams can be constructed. Sometimes, storage ponds are excavated in lowlands adjacent to streams. Control facilities may also be constructed on outlets of lakes, ponds and stream channels to regulate runoff from these natural storages.

Storage works for runoff control can serve some of the varied purposes including water supply for irrigation, domestic, commercial and industrial uses, dilution of polluted water, control of saltwater intrusion in downstream estuary, flood mitigation, water level control for navigation and recreation, preservation of fishery and aquatic resources, as well as electricity generation. The greatest net benefit is usually derived from a judicious combination of reservoir functions in multipurpose developments. However, the main purposes of the developments must be fully justified and always recognized in the planning and design as well as in the operation phase.

The main factors to be taken into consideration in the reservoir development projects include water demand, hydrologic feature, topographic and geologic characteristics, existing land use patterns in the proposed reservoir area, as well as the impacts anticipated to occur from the development.

Flood routing

When floodwater flows through a storage reservoir, retarding basin, or natural stream without additional tributary inputs, the flood peak is more or less decreased. The flood peak reduction is normally called flood attenuation or flood routing. This reduction is due to the available storage between the upstream and downstream sections. Though the reservoir is full and floodwater is released through spillways, some reduction in flood peak can occur due to the availability of additional storage from the rise in upstream water level resulting from the so-called backwater effect. There are several methods for flood routing analysis. The most common method is based on the principle of continuity together with the storage–discharge relationship of the reservoir or stream channel. The continuity equation can be written as

$$S_2-S_1 = I(\mathrm{d}t)-O(\mathrm{d}t) = [I_1+I_2](\mathrm{d}t)/2 - [O_1+O_2](\mathrm{d}t)/2 \qquad (3.1)$$

or in words, the change in storage volume over a given time interval equals the volume of inflow minus the volume of outflow over that time interval. Note that the subscripts 1 and 2 refer to the beginning and end of the time interval, respectively.

From Equation 3.1, the inflows I_1 and I_2 and the initial outflow O_1 as well as the initial storage S_1 are known. So there remain two unknowns O_2 and S_2. With the second equation relating the storage volume with the discharge, these two unknowns can be solved.

Besides the above mentioned flood routing procedure, some other methods are also used. These include the Muskingum method in which the storage is expressed as a function of weighted inflow and outflow, the kinematic routing method in which the continuity equation and a flow equation such as the Manning's equation are used, and the dynamic model in which both continuity and momentum equations are considered. More details on flood routing methods can be found in Linsley & Franzini (1979) and Wanielista (1990).

With these flood routing methods, the effects of storage reservoirs, retarding basins, stream channels and open channel networks on flood attenuation can be analyzed. These models are useful for estimating proper detention or reservoir storage, sizing spillways, pumping stations, drainage channels, as well as evaluating beneficial impact of the reservoir on flood control in the downstream river.

Flood control and drainage

Most lowland areas are downstream portions of the river basins which collect runoff water from the surrounding watersheds. In general, drainage capacity in the lowland is rather poor due to flatness of the area and its low ground surface elevation compared with water level in a nearby river or in the sea. In many lowlands ground surface elevation is even below the sea water level and dikes must be constructed along the river banks and shoreline. With this topographic feature, flooding always occurs when there is heavy rainfall in the watershed area. Sometimes, flash flood occurs through destruction of river banks or dikes which can cause serious damages to life and property.

In coastal areas, flooding can be directly correlated with land subsidence. Sea water level and storm surge level have gradually increased while ground surface is sinking. Increased flooding in most areas is related to changes in land use patterns in the watersheds as well as in the flooded area itself. Destruction of forest lands results in higher rate of surface runoff. Expansion of urban areas also increases the runoff coefficient and is one of the main factors for increased flooding unless adequate drainage facilities are provided.

Many mitigation measures have been applied to control flooding so as to minimize the damages. These include reservoir impoundment to regulate river runoff, construction of a retarding basin to reduce flood magnitude, construction of flood walls along the river banks, diversion of flood water through bypass channels, expansion of discharge capacity of the downstream river to lower flood level, improvement of drainage systems in the protected area, proper management of watershed area, etc. Selection of the most suitable control measure depends on the cause of flooding, hydrologic regime, land use pattern and topographic feature of the area, as well as the available budget. Combination of several flood control methods is normally required, especially in the lowlands where flooding is related to several factors.

Reservoir impoundment will regulate river runoff and thus decrease flood peak. In many reservoir projects, flood control is considered as one of the main purposes in addition to hydropower and irrigation development. For flood control purpose, the impounded reservoir should be located close to the protected area and is operated to cut off the flood peak. In practice, however, selection of the most suitable dam site depends on topographic and geologic features as well as other environmental factors, such as land use, existing valuable environmental resources, socio–economic impact, etc.

Besides the storage reservoir, a retarding basin can be used to reduce flood magnitude. This retarding basin is usually constructed immediately upstream from the protected area. Normally, it is equipped with fixed, ungated outlets which automatically regulate the outflow in accordance with the volume of water in the basin. The outlet can be constructed as a spillway or a sluiceway. The discharge capacity of the outlet for the retarding basin at full storage must not be greater than the maximum discharge that the downstream river can pass without causing severe flood damage. The basin volume must not be less than the flow volume of the design flood minus the volume of water released during the flood. All flow above the safe discharge capacity of the downstream river will be stored in the reservoir, and the stored water is gradually released to recover storage capacity for the next flood.

Construction of a levee or a flood wall to serve as a barrier for protecting land from flood water is one of the oldest and most widely used flood control methods. This flood barrier is usually constructed along river banks or shoreline where overflow often occurs during high discharge or spring tide. Sometimes, the levee must inevitably be constructed across tributary channels which will obstruct drainage of inland storm water to the river or to the sea. The most widely used method to solve this interior drainage problem is to collect water at some low points and pump over the barrier during flooding when gravity flow through outlet gates is impossible. The inland water may also be collected in an open channel on the land side of the barrier and diverted downstream to a point where gravity discharge is possible. Tributary streams are sometimes enclosed in a pressure conduit whose upstream end is at an elevation which permits gravity flow into the main stream at all time (Linsley & Franzini 1979).

Construction of a diversion channel or floodway to bypass flood water is one possible alternative for flood damage reduction. Opportunities for this alternative are usually limited by topography and availability of land area which can be used as the floodway.

Improvement of downstream channel to increase its hydraulic capacity can help reduce flood stage in the upstream river. This can be simply done by removing brush and dredging the deposited sediment. Complete lining and straightening of bends are also undertaken in some places.

Deforestation and urban development in the upstream watershed area are also major causes of flooding in the downstream lowland area. Changing of land use pattern will affect the runoff coefficient. The vegetation cover creates a sort of retarding basin which helps reduce the flood magnitude. Thus, less flood runoff from a well-vegetated area than an urban land bare of vegetation could be expected. Proper watershed management can therefore help reduce flooding problem. This can be done by various means, e.g. reforestation in the bare ground, contour plowing and terracing, water- and soil- conservation measures, etc.

In order to prevent flooding, not only should the facilities such as dikes, reservoirs, or diversion channels be provided, but also the drainage system in the protected area must be designed and operated properly. In the urban area, storm water is usually collected and transported via buried conduits to a point where it can be discharged into a stream or man-made canal. In the lowland area, an open-channel network with a number of pumping stations are commonly used for urban drainage. For an agricultural land, the function of the drainage system includes removal of excess surface water as well as control of groundwater level below the root zone to improve plant growth and reduce accumulation of salts in the top soils.

Groundwater management

As previously mentioned, excessive extraction of groundwater can cause some subsequent problems, such as land subsidence, saltwater intrusion, etc. Pumping of groundwater from each well will produce a drawdown in its vicinity and may affect the pumping of nearby wells. In addition, pollutants entering the aquifer caused by land disposal of wastewater or contamination of a well can be transported at large distances from the original source and affects water quality in other parts of the aquifer. Therefore, it is necessary that exploitation of groundwater must be properly controlled and managed. The management of groundwater resource should be centralized and an appropriate legal and institutional framework is required.

When both surface water and groundwater resources are available, conjunctive use of both is recommended. Each resource should be incorporated in the overall system according to its individual features. Excess surface water in the rainy season can be used for aquifer recharging, while in the dry season, the stored groundwater can be withdrawn to cope with the demand which cannot be adequately supplied by the surface water resource alone.

Groundwater resource management includes determining appropriate locations, rates, and time of pumping, or artificially recharging the aquifer. The quality of pumped water and recharging water must also be considered. The management must be aimed at achieving the specified goals and also satisfy the hydrological and environmental constraints such as specified maximum and minimum groundwater levels, allowable concentrations of certain water quality parameters, limited rate of land subsidence, water demand, available surface water for aquifer recharging, specified length of an intruding sea water wedge, etc. Other constraints including social, political and economic issues must also be taken into consideration.

Usually, the rates of groundwater extraction will be gradually increased due to the growth in population which induce an increase in water demand. Seasonal variations in groundwater utilization are possible depending on the availability of surface water resource. In addition, the seasonal fluctuations of natural replenishment from precipitation and/or streams will serve as an uncontrolled input to the groundwater system. All of these factors will produce an unsteady flow pattern and reserve capacity in the aquifer. These fluctuations must be taken into account in solving the forecasting and management problems. With these fluctuations, a unique optimal management policy for the entire planning period is not possible. Usually, the management of groundwater resource involves determination of the optimal policy for each period based on the state of the aquifer system.

Groundwater development can be made stage by stage as demand arises. A new groundwater well is drilled when actually needed. Therefore, only small investment is required at each stage. This small scale investment makes groundwater development have higher possibility than surface water resource development in which large investment is required for the hydraulic structures at the beginning stage.

Water pollution control

Lowland water resources normally serve multipurpose water uses including being the main sources of domestic, industrial, and agricultural water supply, transportation, aquaculture and fishery, recreation and tourism, etc. Pollution problem in the lowland water will cause great impacts on the existing communities. Therefore, water pollution control is one essential element of the lowland areas.

Public awareness of the consequences of water pollution and the costs of its control should be created in order to get the public support and cooperation in pollution control. To create such awareness, some well–planned public education programs should be prepared and implemented. The public should be convinced that poor water quality will have direct effect on their health and prohibit other beneficial uses. Without strong public support, the water pollution problem is often ignored.

The discharge of all wastewater into the receiving waters must be controlled in a rational manner. In addition to specific wastewater characteristics, some other factors such as discharge location, characteristics of the receiving waters, defined functional uses, and appropriate water quality criteria must be taken into consideration in order to provide adequate protection of the available water resources. For the purposes of pollution control, enforcement by laws are needed. Some regulations and legislation must be established with penalty to those who violate the laws.

In order to prevent intrusion of saltwater into the upstream river, the amount of freshwater flow in the river must be controlled. A storage reservoir may be required for river flow regulation. Control of saltwater intrusion into the coastal groundwater aquifers can be made by reducing the rate of groundwater utilization, i.e. a plan for surface water development is needed to cope with the increasing demand. Artificial recharge of groundwater can help reduce saltwater intrusion in the coastal aquifers.

Water resource systems planning and analysis

Nowadays, greater attention is given to exploitation of natural resources and conservation of the environment, and so public involvement in large development projects has increased. With increased public involvement, the engineers have to change their traditional manner in planning and solving the problems. They have to broaden their perspective and examine a wider range of alternative plans. More detailed information and more systematic approach are required in the planning. Not only should the economic feasibility study be conducted, but also the issues of environmental quality and social welfare are to be considered.

Water resources planning must take into account multiple users, multiple purposes, and multiple objectives. It is impossible to develop a single objective that satisfies all interests, all adversaries, and all political and social viewpoints (Loucks et al. 1981). In the planning stage, a number of reasonable alternatives should be considered; the economic, environmental, social and political impacts from each alternative should be evaluated with adequate details so that the best possible decisions can be made. The systems approach in water resources planning and analysis can help define and evaluate various possible alternatives. Though the systems approach is not restricted to mathematical modeling, models can help evaluate physical and economic consequences of those alternatives of various policies with some different assumptions. The models can be used to identify those management plans or decision variables including engineering designs and operating plans which best meet the objectives.

Two kinds of mathematical models used in water resource systems planning and analysis can be classified, i.e. simulation and optimization models. The simulation model is developed to represent the water resource system characteristics. With some assumed decision variables including system operating policies, the results of objective values can be evaluated. The optimal or near–optimal solutions may be identified by trial–and–error method using the simulation model. However, difficulties often arise when there are a large number of feasible solutions. The optimization model is developed based on some optimization techniques, e.g. linear programming, dynamic programming, quadratic programming, etc. It consists of a set of objective functions and constraint equations which are formulated to incorporate a number of requirements. This model is solved to obtain the optimal solution of the planning problem which is a plan that achieves the maximum or minimum values of the objective functions and satisfies all the constraints. Due to limitation on optimization solution algorithms, the optimization model may not be able to deal with all the complexities and nonlinearities which are easily incorporated in the simulation model.

In applying mathematical models in water resource systems planning and analysis, one should realize the limitation of the models being used. Assumptions used in model formulation and accuracy of the numerical approximation as well as reliability of some input parameters should be recognized. Some input data may be uncertain or controversial, especially those related to future events, which can produce unreliable outputs. Moreover, some qualitative factors such as social and political issues which are also important in the planning and decision–making process are usually ignored in most water resource system models.

Computer tools for water management

With the advancements made in computer technology in the past two decades, a large number of mathematical models have been developed for the purposes of water quantity management. These include modeling studies on rainfall–runoff relationship, water balance, water demand, streamflow and flood routing, groundwater flow and well hydraulics, wave propagation, hydrodynamic circulations in the oceans, seas, coastal areas and estuaries, stratified flows in lakes and reservoirs, reservoir operation simulation, etc.

The design and operation of water resource systems require information on some hydrologic components such as relationship between rainfall and runoff, streamflow at some particular locations, the maximum probable flood at a given return period, etc. Hydrologic models for estimating runoff or flood flows for engineering designs have been developed using some of the different approaches, including statistical analyses based on observed records of adequate length, rational estimates of runoff from rainfall, runoff and soil erosion calculations based on empirical formulae derived from other watersheds with similar characteristics, numerical analyses based on the continuity and momentum equations, etc.

Groundwater flow models are normally formulated based on a number of assumptions, including the assumptions that the porous media are rigid and continuous, the fluid is incompressible, and flow is continuous and irrotational. Most groundwater models have been formulated based on the Darcy's law, which states that the average flow velocity is directly proportional to the hydraulic gradient. The proportionality constant is known as hydraulic conductivity or coefficient of permeability which is a property of the media. The Dupuit assumption has often been employed to obtain simplified formulations. The finite difference method was most widely used numerical procedure in the old models. The recent models are mainly developed by using the finite element method which is recognized as an effective numerical tool for a wide range of engineering problems. Analytical solutions for unconfined and confined flows into groundwater wells can be derived based on the Dupuit assumption and superposition theorem.

Computational methods for wave propagation problems are of great interest for the designers of harbors, offshore structures, and coastal protection works. In general, several phenomena such as breaking of the waves, physical irregularity, viscosity and energy dissipation play a part in water wave propagation problems. However, some assumptions are normally made to simplify the formulations. These models can be of great help to give quantitative information about the phenomenon of wave propagation.

A number of numerical hydrodynamic models have been formulated based on the principles of conservation of mass and momentum. The use of the finite element method in the recently developed models allows for a simple and accurate treatment of irregular boundaries as well as finer resolution in the areas of interest. With the finite element method, the governing partial differential equations are transformed to a set of algebraic equations of which the dependent variables, flow velocities and water depth, are expressed in terms of the values at some nodal points identified in the study domain which is divided into a number of subdomains or elements. With appropriate boundary and initial conditions and with some necessary input data such as bottom topography of the water body, wind magnitude and direction, surface and bottom shear stress coefficients, etc., this set of equations can be solved to obtain the hydrodynamic circulation patterns at each time step. These models have been successfully applied to investigate tidal circulation in the seas, coastal areas, and estuaries.

In an attempt to study the distribution patterns of some water quality parameters in a water body, several mathematical models have been formulated. The water quality models consist of a set of mathematical expressions which represent the effects of some important factors on dispersion and interaction processes of various substances in water. The formulation is based on the principle of conservation of mass. The formulated equations together with the appropriate boundary and initial conditions are solved to obtain the spatial and temporal distributions of the studied water quality parameters. The results obtained from the models are useful in managing and controlling the disposal of polluting substances from industries and other sources located along the coastline and river banks.

Sediment transport in rivers, estuaries, and bays can also be investigated by using mathematical models. There are some different types of sediment transport models. Some models are developed based on empirical formulation and simulation, while others are developed based on the conservation of mass principle.

Besides mathematical models formulated to simulate some physical phenomena mentioned above, new computer softwares, e.g. database and file management systems, the Geographical Information Systems (GIS) and some expert systems, are now very helpful in integrated water management, data processing, planning and design of water resource development projects.

OTHER HYDRO-ENVIRONMENTAL ASPECTS

Erosion and sedimentation

Erosion of surface soils in the upstream watersheds is the main source of sediment in a stream. Small sediment particles can be transported as suspended sediments while larger particles are carried along the stream bottom as bed load. The suspended sediment particles tend to settle to the channel bottom since the specific gravity of these particles is relatively high, while turbulent flow in the stream will counteract the gravitational settling. When flow velocity in the stream decreases as it drains into the sea or a storage reservoir, sedimentation of the suspended sediments will take place. Large particles and most of the bed load are deposited as a delta at the river mouth or head of the reservoir, while smaller particles are deposited farther and some may remain in suspension.

Seasonal variations in sediment load are normally higher than variations in streamflow. In the rainy season, rainwater will erode surface soils and carry along with surface runoff to the receiving stream. In addition, during the periods of high flow, erosion of stream banks usually occurs causing highly turbid water in the stream. In the dry season, water stored in natural storage in the basin including stream banks and groundwater aquifers is gradually released. This water is sediment free that results in clear water in the stream.

Erosion and sedimentation processes in lowland areas have some specific features which are more or less different from other regions. The lowland river is normally of gentle slope, thus flow velocity is rather small which enables deposition of suspended sediments. Suspended sediments play an important role in mass transport processes in estuaries, and they are usually transported to the upper region of estuaries by tidal dynamics and estuarine circulation. In estuaries, the tidal current asymmetry produces sediment transport so that turbidity maxima are formed and travel up– and down–stream according to tidal current (Futawatari & Kusuda 1993).

Conservation and restoration of rivers and floodplains

In the past decade, high attention has been directed to the areas of potentially high conservation value. Among these, the conservation and restoration of rivers and floodplains in lowland areas are of great concern. Ecologists recognize the importance of these areas in determining many ecological functions, both within the river and floodplain themselves and the associated ecosystems (Petts et al. 1992).

The main objective of river and floodplain restoration is the conservation of nature. Morphological dynamics of the river bed are essential for structural diversity which, in turn, affects biodiversity of the river ecosystem. The ecological value of the floodplain depends on the frequency, duration and height of inundations, as well as variations in groundwater levels. Depending on a river's history, the floodplain can be characterized by a pattern of depressions, oxbows, ridges and plains. According to variable sedimentation processes, soil properties vary across the floodplain. Owing to increased transport capacity along the channel, coarser grains are deposited near the river banks leading to rather dry habitats, while silting occurs on the floodplain. With the shifting of the river bed, the succession of different layers will result in heterogenous profiles. Floodplain formation must be regarded as a geological process, which cannot be developed artificially within a short period of time (Kern 1992).

The main items to be considered in the planning of river and floodplain restoration include identification of the objectives of the project and sections of the river and floodplain to be restored, pollution sources and control measures, sediment transport phenomenon, existing land use, water use, and development plans, runoff characteristics, existing river regulation and flood protection facilities, as well as socio–economic and legal aspects. Spawning habitats in all tributaries must be provided in river restoration. Weir and drop structures must allow for fish migration.

In general, the planning for the restoration project will start with a guiding image which is developed as a guide for the final restoration concept which describes the desirable stream properties with regard to the natural potential only while other aspects such as social, economic and political issues are not considered. Then, various restrictions are taken into account and the planners will have to find the optimal feasible solution with regard to the essential ecological rehabilitation of existing conditions.

Eutrophication

Natural waters may be classified, according to their ability to support life, as oligotrophic, mesotrophic, and eutrophic. Oligotrophic waters contain low concentrations of essential nutrients and life forms are present in small numbers. Mesotrophic waters are characterized by the abundance and diversity of life forms at all trophic (food chain) levels. Eutrophic waters characteristically have fewer species present, but the concentration of algae is particularly high. The process of moving from oligotrophic through mesotrophic to eutrophic conditions is called eutrophication (Tchobanoglous & Schroeder 1987). Nitrogen and phosphorus are generally considered to be the main nutrients involved in the eutrophication process, even though they are not the only nutrients required for algal growth. Several factors complicate the attempt to quantify the relationship between the trophic status and the measured nutrient concentrations. A certain fraction of the nutrients, particularly phosphorus, become refractory while passing through successive biological cycles. Also, morphometric and chemical factors affect the availability of nutrients in lakes (Kothandaraman & Evans 1979). Mean depth, basin shape, detention time, and other physical attributes will, more or less, have some effects on eutrophication.

According to Sawyer (1952), any stratified lake with more than 0.3 mg/l of inorganic nitrogen and 0.1 mg/l of inorganic phosphorus at the time of spring overturn can be expected to produce nuisance blooms of algae. Fair et al. (1968) showed an inductive derivation of tolerable amounts of phosphorus based on the fact that 1 mg of phosphorus requires about 130 mg of oxygen in one pass of the phosphorus cycle in lakes. Potential sources of nutrients to lakes are the runoff from watershed areas, domestic and industrial wastewater, and return flow from agricultural lands.

Eutrophication is an important water pollution issue. It causes nuisance problem and restricts the use of water. It results in fluctuation in water quality characteristics, especially the pH value, dissolved oxygen and carbon dioxide concentrations, which affects the survival of aquatic organisms including fish. This phenomenon also deteriorates aesthetic quality and affects biodiversity in the lakes, thus preservation of intrinsic ecological values is at stake. Control of eutrophication can be made by controlling the discharge of limiting–growth nutrients particularly phosphorus. Tertiary treatment processes may be needed for the nutrient–rich wastewater prior to discharging into the receiving waters.

Impacts of reservoir impoundment

Reservoir impoundment in the upstream watershed is an effective method for water resources development and control of floods in lowlands. However, the reservoir impoundment will evidently affect a number of regional and local environmental components. It is necessary that the environmental impacts of the proposed project be assessed in order to identify the probable adverse effects and to find some mitigation measures to avoid or minimize those impacts.

The reservoir impoundment will cause changes in hydrologic regime of the area. The free flowing river will be regulated and the impounded water will be gradually released for various types of water use. The impoundment will also result in equalization of variations in water quality characteristics. The amount of suspended solids and turbidity will be reduced by sedimentation. Some undesirable effects may be introduced due to decaying of the remaining vegetation in the reservoir area. If the nutrient discharge loading into the impounded reservoir is high, the eutrophication may occur. In cold regions, the increase in water depth might develop thermocline and as a result, circulation of water in the reservoir will occur, which may cause significant effects on physical, chemical, and biological processes. Furthermore, the increase in water depth also affects groundwater hydrology in the underlying aquifers. Groundwater table in the top unconfined aquifer will rise, and then affects nutrient contents in the soils.

The reservoir impoundment usually causes losses in forest lands and wildlife habitats. All the forests in the reservoir area must be cleared prior to the impoundment. During the construction and reservoir filling phases, some wild animals are likely to be killed while others have to migrate to the nearby forest areas. However, the impounded reservoir can serve as a water habitat for some birds and amphibians. It also serves as a main water source for wildlife.

The reservoir impoundment will have great impacts on fishery and aquatic ecology as well. Those fish species that prefer standing water will be dominant after the impoundment. The storage dam will inhibit migration of fish to the upstream river which affects spawning of fish and some species will disappear. On the other hand, the impounded reservoir can serve as a significant fishery resource if a properly planned aquaculture system is prepared and implemented.

Land use pattern in the reservoir area will be completely changed. This might cause significant socio-economic impacts. A number of local people situated in the reservoir area will have to evacuate to other places. Well planned resettlement and compensation programs are needed to minimize the social impacts and to avoid conflicts between the local people and the implementing agency.

After the impoundment, the populations of snail and fish which are intermediate hosts of some human parasites might increase, that brings along with spreading of parasitic infections unless some measures against these diseases are practiced at the beginning. In addition, the reservoir impoundment might result in losses of mineral resources, archaeological and historical values of the area, and will have some impacts on recreation, aesthetic and tourism activities of the region.

Due to the relatively great impacts of the reservoir impoundment on the existing environment, planners and decision makers should carefully compare between the benefits to be derived from the impoundment project and adverse impacts or losses caused by the project development. If the reservoir impoundment is considered necessary, some mitigation plans must be properly planned in order to avoid or minimize the adverse impacts which are likely to occur. Some monitoring programs are also needed to detect any changes in environmental quality during the construction phase and in the operation phase, so that a proper corrective program can be immediately undertaken when an adverse impact has occurred.

Sea water rise

Global warming has received high attention worldwide in the past decade. Among the widespread impacts of global warming, a rise in sea level is considered significant since it can cause great subsequent impacts on coastal lowlands of the world. According to the Inter-governmental Panel for Climate Change (IPCC), the rise in sea level of 30 to 110 cm at the end of the 21st century is forecasted under the 'business as usual scenario'. The Coastal Zone Management Subgroup (CZMS) of IPCC suggested all coastal countries to assess the vulnerability to the accelerated sea level rise and make up comprehensive coastal management plans in order to minimize the adverse impacts.

Besides sea level rise, it is anticipated that global warming will result in changes in tidal amplitude, precipitation pattern, magnitude and intensity of typhoon, ocean current, seasonal wind and other climatic condition. These changes will more or less affect other environmental components and water resources of the world particularly in the coastal lowland areas.

Sea level rise will increase risk of flooding on the coastal lowlands. A remarkable extent of shoreline areas will be permanently inundated and most existing infrastructures and facilities will be damaged or not function properly unless some improvements are undertaken.

Sea level rise will have significant impacts on water management in the coastal areas. With higher sea level and higher tidal amplitude, saltwater can intrude farther to the upstream rivers and can cause more damage to agricultural lands and affect water use. Intrusion of saltwater into the coastal groundwater aquifers will also increase which will affect the availability of fresh water supply in the future. The problem on flooding along the river banks will be more severe when flood water from the upstream river enters the downstream estuary during the spring–tide period. Flood protection and drainage facilities in the estuarine and coastal lowland areas will need some improvements which require a lot of budget.

REFERENCES

Bear, J. 1979. *Hydraulics of groundwater.* McGraw–Hill, Israel.
Fair, G.M., J.C. Geyer & D.A. Okun 1966. *Water and wastewater engineering, vol.1: Water supply and wastewater removal.* John Wiley & Sons, New York.
Fair, G.M., J.C. Geyer & D.A. Okun 1968. *Water and wastewater engineering, vol.2: water purification and wastewater treatment and disposal.* John Wiley & Sons, New York.
Futawatari, T. & T. Kusuda 1993. Modeling of suspended sediment transport in a tidal river. In A. J. Mehta (ed.), *Nearshore and estuarine cohesive sediment transport.* Coastal and estuarine studies: 42. American Geophysical Union.
Kern, K. 1992. Restoration of lowland rivers: the German experience. In P.A. Carling & G.E. Petts (eds.), *Lowland floodplain rivers.* John Wiley & Sons, London.
Kothandaraman, V. & R.L. Evans 1979. Nutrient budget analysis for rend lake in Illinois. *J. ASCE., Env. Eng. Div.* 105: 547.
Linsley, R.K. & J.B. Franzini 1979. *Water resources engineering.* McGraw–Hill, Tokyo.
Loucks, D.P., J.R. Stedinger & A.A. Haith 1981. *Water resource systems planning and analysis.* Prentice–Hall, New Jersey.
Petts, G.E., A.R.G. Large, M.T. Greenwood & M.A. Bickerton 1992. Floodplain assessment for restoration and conservation: linking hydrogeomorphology and ecology. In P.A. Carling & G.E. Petts (eds.), *Lowland floodplain rivers.* John Wiley & Sons, London.
Sawyer, C.N. 1952. Phosphorus in lake fertilization. *Sewage and industrial waste.* 24:768.
Tchobanoglous, G. & E.D. Schroeder 1987. *Water quality.* Addison–Wesley, Menlo Park, California.
Wanielista, M.P. 1990. *Hydrology and water quantity control.* John Wiley & Sons, New York.

4 METHODS OF DEVELOPMENT

P. ANKUM
Delft University of Technology, the Netherlands

INTRODUCTION

Lowland development

The development of lowlands is often more difficult and costly than the development of upland floodplains. In lowlands, not only embankments are needed to protect the areas against periodic floods by the sea and/or the rivers, but also the quality of the water creates problems as brackish water may affect the crops. The soils in lowland areas may suffer from acidity after reclamation or may consist of thick layers of peat which offer a poor environment for agriculture. These soils require normally a strict water management regime.

Lowland development is a term which covers a wide field of technical activities. It is used for the development and improvement of land for agricultural, residential, industrial and other purposes. Lowland development requires an integrated approach between hydraulic engineering, hydrology, soil science, agricultural engineering, economy, sociology, etc.

The term "reclamation" is also used in this concept. Reclamation is defined by the International Commission on Irrigation and Drainage (ICID 1967) as: "the act or process of reclaiming swampy, marshy, deteriorated, desert and virgin lands, and making them suitable for cultivation or habitation; also conversion of foreshore into properly drained land for any purpose, either by enclosure and drainage, or by deposition of material thereon".

Development methods

Lowland development can be classified into two methods:
- tidal land reclamation, i.e. reclamation of tidal forelands and coastal marshes.
- impoldering, i.e. reclamation of shallow seas and lakes.

A silting-up of the foreshore is taking place along many sea coasts in deltaic areas because of deposition of sediments. Tidal land development can be carried

out when the foreshore has reached a certain elevation. Typical hydraulic engineering problems are:
  – suitable elevation of tidal foreland for reclamation;
  – silting–up of the outfall channels;
  – the defence of the land against storms;
  – the layout of these progressive impoldered areas.

Reclamation of shallow seas and lakes through impoldering is a very special and expensive type of lowland development, and has been implemented so far in a few countries with serious land problems (e.g. the Netherlands, Japan). Typical hydraulic engineering problems are:
  – the defence of the low–lying land against inundations;
  – seepage of brackish water;
  – pump–lift drainage.

Both methods of lowland development create new land that can be called a "polder". A polder is defined as (Segeren 1982): "a polder is a level area which has originally been subject, permanently or seasonally, to a high water level (groundwater or surface water) and is separated from the surrounding hydro-logical regime, to be able to control the water levels in the polder (groundwater and surface water)".

Thus, a polder shows the following characteristics:
  – a polder does not receive any foreign water from a water course, but there is only water inflow by rain and seepage, or by irrigation intake;
  – a polder has an outlet structure (sluice or pump) that controls the outflow;
  – the groundwater levels and the surface water levels in the polder are in-dependent from the water levels in the adjacent land. These water levels are artificially maintained in order to optimize the objectives of the pol-ders, on basis of land elevation, soil properties, climate, type of land use, etc.

Flood defences and drainage

Lowlands are generally located in deltaic areas which have been formed mor-phologically by river deposits during flooding of the delta. As deltas belong to the most productive and the most densely populated parts of the earth, flood defence is needed in many cases. Flood defence is needed to protect the areas against external floodings, i.e. water from outside the area. Construction of embankments (dikes, levees) is often a part of the required measures. Typical hydraulic engineering problems related to flood defence, are:
  – construction of embankments makes the flood levels to become higher,
  – the discharge of the (internal) drainage to the river is obstructed,
  – the process of gradual land raising by periodic sedimentation is halted.

Swamps and other water–logged areas may have a poor internal drainage system to evacuate rainfall. They require a good land drainage system of sub-surface and/or surface drainage systems. Typical hydraulic engineering pro-

blems related to land drainage, are:
- – subsidence of the drained land,
- – irrigation of the drained areas might become necessary during part of the year.

TIDAL LAND RECLAMATION

Coastal accretion

Sediments are deposited on the tidal lands along many coasts. The sea floor is raised by this process of accretion, and the coast line advances, see Figure 4.1. The tidal lands may become valuable arable lands because of the natural fertility of the deposits, the flatness and the high groundwater tables. The rates of accretion depend on the rate of sediment supply and the degree of exposure of the waves. Accretion is promoted by the presence of a natural vegetation. Vertical rates of accretion of 0.002 – 0.20 m per year might be encountered in tidal lands, with horizontal expansion of 10 to 200 m per year (Wagret 1968).

Only tidal flats well above the level of low water are suitable for technical works to speed–up the accretion. Several methods have been applied to accelerate sedimentation. These methods are based on two aspects (Volker 1993):
- – to allow the suspended silt to settle around the time of slack water after high tide;
- – to dry out the deposit so that the silt adheres to the surface and will not be removed during the next flood period. This requires some sort of a drainage system.

The Schleswig–Holstein method (Figure 4.2) is an example of a method to accelerate the sedimentation by the sea. This method was developed in Germany in the beginning of this century.

Development of tidal lands

An important issue is: at what elevation of the tidal land with respect to the tides, the construction of an embankment could be best undertaken?

The maximum elevation of the tidal lands will be about average high water,

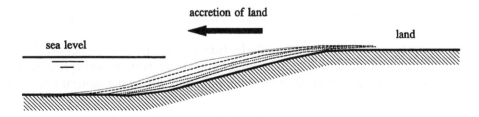

Figure 4.1. The process of accretion.

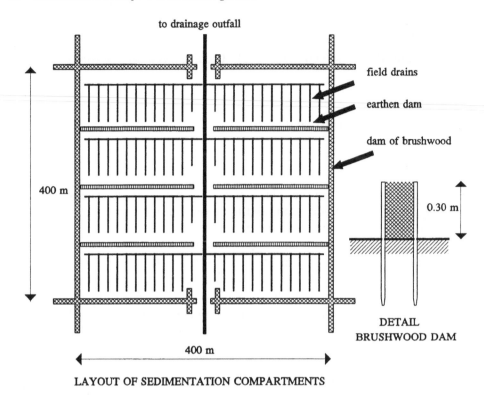

to drainage outfall

field drains

earthen dam

dam of brushwood

0.30 m

DETAIL
BRUSHWOOD DAM

400 m

400 m

LAYOUT OF SEDIMENTATION COMPARTMENTS

Figure 4.2. Schleswig–Holstein system for accretion of tidal lands.

and the question arises whether or not to postpone embankment until the land
has attained that high elevation. Generally, the land elevation is taken between
mean sea level and high tide. However, the pressure of the population can force
an early reclamation. So, the mean sea level is considered in Bangladesh as the
suitable level for reclamation. As a result, the drainage conditions are poor, and
there are difficulties with the construction of the embankments. Moreover,
considerable damage may occur when these embankments breach.

Civil engineering works for tidal land reclamation are related mainly to two
aspects:
- the construction of dikes, to protect the newly gained land against abnor-
mally high seas levels;
- the construction of a land drainage system, and when needed an irrigation
system.

Tidal lands are intersected normally by tidal creeks. The creeks persist during
the building up of the tidal flats. The larger creeks can well be incorporated into
the drainage system after reclamation, because they act as the natural drains of
the area.

Drainage by gravity during the low tide is possible when tidal lands with high elevations have been embanked. A serious problem is often formed by the rapid silting up of the outfall channel outside the tidal drainage sluice, where accretion continues. Sedimentation may even completely obstruct the sluice. The obstruction by sedimentation will be fast when the sluice is located in a tidal creek that is cut off by this sluice. The tidal currents which originally maintained the creek, have now been reduced. The problem can be avoided if the sluice is located on a large tidal creek which is not cut off by the embankment and remains connected with tidal lands. It is often necessary to divert the drainage water to that location.

Another alternative to solve the obstruction of the drainage sluice by sediments, is the construction of a flushing basin. Such a flushing basin is located in front of the drainage sluice and allows an accumulation of the drainage water. Periodic flushing of this basin during low tide will create a strong current that will remove the silt in front of the sluice. The effect is rather local, but can be increased by river training to keep the flow concentrated. Moreover, the accumulation of water in the basin requires water level variations, which have negative effects on the head available for drainage of the reclaimed lands.

Impoldering of tidal lands proceeds by small steps (1000 to 2000 m) and the intermediate periods are quite long. New sea–dikes have to be constructed to protect the newly gained land. The drainage outfall of the older polders becomes longer and longer with progressive impoldering, see Figure 4.3. It means that less head becomes available, while more head would be required because of the subsidence of the older polders. It may then be better to avoid obstruction of the outfalls of the older polders by changing the layout. An alternative is to connect the drainage system of the older polders with the drainage systems further inland.

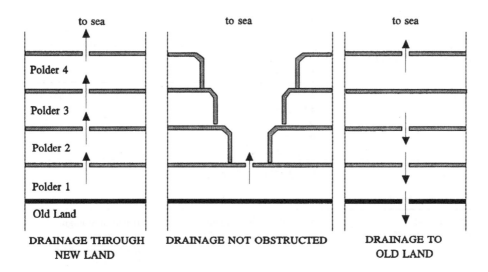

Figure 4.3. Drainage outfall for progressive impoldering.

IMPOLDERING OF SHALLOW SEAS AND LAKES

Design of polders

Impoldering of shallow seas and lakes requires extensive investigations and design work. Experiences with other polders under similar conditions may be of great help. A more or less sequence in the design can be followed, although the actual design is an iterative process for which earlier design assumptions have to be constantly checked. An overall view of the polder design is presented here (Table 4.1). Discussions and more details of the design follows.

The initial design is based on detailed mapping of the new polder area, in respect of the topography, the soil suitability for crop cultivation, the soil permeability, etc.

Table 4.1. Design aspects of polders.

DESIGN ASPECT	TO BE DESIGNED	Supporting design data
LAND USE	● cropping pattern ● size of land holding	o project benefits o drainage requirement o irrigation requirement
LAY OUT	● location of dikes ● location of pumping stations ● location of main drainage system	o project costs o capacity of the drainage systems o capacity of the pumping stations
POLDER WATER LEVEL	● open–water elevation ● polder sections ● water level regulators	o. head at pumping stations o maximum depth of field drainage system
FIELD DRAINAGE SYSTEM	● drain depths ● drain distances ● drain sizes	o layout of farm holdings
MAIN DRAINAGE SYSTEM	● bed widths ● water depths ● bed gradient (=0) ● maximum water level gradients	o minimum freeboard at peak discharge
POLDER OUTLET & INLET STRUCTURES	● pumping stations ● intake sluices ● operation rules	
EXTERNAL WATER SYSTEM	● width of belt canals ● open–water elevation	o irrigation storage o flood storage o negative seepage of the "old"–land

As a first step, the land use and the cropping pattern have to be selected. This leads ultimately to the project benefits. The proposed cropping pattern and the required farm–budget may determine the size of the land holding in the polder. Furthermore, the hydrological data determines the drainage and irrigation requirements for this cropping pattern.

The location of the surrounding polder dikes determines the size of the polder, and thus contributes to the project costs and the benefits. The deeper areas, areas of poor soil suitabilities and areas where high seepage rates can be expected, should be avoided. Also the preliminary layout of the main drainage system can be done in this stage. The canal capacities can also be determined now, as the command areas and the drainage requirements in mm/day or in l/s.ha are known.

The stagnant water level in the main drainage system below the terrain level, i.e. the "polder water level", follows from the optimum agricultural production. Considering also the topography of the polder, this stagnant "polder water level" of the open–water can be designed for the whole polder. Aspects like land subsidence and the oxidation of the soils also play an important role in the design of the polder water level. It might also be possible to design different polder water levels in different sections of the polder. It means that water level regulators, such as gated drop structures, are required between these sections. The stagnant open–water level in the polder determines the required head at the pumping stations. It also determines the maximum depth of the field drainage system.

The design of the field drainage system, whether it consists of open–water trenches or of sub–surface drain–pipes, is based on the above drainage requirements of the crops and the available head above the stagnant open–water level. The design of the field drainage system has an impact on the layout of the main drainage system, and together they determine the ultimate layout of the farm holdings.

The definite layout of the main drainage system is known by now, as well as the definite canal capacities. The water levels in the main drainage system will be most of the time horizontal, so that often the bed gradients also are taken horizontal. The design of the canal dimensions is based on the peak discharge, requiring a sloping water line: lowering at the pumping station and rising at the other end. A check should be made on the minimum freeboard where the water level rises.

The capacities of the pumping stations and the irrigation inlet structures follow from the design of the main drainage system and the water requirements during the irrigation season. Operation rules should be accounted for at the design stage itself.

Pumping stations should be pre–designed at the lower locations in the polder. However, polders with different sections, each with their own stagnant polder water level, may have pumping stations at other locations as well. An example of a polder with three polder sections is presented in Figure 4.4 (Kley & Zuidweg 1969).

In alternative I, the drainage water is discharged from the higher to the lower sections by means of gated drop structures. Only one pumping station, with a capacity of "$Q_1 + Q_2 + Q_3$" is required, but at the highest operation costs, which are related to "$(Q_1 + Q_2 + Q_3) \times H_1$", where H and Q are the required pumping head and discharge of the different polder sections, respectively.

In alternative II, the drainage water is gradually pumped to the higher polder

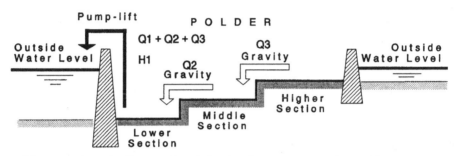

Alternative I: One pumping station (maximum energy requirements).

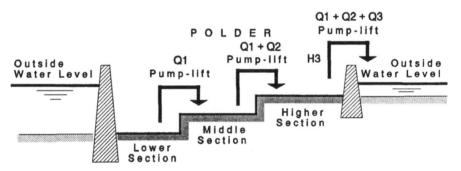

Alternative II: Three pumping stations (minimum energy requirements).

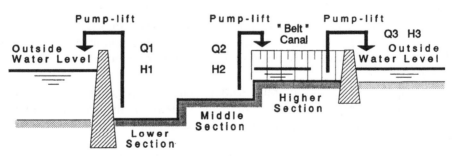

Alternative III: Three pumping stations & high " Belt Canal "

Figure 4.4. Pumping stations in polders with three sections.

sections. The total operation costs will be lower than for the above alternative and will be related to "$H_1 Q_1 + H_2 Q_2 + H_3 Q_3$". Three pumping stations are required here, with capacities related to "Q_1", "$Q_1 + Q_2$" and "$Q_1 + Q_2 + Q_3$", respectively.

In alternative III, the minimum operation costs and the minimum pump capacity costs can be obtained by draining the three polder sections directly to the external water system. The lower and the higher sections are located along the external water system, and can discharge directly. The middle polder section in Figure 4.4 should be connected with the external water system directly, or by means of a "belt canal". Thus, the total operation costs are also at a minimum and are related to "$H_1 Q_1 + H_2 Q_2 + H_3 Q_3$".

The design of an external fresh-water system around the polder is also of importance for the water management of the new polder as water storage can be obtained here. In fact, the proper design of the external drainage system is essential for the water management of the "old"-land, as storm water has to be discharged through this system. Moreover, the wide external water system limits the draw-down effects of the polder on the groundwater levels in the old-land. It is obvious that the pre-selected location of the dike has already determined the size of the external water system, and may have to be adjusted when e.g. more storage is required.

Implementation of polders

Impoldering of open-water areas have to start with the construction of the embankment and the pumping stations. The capacity of the pumping stations is normally based on the agricultural requirements, and is often also sufficient for the initial evacuation of the water from the new polder area. For instance, a pumping capacity of 13 mm/day (1.5 l/s.ha) requires a pumping period of some 8 months for a polder of 3 m deep, when the annual evaporation equals the annual rainfall and seepage.

The land preparation of new clay-polders (Figure 4.5), may start with measures to speed up the ripening of the young soils (Schultz 1983). For instance in the Netherlands, reed-seeds are sown from a plane, to be followed by the de-watering of the soil by digging trenches of ± 0.60 m deep at intervals of some 50 m. One year before cultivation, other trenches at intervals of some 10 m are dug between the existing trenches. Finally, the trench drainage system is replaced by sub-surface drainage pipes after a few years of farming, e.g. at a depth of 1.20 m and at an interval of 50 m for a design discharge of 10 mm/day (1.2 l/s.ha).

Thus, the development of a new polder in the Netherlands is more or less fixed and follows the following sequence (Ankum et al 1988):
1. constructing the enclosing dike with outlets (pumping stations, sluices);
2. dredging main drainage canals when the polder is still submerged;
3. evacuating the water from the polder during 6 to 12 months;
4. sowing reed-seeds on the (soft) mud surface by airplanes. This is to pro-

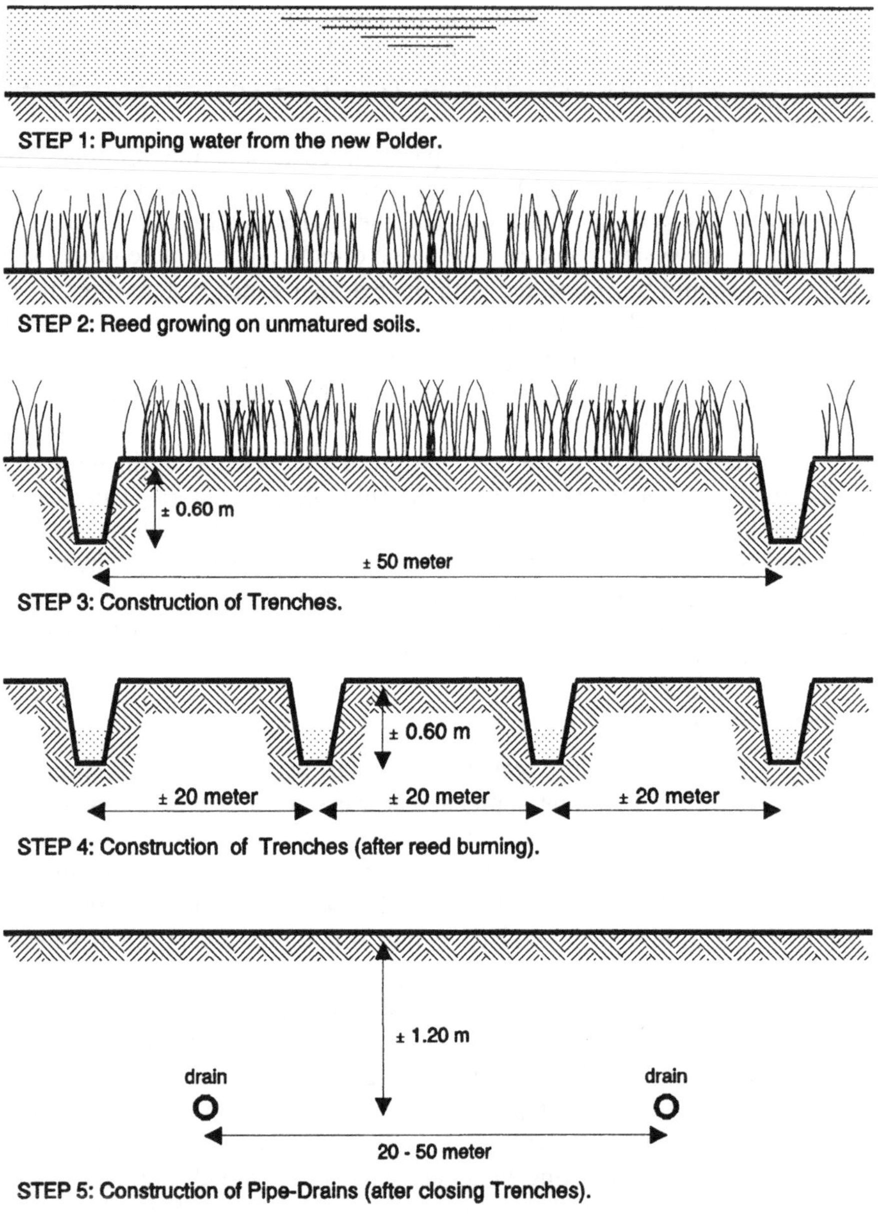

STEP 1: Pumping water from the new Polder.

STEP 2: Reed growing on unmatured soils.

± 0.60 m

± 50 meter

STEP 3: Construction of Trenches.

± 0.60 m

± 20 meter   ± 20 meter   ± 20 meter

STEP 4: Construction of Trenches (after reed burning).

± 1.20 m

drain   drain

O   O

20 - 50 meter

STEP 5: Construction of Pipe-Drains (after closing Trenches).

Figure 4.5. Land preparation of new clay–polders in the Netherlands.

mote the drying process ("soil–ripening") and to prevent other weeds;
5. constructing of the drainage network of main drainage canals, main ditches and plot ditches, as well as the construction of the road system;
6. developing the land by (i) burning the reed vegetation, (ii) construction

of the open field drains or trenches;

7. farming the land for about 5 years with government funds, with a special cropping sequence. For instance, with oilseed in 1st year, wheat in 2nd year, barley in 3rd year, oilseed in 4th year, wheat in 5th year;
8. replacing the open field drains by a subsurface drainage system (pipes) after the soil has ripened and the land subsidence has halted, e.g. after the 5th year;
9. constructing farm buildings and villages, and lay telephone and electricity supply cables and the water mains;
10. allocating land to private individuals on the basis of one to four workers per farm–holding;
11. landscaping, afforest and construct recreation areas;
12. when the polder is ready: transferring of the administration and maintenance functions to e.g. a Polder Board.

WATER MANAGEMENT SYSTEMS IN LOWLANDS

Water balance

Polders, whether they are reclaimed tidal lowlands or reclaimed shallow seas and lakes, can be considered as "catchment areas". The area is separated from the surrounding land by dikes. The water inflow is rainfall and seepage, or irrigation supply that is controlled by an intake structure. The surplus of drainage water has to be discharged artificially, either by gravity through an outlet or by pump–lift (Figure 4.6).

The water balance equation of a polder reads:

$$P + S + I = E_o + ET_c + Q + \frac{\Delta V}{\Delta T} \qquad (4.1)$$

The inflow consists of precipitation, P, seepage from the sub–soil, S, and intake water, I, for irrigation supplies and lockage water from ship–locks. The outflow consists of open–water evaporation, E_o, evapo–transpiration of the vegetation, ET_c, and the pumping and sluicing discharges, Q. The difference between the in– and outflow over the considered period, ΔT, determines the changes in storage, ΔV, in the soil and in the open–water. This time interval, ΔT, can be taken as e.g. one year for determination of the annual pumping costs, or as one day or even shorter for the determination of storm–drainage requirements of the canals and the pumping stations.

The assessment of the seepage to new polders is quite difficult as the seepage depends on the permeability of the sub–soil, the thickness of the aquifers and on several other geo–hydrological parameters. The seepage can be divided into (i) a seepage through the embankment, and (ii) a seepage from the deeper soil

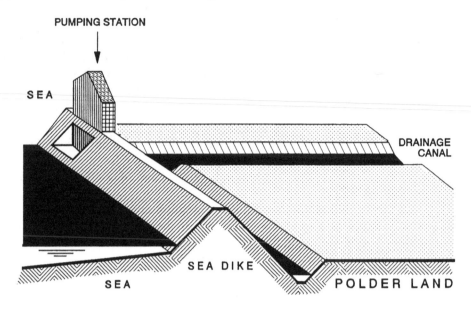

Figure 4.6. Water management system of a polder.

layers. The total seepage of existing polders can be calculated by means of the water balance equation. The typical seepage value of polders in the Netherlands is 1 mm/day (0.1 l/s.ha), but values as high as 13 – 16 mm/day (1.5 – 1.9 l/s.ha) are also encountered (Schultz 1992).

Many polders require a regular intake of fresh water during a part of the growing season. For instance, grass–polders in the Netherlands may require an irrigation supply upto 0.6 l/s.ha during dry years. Some 1.5 l/s.ha is required in the paddy–polders in Indonesia. The quality of deep seepage water is often poor, as it can be brackish or even salty. Thus, the salt water balance of the polder should be checked. Polders in (semi) arid climates may require a considerable irrigation supply together with a good drainage system, to control the saline conditions of the root zone of the crops.

Water storage in open–water channels depends on the percentage of open–water in the polder and the allowable rise in water level. For instance, a modern polder may have 1% of open–water, with an allowable rise of water level of 0.20 m. The rate of open–water can also be expressed in a water depth over the whole polder. So, the polder in the above example has an open–water storage of 0.01 × 0.20 = 0.002 m = 2 mm.

Water storage in the soil depends on factors like the void–ratio of the soil and the allowable rise of the groundwater table (Luijendijk 1982). So, polders with deeper open–water level during the dry season will store more water in the soil during heavy rainfall (Table 4.2).

Table 4.2. Typical values of water storage in the soil.

Open–Water Level below terrain level	Water storage in soil		
	peat	light clay	heavy clay
0.20 m	20 mm	4 mm	2 mm
0.40 m	45 mm	12 mm	6 mm
0.60 m	75 mm	25 mm	12 mm
0.80 m	105 mm	40 mm	20 mm
1.00 m		60 mm	28 mm
1.20 m		85 mm	38 mm
1.40 m		115 mm	50 mm

The water balance with a time interval of $\Delta T = 1$ day can be used for a rough estimate of the required pump capacity (Kley & Zuidweg 1969). As an example (Table 4.3), a polder may have an open–water ratio of 10% and the ground-water level is initially 0.10 m above the stagnant open–water level. The water storage in the soil is 7 mm per 0.10 m groundwater rise. A seepage flow to the polder amounts to 1 mm/day (0.1 l/s.ha). The design rainfall may consist of a three–day storm of 18 mm/day, 33 mm/day and 12 mm/day, respectively. It is assumed that the rainfall of one day percolates during 3 days, i.e. during the day of rainfall and the two following days. It follows from the calculations in Table 4.3 that the average groundwater level will rise by 0.40 m, and the open–water level by 0.14 m for an assumed pump capacity of 14 mm/day (1.6 l/s.ha).

Table 4.3. Water balance of a polder for the design rainfall.

		Day no.1	Day no.2	Day no.3	Day no.4	Day no.5	Day no.6
Rainfall	mm/day	0	18	33	12	0	0
Percolation (Rain–day1)	mm/day		6	6	6		
Percolation (Rain–day2)	mm/day			11	11	11	
Percolation (Rain–day3)	mm/day				4	4	4
Groundwater discharge	mm/day	0	6	17	21	15	4
Seepage	mm/day	1	1	1	1	1	1
Discharge to open–water	mm/day	1	7	18	22	16	5
Pumping	mm/day	−1	−7	−14	−14	−14	−14
Open–water storage	mm	0	0	4	12	14	5
Open–water elevation	m+MSL	0.00	0.00	0.04	0.12	0.14	0.05
Water storage in soil	mm	0	12	28	19	4	0
Groundwater elevation	m+MSL	0.10	0.27	0.50	0.37	0.16	0.10

Polder water level

The water management system of a polder consists of a vast network of open–water drainage channels.

- field drainage system, to maintain the groundwater table under the root zone;
- main drainage system, to divert the drainage water from the field drainage system to the outlet;
- sluices and/or pumping stations, to evacuate the water from the main drainage system to the surrounding canals;
- the external water system of belt canals, surrounding the polder.

The main drainage system flows mainly because of rainfall. So most of the time it contains stagnant water. The water level of this stagnant water must be carefully controlled. A lower or a higher water level might have negative effects on the crops and on land subsidence. This target water level of the open–water drainage system of a polder can be called the "polder water level". Thus, water management in a polder is mainly determined by "water level control" in the drainage system, and to a lesser extent by "discharge control" like in irrigation schemes.

The polder water level is often established on agricultural considerations to obtain the highest economic return of the crops (Kley 1969). The relevant parameters are:

- type of the soil,
- proposed land use, and the yield curve for different open–water levels, see Figure 4.7 for example,
- the elevation of the terrain.

But the polder water level has also to meet interests of navigation, recreation and nature preservation. Moreover, it should prevent or reduce the (irreversible) subsidence of the soil.

Generally, peat soils require high polder water levels of some 0.30 – 0.60 m below terrain level to prevent the subsidence of the soil. Thus, grass and other crops only with shallow roots can be cultivated. Clayey soils permit deep polder water levels of some 1.20 – 1.80 m below terrain. It means that dryland crops with a thick root zone can be cultivated. Even greater depths of 1.50 – 2.00 m below terrain have to be selected in polders with brackish seepage and with low rainfall. Capillary rise of the brackish water can be avoided by over–irrigating the crops and allowing a percolation flow to the drains.

Polders may be divided into sections having different polder water levels, depending on the topography and on the soil conditions. Often a "summer" polder water level and a deeper "winter" polder water level is applied, as to provide for more drainage of the soils in the wet winter season.

An example of the determination of the polder water level based on the highest crop yields is given here. A certain polder has a size of 1200 ha. A lower area at 0.00 m+MSL of 900 ha of peat soils will be used for grass, a higher area at 1.00 m+MSL of 300 ha of clayey soils will be used for crop

Percentage of the optimum yield

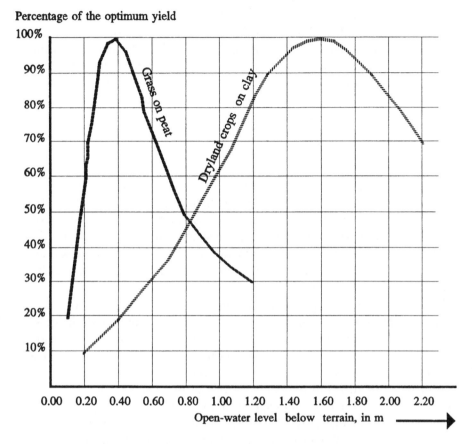

Figure 4.7. Example of crop yields for different open–water levels.

cultivation. The yield curves for different (stagnant) open–water levels in the channels below terrain level, are presented in Figure 4.7. The economic value of arable land is twice as high as the value of grassland. Thus, the annual yield of arable land is $ 2X, against an annual yield of grassland of $ X. The calculation is done by means of Table 4.4.

Table 4.4. Example: Calculation of the polder water level.

Polder water level	Grass land (900 ha)		Arable land (300 ha)		Total polder (1200 ha)
m–MSL	Water level m–terrain	yield in $	Water level m–terrain	yield in $	yield in $
0.60	0.60	0.75 × 900X	1.60	1.00 × 600X	1275X
0.50	0.50	0.90 × 900X	1.50	0.98 × 600X	1398X
0.40	0.40	1.00 × 900X	1.40	0.95 × 600X	1470X
0.30	0.30	0.93 × 900X	1.30	0.90 × 600X	1377X

Each assumed open–water level gives a certain relative yield of the grassland and of the arable land. It shows that the total annual yield of $ 1470X$ is maximum for a polder water level of 0.40 m+MSL. As an alternative, two different polder water levels can be used: 0.40 m+MSL for the grassland and 0.60 m+MSL for the arable land. The total annual yield would be $ 900X + $ 600X = $ 1500X$. The extra construction costs would include a gated drop structure between the two polder sections.

Field drainage system

The field or land drainage system collects the excess water from the land. Two principal types of field drainage may be distinguished: (i) a surface drainage system of open–water trenches, (ii) a sub–surface drainage system of pipe drains. The function of the field drainage system is two–fold: (i) to speed up the soil–ripening process during the reclamation period, when the very soft and wet soils are transformed into normal soils, and (ii) to maintain adequate drainage conditions after reclamation.

The design of the field drainage system in a polder depends on the future land use. Paddy is grown with high groundwater levels, and the drainage system has mainly a function for the surface runoff. Grass can grow well in areas with shallow groundwater levels, but dryland crops require deep groundwater levels. Even deeper groundwater levels might be required when the groundwater is saline or brackish to avoid the capillary rise of the water to the root zone.

Modern criteria for field drainage systems with dryland crops are related to a groundwater level that may be reached during a certain duration and with a certain return period. For example, the groundwater level may reach upto e.g. 1.00 m below terrain level during 2 days with a frequency of once per year, and upto e.g. 0.75 m during one day with a frequency once per 10 years.

The design of the field drainage system can be based on the Hooghoudt formula (Figure 4.8). The formula describes the steady flow to pipe drains and open–water trenches, and incorporates the radial flow by means of a reduced value, d, of the aquifer thickness, D. The distance, L, between the drains is calculated by iteration. The Hooghoudt formula reads (ILRI 1974, Smedema & Rycroft 1983):

$$q = \frac{8\ K_b\ d\ h\ +\ 4\ K_a\ h^2}{L^2} \tag{4.2}$$

where d follows from:

$$\frac{L}{D} \leq 4 : \quad d = \frac{\pi\ L}{8\ \ln(\frac{L}{u})} \tag{4.3a}$$

or from:

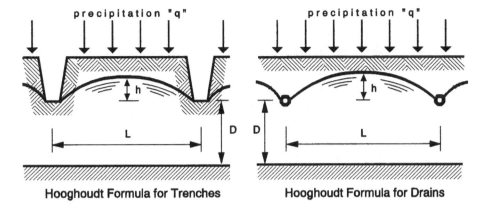

precipitation "q"   precipitation "q"

Hooghoudt Formula for Trenches   Hooghoudt Formula for Drains

Figure 4.8. Drainage formula of Hooghoudt.

$$\frac{L}{D} \geq 4 : \quad d = \frac{\pi\,D\,L}{\pi\,L + [\,8\,D\,\ln(\frac{D}{u})\,]}$$

(4.3b)

where: q is the (steady) precipitation on the terrain, in m/day; K_b is the permeability below the drainage basis, in m/day; K_a is the permeability above the drainage basis, in m/day; typical permeabilities of different soils are presented in Table 4.5; h is the maximum head of groundwater above drainage basis, in m; L is the distance between the drains, in m; D is the thickness of the permeable soil below the drainage basis, in m; u is the wet perimeter of the drain, in m. For a drain–pipe with radius, r: $u = \pi\,r$ (= ½ circumference!); for an almost dry trench with bed width, b: $u = b$; and for a trench with a bed width, b, and a water depth, y: $u = b + 1.4\,y$.

Table 4.5. Typical permeabilities of soils.

Type of soil	Permeability K in m/day
frictured clay in new polders	10 – 100
coarse sand with gravel	10 – 25
middle fine sand	1 – 5
very fine sand	0.2 – 0.5
sandy clay	0.10
peat	0.01
heavy clay	0.01
un–ripened clay	10^{-5}

Pipe drains are normally corrugated pipes, which are easier to handle by pipe–laying machines than smooth pipes. The discharge through corrugated pipes can be calculated by the Manning–Strickler formula with a roughness coefficient of $k \approx 33$ m$^{1/3}$/s. However, the non–uniform flow condition of a gradual filling of the pipe with drainage water is often compensated by increasing the roughness coefficient to $k \approx 57$ m$^{1/3}$/s. Thus, the discharge formula for full–flowing corrugated pipes under non–uniform flow reads (ILRI 1974):

$$Q = 38 \ (2 \ r)^{2.67} \ s^{0.5} \tag{4.4}$$

where Q is the pipe discharge at the outlet, in m^3/s; $2r$ is the diameter of the pipe, in m; and s is the slope of the (energy) gradient.

Main drainage system

The function of the main drainage system is to collect the water from the field drainage system, to transport the water to the outlet, and to evacuate the water from the area.

The main drainage system of a typical new polder in the Netherlands consists of: (i) canals, and (ii) main ditches and plot ditches. The canals are often navigable and have (stagnant) water depths of some 1.50 – 2.50 m. The gradient of the canal bed is normally horizontal. The main ditches and plot ditches run along the plots, e.g. the rectangular fields of about 500 m × 1200 m.

The required capacity of the drainage system depends not only on the rainfall but also on the seepage, e.g. 1 mm/day (0.1 l/s.ha) or higher, and on lockage water from polder shiplocks, e.g. 1 mm/day (0.1 l/s.ha). Furthermore, the required capacity depends also on the percentage of open–water, which is kept low in the modern polders (1 – 2%) with electric pumping stations, but might be much larger (5 – 15%) in older peat polders with pumping by windmills.

Table 4.6 presents some typical values of canal capacities in the main drainage system for moderate climates (Luijendijk & Schultz 1982). These discharges are of the order of 13 mm/day (1.5 l/s.ha) for the smaller ditches and of 11 mm/day (1.3 l/s.ha) for the larger canals. For urban polder areas, values for the main system of 20 mm/day (2.3 l/s.ha) and more are often applied.

Table 4.6. Design parameters of main drainage systems for moderate climates.

Polder type	Soils	Land use	Open water %	Polder water level m–terrain	Canal capacity mm/day
old polders	peat	grass	5 – 10	0.20 – 0.50	8 – 12
old polders	clay	grass	3 – 10	0.40 – 0.70	8 – 12
old polders	clay	crops	5 – 10	0.80 – 1.00	8 – 12
new polders	clay	crops	1 – 2	1.40 – 1.50	11 – 14
urban polders	– –		3 – 8	1.50 – 1.80	15 – 30

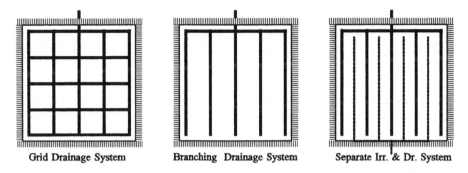

| Grid Drainage System | Branching Drainage System | Separate Irr. & Dr. System |

Figure 4.9. Pattern of main drainage system in polders.

The main drainage system of a polder can be designed in a grid pattern, or in a branching pattern (Figure 4.9). Moreover, the question may arise in polders with irrigation water needs, whether separate irrigation and drainage systems are constructed, or that the drainage system is also applied for irrigation supplies as well. A grid pattern can be applicable for flat polders. It has the advantage that the canals can be used as boundaries between land holdings. Examples of grid patterns can be found in the polders of the Netherlands. A branching pattern of the main drainage system might be applied for sloping polders. A branching drainage system is also applied in polders with sulphate–acid problems to flush the acid water. Examples can be found in Indonesia.

Irrigation supply to the crops with a grid drainage system is possible by means of capillary rise of the groundwater. The appropriate groundwater level is maintained by bank–infiltration from the open–water system. The other possibility for irrigation in a grid drainage system is by means of pumping by the individual farmers on an "on–demand" supply basis, where the open–water level in the polder is maintained by subsequent intake into the polder. A branching pattern must be applied in polders with a separate irrigation and drainage systems, such as in Japan, Korea and Taiwan. The water levels in the drainage system are kept sufficiently below terrain level, while the level of the (gravity) irrigation water is kept at a higher level.

The present design procedure of the main drainage system with open–water channels, is still rather empirical and often based on steady and uniform flow only. The Strickler–Manning formula is normally used in the design:

$$Q = k A R^{2/3} s^{1/2}, \text{ and } Q = v A \tag{4.5}$$

with the wet cross–sectional area, A:

$$A = (b + ym)y \tag{4.6}$$

and the hydraulic radius, R:

$$R = \frac{A}{b + 2y\sqrt{1+m^2}} \tag{4.7}$$

where Q is the discharge, in m³/s; v is the velocity, in m/s; A is the wet cross–

sectional area, in m²; R is the hydraulic radius, in m; s is the water level (energy) gradient; b is the bed width, in m; y is the water depth, in m; m is the side slope $(1_{Vert} : m_{Hor})$; and k is the (Strickler–Manning) roughness coefficient, in $m^{1/3}/s$.

The Strickler–Manning formula describes only the relation between the parameters. In general, the design discharge, Q, is known, while also an assumption can be made on the roughness coefficient, k, and the side slope, m, of the channel. The Strickler–Manning roughness coefficient, k, is independent of the water depth, but depends strongly on the maintenance conditions. A well-maintained channel may have a roughness coefficient of $k = 35 - 40$ $m^{1/3}/s$. But, a channel full with weeds may have roughness coefficients as low as $k = 5 - 10$ $m^{1/3}/s$. The side slope "$1_{Vert} : m_{Hor}$" depends on the principles of soil mechanics, but it is often selected on basis of regional experiences as a function of the design discharge. For example, $m = 1$ for $Q < 0.5$ m³/s, $m = 1.5$ for $0.5 < Q < 4$ m³/s, and $m = 2$ for $Q > 4$ m³/s.

The design of the channel requires three parameters, i.e. the bed width, b, the design water depth, y, and the gradient of the energy line (water line), s, while only one equation is available. It means that two other conditions should be set.

The first condition is related to the water level gradient, s. In sloping polders, the criterion of the tractive force will be selected.

$$T = \rho\, g\, y\, s \text{ and } \quad T \le T_{cr} \tag{4.8}$$

where T is the tractive force on the bed in N/m², T_{cr} is the critical tractive force for that soil in N/m², $\rho = 1000$ kg/m³ is the density of water, $g = 9.8$ m/s² is the acceleration of gravity, y is the water depth in m, s is the water level (energy) gradient. The critical tractive force for that soil should be investigated before the detailed design starts, but as a first estimate, $T_{cr} = 3 - 5$ N/m² can be taken.

The first condition in flat polders, however, is related to the water management practice of "water level" control to maintain the polder water level constant. However, a water level gradient will be needed during discharges. It means that during periods of the design–discharge, the actual canal levels in the upper part of the polder will exceed the polder water level (+0.10 to +0.20 m), while near the outlet the water level may be lowered well below the polder water level (–0.30 to –0.40 m). In the Netherlands, for instance, the velocity during the design discharge is set at a maximum of $v = 0.25$ m/s. This makes the gradient of the water level during the design discharge to be flat, of the order of $0.05 - 0.10$ m per kilometer $(0.05 - 0.10 \times 10^{-3})$. It means that the water level deviations from the (stagnant) polder water level are only small for changing discharges, also in long channel reaches. It should be noted that the designer often ignores the very flat gradient of the bed and that the bed is constructed horizontal. In the design of the water levels, however, the preliminary design calculations are often simplified by assuming a uniform flow. It is also assumed that the bed is indeed constructed according to the flat slope thereby the bed gradient equals the actual energy gradient.

The second condition is related to the relation "*b/y*", between the bed width, *b*, and the water depth, *y*. Large values of the "*b/y*"–relation have the advantages that:

- a wider canal leads to less water level variations for changing discharges;
- the open–water ratio increases, so the capacity of the pumping station might be reduced by a larger open–water storage;
- the depth of the channels is restricted, so less cutting through any horizontal (impermeable) layers to limit the seepage.

There are mainly two types of structures in the drainage system of a polder:

- water level regulators and gated drop structures, such as overflow weirs and undershot gates, to maintain the target polder water level;
- road crossings, such as bridges and culverts.

Road crossings, but also often the water level regulators should have preferably a low head loss during the design discharge, as to prevent large variations in water levels in the tail–end of the drainage system between stagnant water level and the higher storm water level. This can be achieved by a submerged (sub-critical) flow through the structure during the maximum discharge. Moreover, it means that for gated structures the gates should not obstruct the flow. Thus, the head loss over the above structures is based on the formula for sub–critical flow through structures. This formula can be simplified by considering the entrance and exit losses only:

$$\Delta H = \alpha \frac{v^2}{2 g}, \text{ and } v = \frac{Q}{A} \qquad (4.9)$$

where ΔH is the (energy) head loss over the structure in m, α is the entrance-and–exit coefficient (e.g. $\alpha = 1.0$), v is the velocity of the water in the structure in m/s, Q is the discharge in m³/s, A is the wet cross–sectional area of the structure in m², $g = 9.8$ m/s² is the acceleration of gravity.

Drainage outlets

Drainage water can be evacuated from a polder either by gravity through a tidal outlet in the dike during low tide, when a polder or the external water system of belt canals is surrounded by open water in which tides are active, or by means of pump lift. Tidal outlets can be constructed as a gated culvert in the dike or as a sluice when ships have to pass. Both structures should be designed at the smallest dimensions, to limit the costs, by aiming at:

- overflow, instead of orifice flow,
- free flow, instead of submerged flow.

The discharge formula of a sluice and a culvert, both with a free water level, reads:

- for free flow:

$$Q = 1.7 \ b \ H^{1.5} \qquad\qquad \text{valid for } z > \frac{1}{3} H \quad (4.10)$$

● and for submerged flow:

$$Q = \mu \, b \, y \, \sqrt{2 \, g \, z} = 4.4 \, \mu \, b \, y \, \sqrt{z} \text{ valid for } z < \frac{1}{3} H \qquad (4.11)$$

where Q is the sluice discharge in m³/s, b is the width of the sluice in m, H is the upstream energy head above the sill in m, y is the (outer) water level above the sill in m, z is the (energy) head loss over the structure in m, μ is a coefficient ($\mu \approx 1.0$).

The design calculations of a tidal outlet have to be done in time–steps. For each time step, the outfall volume should be calculated for that specific tidal water level. This outfall volume decreases the volume of water in the polder, so the open–water level in the polder decreases. This new water level is the input for the next time–step.

As an example, the width of a tidal outlet structure is designed. This tidal sluice drains a polder area of 6000 ha. The open–water system has an area of 125 ha, and is supplied by a field drainage system at a rate of 10 mm/day. The sill elevation is located at the bed level of the drainage canal at 1.20 m–MSL. The tide in the sea is a "single–daily" tide, i.e. a cycle of 12½ hours high water and 12½ hours low water (Figure 4.10). The water level in the drainage system after flushing must be again at the target water level at mean sea level (MSL). The water level will rise during the period that the tidal gates are closed, e.g. during about 10 hours, thus 10/24 × 6000 / 125 × 0.010 = 0.20 m. The total volume that have to be discharged during one sluicing–period amounts to 25/24 × 0.010 × 6000 × 10⁴ = 625,000 m³.

The flow will be mainly free flow. Submerged flow will occur only for $z < 1/3\ H$, i.e. for the higher outer water levels:

– during the beginning of the sluicing period for an inner water level of 0.20 m+MSL, when $z < 1/3 \times 1.40 = 0.47$ m, thus for an outer water level of $y > 0.93$ m (> 0.27 m–MSL),

– at the end of the sluicing period for an inner water level of 0.00 m+MSL, when $z < 1/3 \times 1.20 = 0.40$ m, thus for an outer water level of $y > 0.80$ m (> 0.40 m–MSL).

The calculation is made by means of a simple computer program or by hand

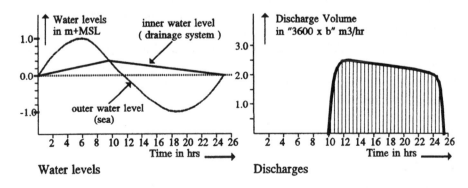

Figure 4.10. Water levels and water volumes of a tidal outlet.

(Table 4.7). The time–step may be taken at one hour. The total volume of outflowing discharge depends on the sluice width, b, and amounts to $34.28 \times b \times 3600 = 123,408 \times b$ m^3. It was already known that the total volume equals to 625,000 m^3, so that $123,408 \times b = 625,000$. It means that the minimum required width, b, of the sluice amounts to $b = 5.06$ m.

Pumped drainage is required from low–lying polders where drainage by gravity is impossible. Different types of pumps are available, such as Archimedean screw pumps, rotodynamic pumps (Smedema & Rycroft 1983). The annual energy costs for pumping depends on the price per kWh and the running hours per year, as well as on the power consumption. The power consumption of pumped drainage may be calculated as:

$$E = \frac{9.8 \; Q \; H}{\eta} \tag{4.12}$$

where: E is the power consumption in kW, Q is the discharge in m^3/s, H is the head in m, η is the pump efficiency ($\eta \approx 0.65$).

Table 4.7. Design of a tidal outlet.

Timestep in hrs	Water levels				Calculation			
	Inside "H"		Outside "y"		Head loss "z"	Type of Flow		Outfall Volume
	m+MSL	m	m+MSL	m	m			m^3
11–12	+0.19	1.39	−0.02	1.18	0.21	submerged		2.38×b×3600
12–13	+0.17	1.37	−0.10	1.10	0.27	submerged		2.51×b×3600
13–14	+0.16	1.36	−0.38	0.82	0.54	free flow		2.70×b×3600
14–15	+0.14	1.34	−0.56	0.64	0.70	free flow		2.64×b×3600
15–16	+0.13	1.33	−0.80	0.40	0.93	free flow		2.61×b×3600
16–17	+0.11	1.31	−0.88	0.32	0.99	free flow		2.55×b×3600
17–18	+0.10	1.30	−0.96	0.24	1.06	free flow		2.52×b×3600
18–19	+0.08	1.28	−1.00	0.20	1.08	free flow		2.46×b×3600
19–20	+0.07	1.27	−0.96	0.24	1.03	free flow		2.43×b×3600
20–21	+0.05	1.25	−0.92	0.28	0.97	free flow		2.38×b×3600
21–22	+0.03	1.24	−0.82	0.38	0.86	free flow		2.35×b×3600
22–23	+0.02	1.22	−0.70	0.50	0.72	free flow		2.29×b×3600
23–24	+0.01	1.21	−0.56	0.64	0.57	free flow		2.26×b×3600
24–25	+0.00	1.20	−0.33	0.87	0.33	submerged		2.20×b×3600

Total Volume of outflowing discharge: 34.28×b×3600

Remark: free flow: $Q = 1.7 \; b \; H^{3/2}$, for $z > 1/3 \; H$
submerged: $Q = 4.4 \; b \; y \; \sqrt{z}$, for $z < 1/3 \; H$

External water system of the polder

The polders evacuate the drainage water from the polder canals to the sur–
rounding water. This open–water can be the sea. More often the open–water
surrounding the polders is fresh, and consists of an extensive system of channels
and lakes (Wagret 1968).

The surrounding fresh water system of a polder may have an important
function on the water management in the polder and at the surrounding "old"
land:

– discharging the drainage flow of the individual polders and the old–land
  to the sea;
– supply of fresh–water to the polder;
– the percolation control in the surrounding old–land for deep polder water
  level.

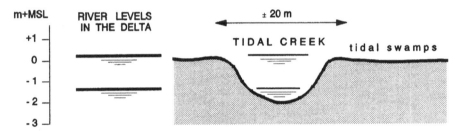

BEFORE YEAR 1300: Open (tidal) creeks without dikes.

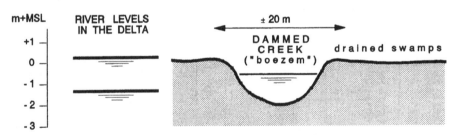

AROUND YEAR 1300: Creeks are dammed from the Sea.

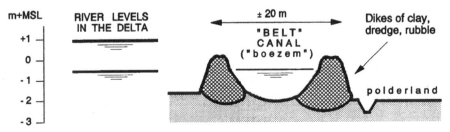

AT PRESENT: Subsidence of the polderland.

Figure 4.11. The transfer of tidal creeks to "belt canals" in the Netherlands.

The old tidal creeks can form the external open–water system in the modern polder landscape, like in the western part of the Netherlands (Luijendijk et al 1988, Ven 1993). Here, groups of farmers made polders some 1200 years ago by constructing embankments along the creeks (Figure 4.11). The creeks were closed off from the sea by dams, and their water level was maintained some 0.50 m below mean sea level to allow tidal flushing to the sea. The land sub-sided gradually, so windmills and later pumps had to be used to drain the polders on the external open–water system of higher–lying canals, lakes or former rivers.

The old system of tidal creeks still exists in the Netherlands, and a de-centralized water management became possible (Figure 4.12). The water management of the external storage system of belt canals is done by a higher authority than that of the individual polders. This authority is charged to maintain a target water level in the system of belt canals by evacuating water into the sea by means of a sluice or by a pumping station. The individual polders pump their drainage water into the surrounding "belt" canal system. However, pumping into the belt canal system might be prohibited when a certain maximum belt water level is reached, to protect the polder embankments against overtopping.

The water balance equation of the belt canal system expressed as a water depth, reads:

$$P\,\Delta T \; + \; R\,\Delta T \; + \; I\,\Delta T \; = \; Q\,\Delta T \; + \; \Delta V \qquad (4.13)$$

where for $\Delta T = 1$ day: P is the precipitation directly on the belt canal system in mm, R is the gravity runoff from the non–polder areas in mm, I is the pumping discharge from the polders in mm, Q is the discharge from belt system in mm, ΔV is the storage in the belt canal system in mm.

An example of the water balance of belt canal system in the Netherlands is presented here. Some 60 polders pump their drainage water individually on a belt canal system. The total polder area of 25,000 ha, and the installed pump capacity is 50 m^3/s. Also some higher urban areas of 8500 ha with a runoff coefficient of 40% drain by gravity to the belt canal system. The surface area of the belt canal system is 680 ha, i.e. some 2% of the total surface area. The

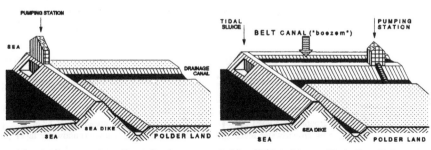

Polder discharging directly to the Sea.   Polder discharging to "Belt Canal" System.

Figure 4.12. Relation between polder, belt canal and sea.

Table 4.8. Example of a water balance of a belt canal system.

Parameters of water balance	water depth in belt system
o Precipitation, P, directly on belt canal system	40 mm
o Runoff, R, from the urban areas: 0.40 × 40 × 8500/680	200 mm
o Discharge, Q, from belt system: 52 × 24 × 3600 × 10^{-1} / 680	−660 mm
o Storage, ΔV, in the belt canal system	−150 mm+
o the pumping discharge, I, from the polders	−570 mm

installed pump capacity from the belt canal system is 52 m³/s. The design rainfall with a frequency of 1 : 10 years amounts to 40 mm/day. The polders are allowed to pump on the belt canal system as long as the water level rise in the belt canal system is less than 0.15 m. Table 4.8 presents the different terms of the water balance. It is calculated in the water balance that the polders may pump 570 mm (belt) water depth on the belt canal system, i.e. 0.570 × 680 × 10^4 = 3876 × 10^3 m³ during ΔT = 1 day. This means a continuous pumping discharge from the combined 60 polders of 3876 × 10^3 / (24 × 3600) = 45 m³/s. This means that all pumping stations may discharge on the belt canal system during only 90% of the day. Furthermore, a (polder) water depth of 3876 × 10^3 / (25000 × 10^4) = 0.0016 m = 16 mm has been pumped. It means that 40 − 16 = 24 mm water depth is stored in the polders, either as the soil–storage, or as open–water storage.

The external water system of the polder may also have a function on water storage, either during periods of water surplus or during periods of water shortages. Specially fresh water lakes with large surface areas and some allowable level variations are suitable for storage.

An additional function of the surrounding open–water was only understood since 1940, after the groundwater problems with the North–east Polder in the Netherlands (Volker 1991). The North–east polder appeared to have a serious draining effect on the adjacent old–land, as this polder lowers the groundwater table in a vast area (Figure 4.13). This effect was prevented at the next IJsselmeer polders in the Netherlands by constructing a wide lake of 1 – 5 km between the polder and the old–land.

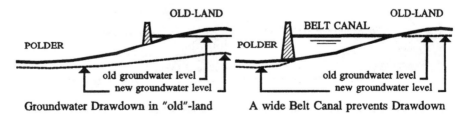

Figure 4.13. A belt canal prevents the groundwater draw–down in the old–land.

REFERENCES

Ankum, P., K. Koga, W.A. Segeren & J. Luijendijk. 1988. Lessons from 1200 years impoldering in the Netherlands. *Proc. int. symposium on shallow sea and low land, Saga*: 101–108. Saga University, Japan.

ICID. 1967. *Multilingual technical dictionary on irrigation and drainage.* International commission on irrigation and drainage. New Delhi.

ILRI. 1974. *Drainage principles and Applications.* Publication No. 28, International Institute for land reclamation and improvement (ILRI), Wageningen, the Netherlands.

Kley, J. van der, & H.J. Zuidweg. 1969. *Polders en dijken* (in Dutch). Agon Elsevier, Amsterdam.

Luijendijk, J. & E. Schultz. 1982. Waterbeheersingssysteem van een polder. *PT Civiele Techniek Nr.9/82*: 21–29 (in Dutch). NIRIA, the Hague, the Netherlands

Luijendijk, J., E. Schultz & W.A. Segeren. 1988. Polders. *Developments in hydraulic engineering,* volume 5 (P. Novak ed.). Elsevier, London/New York

Schultz, E. 1983. From natural to reclaimed land. *Final report, int. symposium polders of the world.* International institute for land reclamation and improvement (ILRI), Wageningen, the Netherlands.

Schultz, E. 1992. *Waterbeheersing van de Nederlandse droogmakerijen* (in Dutch). Dissertation, University of Technology, Delft, the Netherlands.

Segeren, W.A. 1982. Introduction to polders of the world. *Final report, int. symposium polders of the world.* International institute for land reclamation and improvement (ILRI), Wageningen, the Netherlands.

Smedema, L.K. & D.W. Rycroft. 1983. *Land Drainage.* Batsford, London.

Ven, G.P. van de (ed.). 1993. *Man–made lowlands, history of water management in the Netherlands.* ICID, the Hague, the Netherlands

Volker, A. 1991. *Land development and land reclamation.* Lecture notes, International institute for hydraulic and environmental engineering (IHE), Delft, the Netherlands.

Volker, A. (ed.). 1993. *Hydrology and water management of deltaic areas.* Ministry of transport, public works and water management, CUR–report 93–5, the Netherlands.

Wagret, P. 1968. *Polderlands.* Methuen, London.

5 COASTAL PROTECTION AND REVETMENTS

Y. HOSOKAWA
Port and Harbour Research Institute, Yokosuka, Japan

INTRODUCTION

Planning

Coastal protection facilities such as seawalls and revetments are built along existing beaches and waterlines of land to protect the land areas from high waves or high water during a storm. Accordingly, these facilities and protection systems should be designed and built to be stable and safe enough against external forces acting on them. At the same time, the areal integration of each facility and system component is also required for planning the effective development of coastal protection. Isolated local protection with hard facility might sometimes cause serious effects of erosion or overtopping to the neighboring coasts. Dynamics of the external forces and internal responses should be analyzed carefully for planning coastal protection.

External natural conditions to be taken into account shall be;
a) Waves and wave forces;
b) Tides and anomalous sea level;
c) Currents and current forces;
d) Estuarine hydraulics and littoral drift;
e) Soil conditions;
f) Earthquake and seismic forces;
g) Buoyancy; and
h) Others.

These conditions shall be chosen with careful consideration for environmental, service and construction conditions of the facility as well as characteristics of the materials and the social requirements.

Safety levels attained by the protection facilities are usually evaluated from the view point of potential capability against inundation and strength of facilities against the affecting external forces. Associated systems such as drainage system and building setbacks are also needed for protection against inundation damages. For some coastal lowlands, other important factors such as irrigation capacity, aquacultural availability or tourism utilization might also be considered in the design. Historical background and traditional knowledge of each coastal community for coping up with the cyclones or extreme climatic events at the place, may also be appreciated and useful in addition to such economic considerations as the cost–effectiveness analysis.

In this chapter, the dynamics of the external forces are reviewed at first. Then, a design methodology is explained for the revetments. The design still has some technically unclear parts and relies to some extent on empirical estimations. This design methodology is based on the Japanese Standards (e.g. OCDI 1991) developed by Ministry of Transport for coastal protection of the Japanese coasts. Japan has a long history of severe attacks of typhoons with heavy damages induced by them. Coastal areas have been intensively used and the population density in the coastal flats is extremely high.

Design flow

The main external forces considered for the design of the coastal revetments change with the facing depth. Hard revetment is common along the coast facing in the deeper sea for protection against a) incident waves, b) high tide, and c) storm surges. Simpler and light banking can be seen on the land areas above the high water level. Foundation characteristics, including the bearing capacity, are also important factors for soft ground areas such as deltaic lowlands at the mouth of a river. Earth pressures from the soils behind and overtopping into the adjacent land areas are the two characteristic factors for revetments. The above four conditions or factors; namely incident waves, high tide, storm surges and bearing capacity; are the main considerations for designing the cross–section of the revetments in order to secure the desired safety levels for stability and overtopping rate.

EXTERNAL FORCES

Waves

Definition of waves
Waves to be used in the design of the protective facilities shall be determined from data with the help of appropriate statistical analysis. Transformation and deformation of waves at shallow coastal sea shall also be considered. Procedure for determining the design waves is shown in Figure 5.1.

Various waves are used in the design as defined below:
a) Significant wave ($H_{1/3}$, $T_{1/3}$): A hypothetical wave having wave height and period which are equal, respectively, to average values of the wave height and period of the largest one–third of all waves in the train as counted in the order of greater wave height.

Wave Data

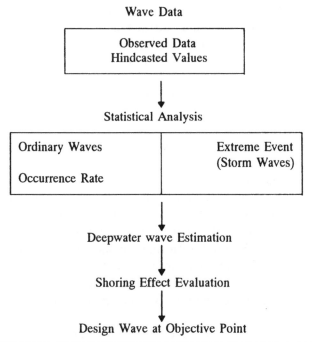

Figure 5.1. Procedure to determine the design waves.

b) Highest wave (H_{max}, T_{max}): Wave of the maximum wave height in the wave train.

c) Deepwater wave (H_0, T_0): Wave at a point where the water depth is equal to half of the wave–length or greater, to be expressed in terms of the parameters of the significant wave.

d) Equivalent deepwater wave height (H'_0): A hypothetical wave with the height of deepwater wave corrected for the effects of submarine topographic changes such as refraction and diffraction of wave, but excluding wave shoring and breaking, to be expressed in terms of the significant wave height. Equivalent deepwater wave height (H'_0) can be expressed as:

$$H'_0 = Kr.Kd.(H_{1/3})_0 \tag{5.1}$$

where $(H_{1/3})_0$ is the deepwater significant wave height. *Kr* and *Kd* are the coefficients of wave refraction and diffraction, respectively. Equivalent deepwater wave height is useful for the estimation of wave shoring.

Parameters of the deepwater waves shall be determined from the following:
a) observed values of waves over a considerably long period,
b) hindcasted values based on the meteorological data over a long period. This value should be obtained through appropriate statistical processing and validated by observed data.
c) hindcasting values based on the hypothetical cyclones/typhoons.

Wave deformation in shallow area
When deepwater waves approach from offshore to the shore line, they begin to feel the friction of the sea bottom to refract and begin to stop progressing by islands or structures to diffract. When these waves reach into more shallower area, they deform rapidly. Shoring transformation should be considered. When water depth becomes shallower than around three times of the equivalent deepwater wave height ($d < 3.H'_0$), wave breaking and attenuation of wave height occur. Dissipated wave energy is converted to littoral current or wave–set up.

Refraction due to local change of wave celerity with change of water depth, alters the wave height and direction.

Reflection of the incident wave energy by a structure increases wave height in front of the structure. Superposition of the energy of incident waves and that of reflected waves gives the local wave energy of the adjacent area. Wave height is proportional to the square root of wave energy.

Shoring changes the local wave height at a certain depth. Local wave height ($H_{1/3}$) at a certain depth (h) is expressed as the product of the shoring coefficient (Ks) and H'_0 for $h/L_0 > 0.2$. Shoring coefficient (Ks) varies with wave steepness (H'_0/L_0), and relative depth (h/L_0) as shown in Figure 5.2 (Goda et al. 1975). Above the dotted line with the legend "wave breaking" in the figure, the attenuation of $H_{1/3}$ exceeds 2%. In such shallower zones, wave breaking should be taken into account.

Breaking occurs when waves progress into much shallower areas near the coast. Oscillatory wave motion before breaking changes into strong turbulent waves with air entrainment after breaking. Limiting the breaker height of regular waves is shown in Figure 5.3 (Goda 1970) as an average relation.

Tide

Water levels change periodically, semidiurnal, diurnal or longer periods, due to astronomical effects.   This tidal fluctuation of water level reaches the comparative magnitude as incident wave height along many coastal lines. In some areas, meteorological tide and/or seiche are also significant for water level fluctuations. High water levels in front of the seawall will easily increase the

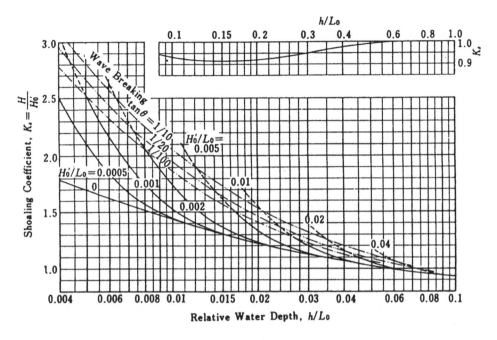

Figure 5.2. Diagram of nonlinear wave shoring (relative water depth vs. shoring coefficient).

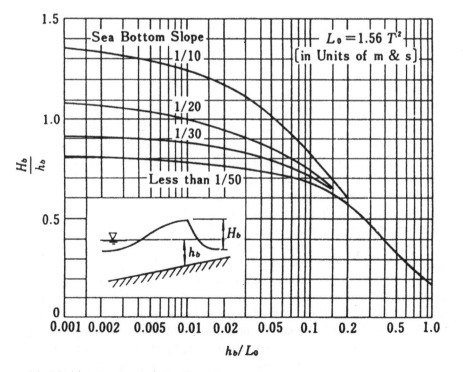

Figure 5.3. Limiting breaker height of regular waves.

possibility of wave overtopping and the safety factor against stability reduces. Such extremely high water levels as those exceeding the crest height of the protection facilities means serious damages on the land and severe inundation.

Storm surge

Storm surge and tsunami attack are extreme events which give high water above the astronomical tidal level to cause coastal disasters. Increase of water level along the coast by a storm is generated mainly through: a) suction due to the low atmospheric pressures, b) water set up near the shore due to the wind blowing on the water surface, and c) river discharge to the coast due to heavy precipitation on the catchment.

Magnitude of the uplifting of water level due to a storm can be estimated by the numerical simulation or by the observed water level data from past storm attacks. Records of runup height on the coastal land are often referred for the coastal protection planning. A storm surge simulation model usually includes the effects of above a) suction and b) set up. Assigning the depth of the central low pressure and the magnitude of storm radius, a hypothetical storm runs in the model field from the outer sea area through the coastal line. If the wind field can be given on the sea surface, wind waves due to a storm passing can also calculated by a wind wave model.

DESIGN OF REVETMENT

Design criteria

The main body of the protection facilities must be strong enough to withstand the storm attacks. Stability should be checked for the following factors in general: a) sliding of the wall, b) bearing capacity of the foundation, c) overturning of the wall, d) circular slip and settlement. Weight of the revetment, earth pressure with residual water pressure, and buoyancy also need to be considered as well as acting wave forces.

Overtopping is another important criterion. The crown height of the protection facility must be high enough to prevent the intrusion of sea water onto the land by overtopping. Wave overtopping is primarily controlled by the absolute heights (water level + wave height) of waves relative to the crown elevation of the wall. The mean rate of overtopping water volume per unit length during the occurrence of storm waves is a typical index, denoted as q (m^3/m.s). For vertical revetments, design curves or diagrams of q are proposed by Goda et al. (1975), some of which are shown in Figure 5.4 for a sea bottom slope of 1/30. Overtopping rate is sensitive to relative depth (h/H'_0), relative crest height (hc/H'_0), and wave steepness (H'_0/L_0). Wind is also an important factor for estimating overtopping, although the effect cannot be estimated quantitatively

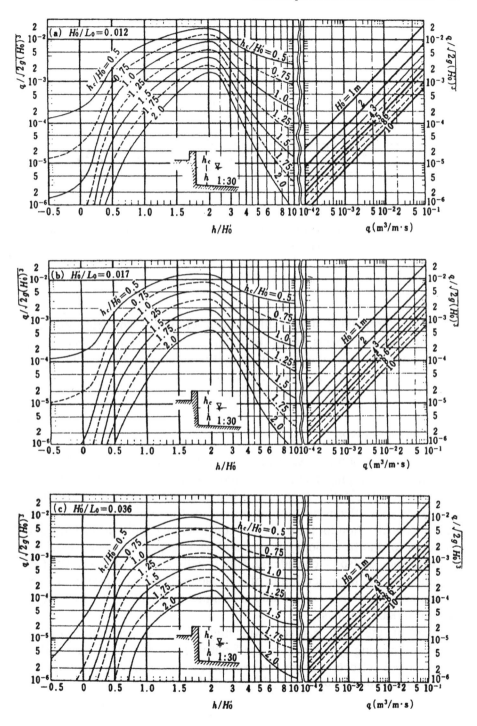

Figure 5.4. Design diagrams of wave overtopping rate of vertical revetments on a sea bottom slope of 1/30 (Goda et al. 1975).

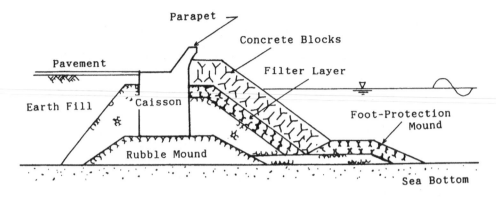

Figure 5.5. Idealized typical section of a seawall built with concrete block mound of the energy–dissipating type (Goda 1985).

as of now.

During a storm attack, drag and lift forces act toward armouring stones settled at the foot of revetments or on the outer surface of slope mound due to the wave motion. Necessary minimum weight is proportional to the third power of the acting wave height. As wave forces decrease with depth, less weight is required for the armouring stone at a deeper place.

Typical cross section

Typical cross section of a seawall in Japan is shown in Figure 5.5. This wall is made of a) sloping mounds of rubble stones, b) concrete blocks of the wave–energy–dissipating type and c) heavy caisson with parapet on the top.

The setting of tolerance limits according to structural type may be too crude without consideration for the particular construction conditions including surface armouring of the front slope, crown, and back slope. Drainage capacity at the backyard should also be considered. But, usually a value of 0.005–0.2 $(m^3/m.s)$ is applied for the tolerance limits of q in Japan. If the safe passage of vehicles is to be guaranteed at all times, tolerable limits may have to be decreased to the order of 0.0001 $(m^3/m.s)$.

Setting of the high parapet, widening of the crown, and installing of energy–dissipating type blocks at the front slope often reduce the overtopping, though these effects cannot be estimated quantitatively without careful experiments with appropriate hydraulic models.

REFERENCES

Goda, Y. 1970. *Deformation of irregular waves due to depth-controlled wave breaking.* Report of the Port and Harbour Res. Instt. 9(3):59–106 (in Japanese).

Goda, Y., Y. Kishira & Y. Kamayama 1975. *Laboratory investigation on the overtopping rate of seawalls by irregular waves.* Report of the Port and Harbour Res. Instt. 14(4):3–44 (in Japanese).

Goda, Y. 1985. Random seas and design of maritime structures. University of Tokyo Press, 323p. (Above two reports of the same author are summarized in this book in English).

OCDI 1991. *Technical standards for port and harbour facilities in Japan.* The Overseas Coastal Area Development Institute of Japan, 438p.

6 IMPROVEMENT TECHNIQUES FOR SOFT AND SUBSIDING GROUND

D. T. BERGADO[*] and N. MIURA[**]
[*]*Asian Institute of Technology, Bangkok, Thailand*
[**]*Saga University, Saga, Japan*

INTRODUCTION

General

Various soil/ground improvement methods are presented that have been tested to provide improvement in soil strength, mitigation of total and differential settlements, shorten construction time, reduce construction costs, and other characteristics which may have impact on their utilization to specific projects on soft ground. In general, the term soft ground includes soft clay soils, soils with large fractions of fine particles such as silts, clayey soils which have high moisture content, peat foundations, and loose sand deposits near or under water table (Kamon & Bergado 1991). For clayey soils, the softness of the ground can be assessed by its undrained strength, S_u, or by its unconfined compressive strength, q_u. On the other hand, the SPT N–values are utilized to ascertain the consistency of the ground and its relative density. Table 6.1 outlines the identification of soft ground according to the types of the structures using the aforementioned assessment methods. Considering such factors as the significance of the structure, applied loading, site conditions, period of construction, etc. it becomes important to select appropriate method suitable for specific soil types as tabulated in Table 6.2. For soft and cohesive soils in subsiding environments, ground improvement by reinforcement (i.e. sand compaction piles), by admixtures (i.e. deep mixing method) and by dewatering (i.e. vertical drains) are applicable. For loose sand deposits, various in–situ compaction methods are applicable such as dynamic compaction, resonance compaction, vibroflotation, sand compaction piles, etc.

A chart showing the procedures for modelling and selecting shallow ground improvement techniques has been presented by Kamon & Bergado (1991) as given in Figure 6.1. Figure 6.2 shows the selection procedures for deep ground improvement techniques. For infrastructures on embankment fill on soft ground, soil/ground improvement is not limited to portions below ground but

Table 6.1. Outline for identification of soft ground.

Structure	Soil condition	N-value (SPT)	q_u (kPa)	q_e (kPa)	Water content (%)
Road	A: Very soft	Less than 2	Less than 25	Less than 125	
	B: Soft	2 to 4	25 to 50	125 to 250	
	C: Moderate	4 to 8	50 to 100	250 to 500	
Express highway	A: Peat soil	Less than 4	Less than 50		More than 100
	B: Clayey soil	Less than 4	Less than 50		More than 50
	C: Sandy soil	Less than 10	–		More than 30
Railway	(Thickness of layer)				
	More than 2m	0			
	More than 5m	Less than 2			
	More than 10m	Less than 4			
Bullet train	A	Less than 2		Less than 200	
	B	2 to 5		200 to 500	
River dyke	A: Clayey soil	Less than 3	Less than 60		More than 40
	B: Sandy soil	Less than 10			
Fill dam		Less than 20			

Table 6.2. Applicability of ground improvement for different soil types.

Improvement mechanism	Reinforcement	Admixtures or grouting	Compaction	Dewatering
Improving period	Depending on the life of inclusion	Relatively short term	Long–term	Long–term
Organic soil Volcanic clay soil Highly plastic soil Lowly plastic soil Silty soil Sandy soil Gravel soil	↕	↕	↕	↕
Improved state of soil	Interaction between soil and inclusion (No change in soil state)	Cementation	High density by decreasing void ratio (change in soil state)	

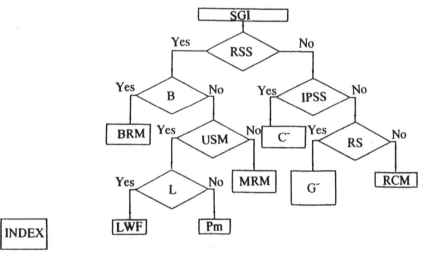

INDEX

SGI: Shallow Ground Improvement   RSS: Do you Replace the Soft Soils?
B: Do you use Blast?
IPSS: Do you Improve Properties of Soft Soils?
BRM: Blasting Replacement Method   USM: Do you Use Special Material?
C ¯: Cement and/or Lime Stabilization Method
RS: Do you Reinforce the Soils?   L: Is it Light?
MRM: Mechanically Replacement Method
G ¯: Geotextile – Sheet, Net and/or Grid Reinforcing Method
RCM: Roller Compacted Method
LWF: Light Weight Fill Method
Pm: Pre–mixed Soil Method

Figure 6.1. Selection flow of shallow ground improvement technique.

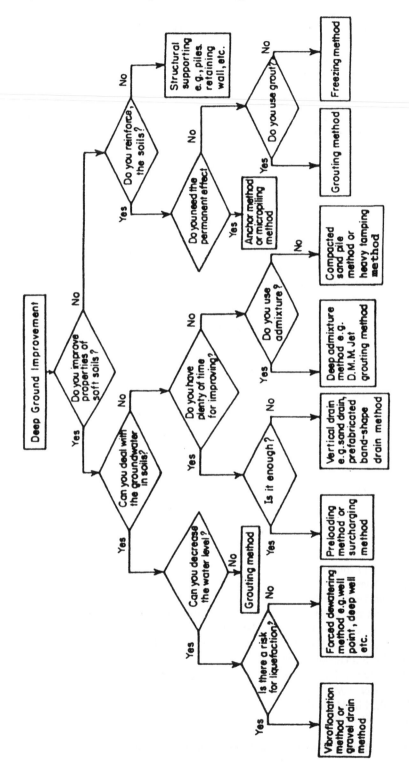

Figure 6.2. Selection flow of deep ground improvement technique.

also includes improvement of fill soils by reinforcing with grids and/or geotextiles as well as by the use of lightweight materials such as expanded polystyrene and other related products. In this chapter, the discussions are limited to ground improvement techniques of soft ground using vertical band drains, granular piles, and lime/cement mixing methods.

The existing conditions

Associated with the low strength and high compressibility of the soft ground are the problems of low bearing capacity and unstable embankment and excavation slopes. Thick deposits of soft clay consolidate and cause large settlements when loaded. Saturated loose sands may liquefy under seismic loads during earthquakes. Loadings may also result directly from engineering structures (e.g. fill embankments) or due to increases in effective stresses caused by piezometric drawdown due to excessive extraction of groundwater causing ground subsidence. Ground subsidence studies on soft Bangkok clay in the Chao Phraya Plain of Thailand revealed variable amounts of surface settlements causing severe cracking of pavements. Differential ground subsidence has been attributed to the variable occurrences of silts and fine sand lenses with compressible soft clay layer. Further, the areas with the greatest amount of subsidence often coincided with the presence of weak pockets within the soft clay layer.

GROUND IMPROVEMENT BY VERTICAL DRAINAGE

General

Vertical drains are artificial drainage paths created to accelerate consolidation of soft clay grounds. Early applications of vertical drains utilized sand drains which are formed by infilling sand into a hole in the soft ground. There are two categories of installation methods, namely: displacement and non-displacement types. In the displacement type, closed–end mandrel is pushed into the soft ground with resulting displacements in both vertical and lateral directions. Whereas, the non–displacement type installation requires pre-drilling the hole by means of power auger or water jets and is considered to have less disturbing effects on soft clay. Non–displacement sand drains were tested during the 1984 Second Bangkok International Airport site study.

Prefabricated band drains

A prefabricated band drain consists of a synthetic filter jacket surrounding a plastic core having the following characteristics:

a) Ability to permit porewater from the soil to seep into the drain
b) Transmit the collected porewater along the length of the drain

The jacket material consists of non–woven polyester or polypropylene geotextiles or synthetic paper that functions as physical barrier separating the flow channel from the surrounding soft clay soils and a filter to limit the passage of fine particles into the core to prevent clogging.

The plastic core serves two vital functions, namely: to support the filter jacket and to provide longitudinal flow paths along the drain even at large lateral pressures. Some details of various drain cores are shown in Figure 6.3.

Consolidation with vertical drains

Barron (1948) presented solution to the problem of consolidation of soil cylinder with central drain. Assuming equal vertical strain, the differential equation governing the consolidation process is given as:

$$\frac{\partial U}{\partial t} = C_h[(\frac{\partial^2 U}{\partial r^2}) + \frac{1}{r}(\frac{\partial U}{\partial r})]$$
(6.1)

where U is the average excess pore pressure at any point and at any given time; r is the radial distance of the considered point from the center of the drained cylinder; t is the time after an instantaneous increase in the total vertical stress, and C_h is the horizontal coefficient of consolidation.

Hansbo (1979) modified the equations developed by Barron (1948) for the solution of the governing differential equation by incorporating the physical dimensions and characteristics of the band drain. The modified general expression for the average degree of consolidation is given as follows:

$$U_h = 1-\exp(\frac{-8T_h}{F})$$
(6.2)

and

$$T_h = \frac{C_h t}{D_e^2}$$
(6.3)

and

$$F = F(m)+F_s+F_r$$
(6.4)

where D_e is the equivalent diameter of the soil cylinder being drained; C_h is the coefficient of horizontal consolidation; F is the factor expressing the additive effect due to the spacing of the drains, $F(m)$; smear effect, F_s; and well–

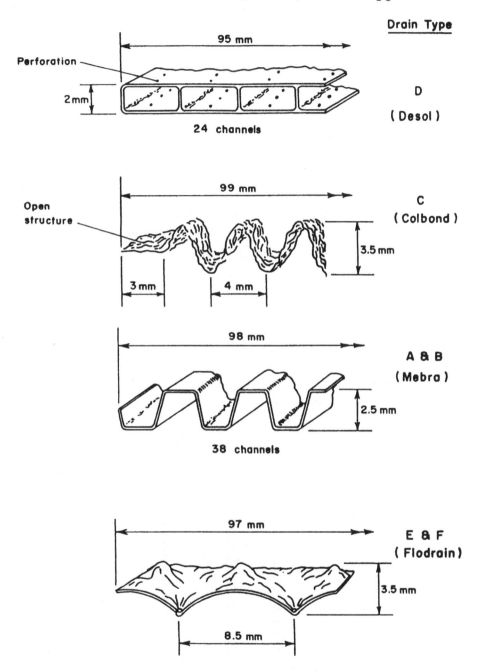

Figure 6.3. Geometrical shapes of various drain cores.

resistance, F_r. For typical values of the spacing ratio, n, of 20 or more, the spacing factor simplifies to:

$$F(m) = \ln(m) - \frac{3}{4}$$

(6.5)

and

$$m = \frac{D_e}{d_w}$$

(6.6)

where d_w is the equivalent diameter of the band drain. To account for the effects of soil disturbance (smear) during installation, a zone of disturbance with a reduced permeability is assumed around the vicinity of the drain as shown in Figure 6.4 and the smear factor is given as:

$$F_s = [(\frac{K_h}{K_s} - 1) \ln(\frac{d_s}{d_w})]$$

(6.7)

where d_s is the diameter of the disturbed zone; and K_h is the coefficient of permeability in the horizontal direction of the undisturbed soil, and K_s is the coefficient of permeability in the disturbed zone. Since the band drain has limited discharge capacity, a drain resistance factor, F_r, was developed for flow along the longitudinal axis of the drain as follows:

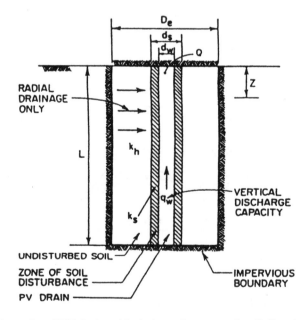

Figure 6.4. Schematic of PV drain with drain resistance and soil disturbance.

$$F_r = \pi z (L-z) \frac{K_h}{q_w} \tag{6.8}$$

where z is the distance from the drainage end of the drain; L is the length of the drain when drainage occurs at one end only; L is half the length of the drain when drainage occurs at both ends; and q_w is the discharge capacity of the drain at hydraulic gradient of 1.

To obtain the extent of disturbances around the drain, the following relationships are recommended:

$$d_s = 2d_m \tag{6.9}$$

and

$$K_s = K_v \tag{6.10}$$

where d_s is the diameter of the disturbed zone; d_m is the equivalent diameter of the mandrel; and K_v is the vertical permeability of the undisturbed soil.

Assuming the aforementioned relationships, the time, t, to obtain a given degree of consolidation at an assumed spacing of drains, is given as follows:

$$t = (\frac{D_e^2}{8C_h})(F(m)+F_s+F_r)\ln(\frac{1}{(1-U_h)}) \tag{6.11}$$

The greatest potential effect on consolidation time is due to the variation of C_h and D_e. C_h can easily vary by a factor of 10 and D_e by a factor of 2 to 3. The term D_e reflects the effects of drain spacing. For convenience on the part of the users in designing vertical drain scheme, a design graph devised by the authors is shown in Figure 6.5. This design graph incorporates both the effects of smear and well–resistance.

Case records on soft and subsiding Bangkok clay

In 1983, three test embankments were designed, constructed, and monitored at the proposed site of the Second Bangkok International Airport at Nong Ngu Hao, Thailand (Dept. of Aviation 1984). Non–displacement sand drains of 260 mm diameter at 2.0 m spacing were installed by non–displacement jet bailer method under the test sections down to 14.5 m. Surcharge loading up to 4.2 m high was used at one test section (Moh & Woo 1987) and vacuum preloading at the other section (Woo et al. 1989). The soil profile and piezometric pressure distribution are shown in Figure 6.6 indicating drawdown in the underlying layers. Subsequently, it was found that more than 50 percent of settlement occurred between 3 to 7 m depth and that the clay layer between 11

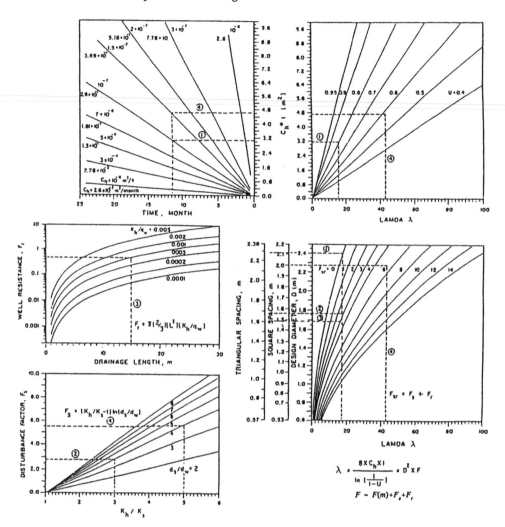

Figure 6.5. Design chart for flodrain.

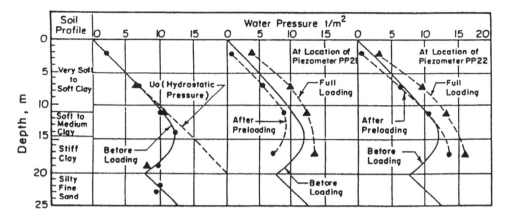

Figure 6.6. Porewater pressure changes before and after preloading (Moh & Woo 1987).

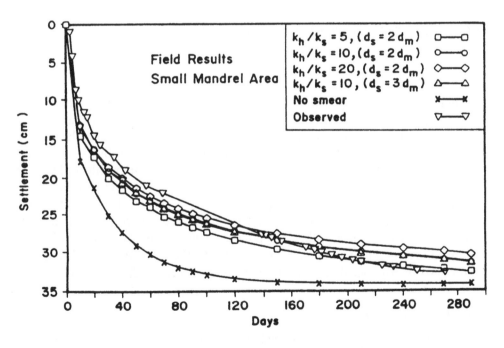

Figure 6.7. Comparison between predicted and observed settlements for test embankment on vertical drains.

to 15 m depth had little contribution to the total settlement.

A 4.0 m high test embankment for preloading was constructed inside AIT campus on an improved ground with prefabricated (Mebra) drains. The drains were installed in triangular pattern at 1.5 m spacing down to 8.0 m depth. The details of the test embankment are described in Bergado et al. (1990). Figure 6.7 shows the comparison between the predicted and the observed settlements by different methods using back–analyzed parameters (Bergado et al. 1990). The Asaoka (1978) and Skempton & Bjerrum (1957) methods yielded good predictions. From back–analysis, values of K_h/K_s = 10 and d_s/d_w = 2.5 were found to be appropriate. In addition, values of $C_h(field)/C_h(lab)$ = 4 and $C_v(field)/C_v(lab)$ = 12 were also obtained.

Bergado et al. (1991) installed band drains (Alidrains) at 1.2 m spacing in square pattern to 8.0 m depth in AIT campus. A small mandrel was used in one–half of the site and a large mandrel in the other half. The height of the test embankment was 5 m. The time–settlement relationship is shown in Figure 6.8 for the small mandrel area. The results of the back–analysis confirmed the proposal of Hansbo (1987) that $d_s = 2d_m$ and $K_s = K_v$ in the smeared zone.

GROUND IMPROVEMENT BY GRANULAR PILES

General

Granular piles are composed of compacted sand or gravel inserted into the soft clay foundation by partial or full displacement methods. This term also refers to stone columns. When the composite ground improved by granular piles is loaded, the piles deform by bulging laterally into the subsoil strata and distribute the stresses at the upper portion of the subsoil profile rather than transferring the stresses into deeper and stronger layers. The strength and bearing capacity of the composite ground is increased and the compressibility is reduced. In addition, less stress concentration is developed on the granular piles. Moreover, since the component material is granular with high permeability, granular piles can also accelerate the consolidation settlements, and consequently, accelerate the strength gain of surrounding clay subsoil.

Granular pile construction

Various methods of granular pile installation have been used all over the world depending on their proven applicability and availability of equipments. The following methods will be briefly described, namely: vibro–compaction, vibro–replacement, vibro–compozer, and cased–borehole methods.

The vibro–compaction method is used to improve the density of cohesionless, granular soils using a vibroflot which sinks into the ground under its own

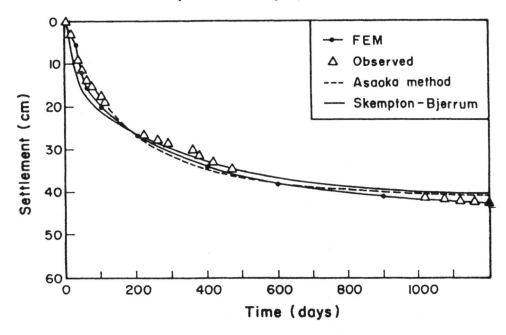

Figure 6.8. Time–settlement relationships at Alidrain test embankment: small mandrel area.

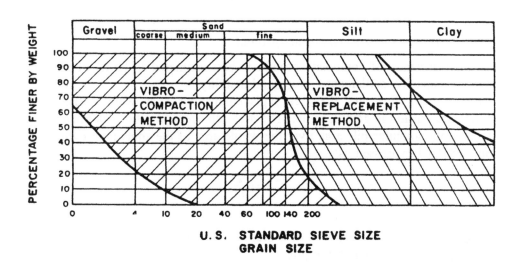

Figure 6.9. Range of soils suitable for vibro–compaction and vibro–replacement methods (Baumann & Bauer 1974).

weight and with the assistance of water and vibration (Baumann & Bauer 1974). After reaching the predetermined depth, the vibroflot is then withdrawn gradually from the ground with subsequent addition of granular backfill thereby causing compaction. The range of grain size distribution of soils suitable for this method is shown in Figure 6.9.

The vibro–replacement method is used to improve soils with more than 18% passing No. 200 U.S. standard sieve size. The equipment used is similar to that for vibro–compaction. The method can be carried out either with wet or dry process. In the wet process, the hole is formed by jetting the vibroflot with water while in the dry process, jetting is done by compressed air. The wet process is generally suited for unstable borehole and with high groundwater table. The grain size range of suitable soils for this treatment is also shown in Figure 6.9.

The vibro–compozer method has been popularized in Japan for stabilizing soft clays with high groundwater level using sand materials. The resulting pile is usually termed as sand compaction pile. The sand compaction piles are constructed by driving the casing pipe to the desired depth using vibratory hammer located at the top of the pipe. The casing is filled with specified volume of sand and the casing is then repeatedly extracted and partially redriven using the vibratory hammer starting from the bottom.

In the cased–borehole method, the piles are constructed by ramming granular materials in the prebored holes in stages using heavy falling weight (usually 15 to 20 kN) from a height of 1.0 to 1.5 m (Ranjan 1989).

Engineering behavior of composite ground

The performance of composite ground is best estimated in terms of ultimate bearing capacity, settlement and general stability.

Figure 6.10 illustrates the area replacement factor as well as the stress concentration in the granular pile. The area replacement ratio is defined as the ratio of the granular pile area over the whole area of the equivalent cylindrical unit cell and expressed as:

$$a_s = \frac{A_s}{A_s + A_c} \qquad (6.12)$$

where A_s is the horizontal area of a granular pile and A_c is the horizontal area of the ground surrounding the pile.

When the composite ground is loaded, the vertical deformation of the granular pile and the surrounding soil is approximately the same causing the occurrence of stress concentration in the granular pile which is stiffer than the surrounding cohesive or loose cohesionless soil. The distribution of vertical stress within the unit cell can be expressed by a stress concentration factor, n, defined as

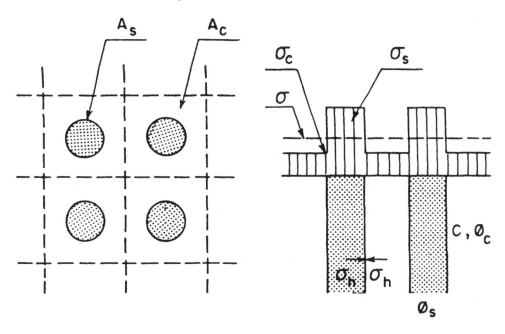

Figure 6.10. Diagram of composite ground.

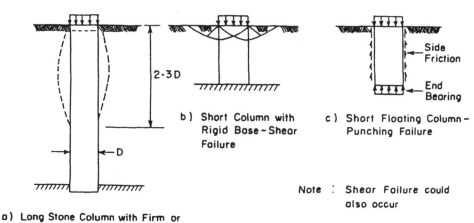

b) Short Column with
   Rigid Base - Shear
   Failure

c) Short Floating Column -
   Punching Failure

Note : Shear Failure could
       also occur

a) Long Stone Column with Firm or
   Floating Support - Bulging Failure

Figure 6.11. Failure mechanisms of a single granular pile in a homogenous soft layer
            (Barksdale & Bachus 1983).

follows:

$$n = \frac{\sigma_s}{\sigma_c} \qquad (6.13)$$

where σ_s is the vertical stress in the granular pile and σ_c is the vertical stress in the surrounding soil. The variation of stress concentration factor may range from 2 to 5 as compiled by Barksdale & Bachus (1983).

Failure mechanisms

The failure mechanisms of single and isolated granular piles are illustrated in Figure 6.11a,b,c, respectively, indicating the possible failures as: a) bulging, b) general shear, and c) punching. If the granular piles are long enough as commonly used in actual practice, punching failure may not occur. Usually, a foundation pad is provided on top of the composite ground such that general shear failure may not happen. Thus, the most probable failure mechanism is the bulging failure.

Ultimate bearing capacity

For single, isolated granular piles, the most probable failure mechanism is the bulging failure. The lateral confining stress which supports the granular pile is usually taken as the ultimate passive resistance which the surrounding soil can mobilize as the pile bulges outward. Various approaches have been developed in predicting the ultimate bearing capacity of single and isolated granular pile based on bulging failure mechanism by Vesic (1972), Hughes & Withers (1974), Hughes et al. (1975), Madhav et al. (1979), etc. Bergado & Lam (1987) conducted full scale testing of single and isolated granular piles. The results of the actual field tests are plotted in Figure 6.12 along with the aforementioned theoretical predictions.

The most common method for estimating the ultimate bearing capacity of granular pile groups assumed full strength mobilization and that the angle of friction in the surrounding cohesive soils as well as the cohesion in the granular pile material are negligible. For firm and stronger cohesive soils, Barksdale & Bachus (1983) suggested the expressions for ultimate bearing capacity of granular pile groups assuming the failure surface with two straight rupture lines and the loading by rigid foundation. For the case of very soft to soft cohesive soils, the pile group capacity is predicted using the capacity of a single and isolated pile located within the group and to be multiplied by the number of piles (Barksdale & Bachus 1983). Thus, the Terzaghi bearing capacity equation can be used to obtain the ultimate bearing capacity using a modified bearing capacity factor as follows:

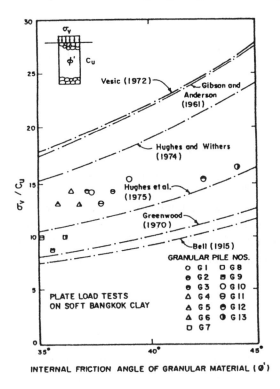

Figure 6.12. Relationship between internal friction angle of granular material, strength of surrounding clay and ultimate bearing capacity of single granular pile (Bergado & Lam 1987).

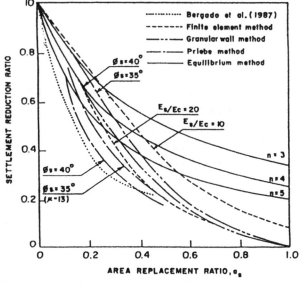

Figure 6.13. Comparison of estimating settlement reduction of improved ground (Aboshi & Suematsu 1985).

$$q_{ult} = cN_c'$$

(6.14)

where N_c' is the composite bearing capacity factor. For soft Bangkok clay, N_c' ranges from 15 to 18 using an initial pile diameter of 254 mm with gravel pile material compacted at 1.0 m lifts using 0.16 ton hammer and dropping 0.60 m (Bergado & Lam 1987).

Settlement of composite ground

In estimating the settlement of the composite ground, an infinitely wide, loaded area is assumed to be reinforced with granular piles having constant diameter and spacing. For this loading condition and geometry, the unit cell idealization is valid. The model of a unit cell loaded by a rigid plate is analogous to one–dimensional consolidation condition. To estimate the settlement of composite ground, a settlement reduction ratio, R, is devised as follows:

$$R = \frac{S_t}{S_o}$$

(6.15)

where S_t is the settlement of composite ground and S_o is the settlement of unimproved ground assuming one–dimensional consolidation. The settlement reduction ratio, R, is also expressed as function of the area replacement ratio, a_s, angle of the internal friction of granular pile materials, ϕ_s, stress concentration factor, n, and modulus ratio of granular pile material and surrounding soil, E_s/E_c. Figure 6.13 shows the relationship between the settlement reduction ratio and the aforementioned parameters together with the results of the work by Bergado & Lam (1987) on soft Bangkok clay.

Slope stability of composite ground

The average shear strength method is widely used for stability analysis of slopes and embankments constructed over composite ground improved with granular piles (Aboshi et al. 1979). The method considers the weighted average material properties of the materials within the unit cell (Figure 6.14). The soil having the fictitious weighted material properties is then used in stability analysis.

The shear strength of the granular pile and the surrounding cohesive soil are given as follows:

$$\tau_s = (\sigma_z^s \cos^2\beta)\tan\phi_s$$

and   (6.16)

$$\tau_c = c + (\sigma_z^c \cos^2\beta)\tan\phi_c$$

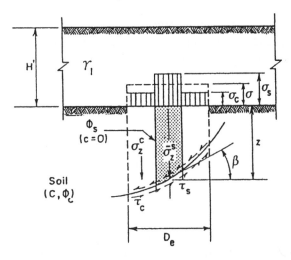

Figure 6.14. Average stress method of stability analysis (Barksdale & Bachus 1983).

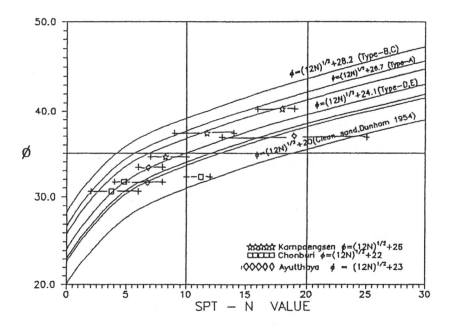

Figure 6.15. Internal friction angle (ϕ) vs SPT N-value.

where σ_z^s is the effective vertical stress in the granular pile and σ_z^c is the total vertical stress in the surrounding soil. The rest of the terms are defined in Figure 6.14. The average weighted shear strength, τ, within the area tributary to the unit cell is given as:

$$\tau = (1-a_s)\tau_c + a_s\tau_s \qquad (6.17)$$

The weighted average unit weight within the composite ground to be used in calculating the driving moment, is given as:

$$\gamma_{avg} = \gamma_s a_s - \gamma_c a_c \qquad (6.18)$$

where γ_s and γ_c are the saturated unit weight of the granular pile materials and the surrounding soil, respectively. Thus, the weighted shear strength and unit weight are calculated for each row of granular piles for use in the stability analysis.

Performance of granular piles on soft Bangkok clay

Bergado et al. (1984) conducted full scale test embankment loading on soft Bangkok clay improved by granular piles. The cased borehole method was employed in the construction of granular piles. The compaction was done by dropping 1.6 kN hammer at 0.6 m falling height for each 1.0 m lifts. Each layer was compacted with 15 blows. The 300 mm diameter granular piles were arranged in triangular pattern at 1.50 m center to center spacing having lengths of 8.0 m fully penetrating the soft clay layer.

Based on full scale plate load tests (Bergado et al. 1984, Bergado & Lam 1987), the bearing capacity of composite ground, improved with granular piles, increased by up to 4 times, the total settlements reduced by at least 30% and the slope stability safety factor increased by at least 25%. The comparative study of the performance of test embankments on granular piles and on vertical band drains indicated that the former settled about 40% less than the latter (Bergado et al. 1991). This confirms the notion that granular piles function more as reinforcement to the clay than as drains.

Using the finite element method (FEM), Bergado & Long (1992) analyzed the consolidation of composite ground due to embankment load by 2–D model using FEM program CON2D (Duncan et al. 1981) and by axi–symmetric model using FEM program CONSAX (D'Orazio & Duncan 1982). The prediction methods of Asaoka (1978) and Skempton & Bjerrum (1957) were also used. The results of FEM and Asaoka (1978) methods agreed with the observed data. The results of Skempton & Bjerrum (1957) method overestimated the results.

Future outlook of granular piles

There is no doubt that granular piles are more economical compared to precast reinforced concrete piles in supporting lightly–loaded structures on soft and subsiding ground. However, there is a need to reduce the local construction costs. Recently, Aoyama et al. (1990) presented results of prototype tests on granular piles with low replacement ratio ranging from 20 to 40 percent. For projects where certain degree of deformation is allowed, this technique is more economical. Another cost saving exercise is to use locally–available but low–quality construction materials which need to be investigated regarding its suitability for granular pile construction. Bergado et al. (1993) conducted model tests to study the suitability of locally–available clayey to silty sands as construction materials for sand compaction piles. The correlations between internal friction angle and cone resistance of the sand materials generally agreed with the graphical correlation suggested by Robertson & Campanella (1983). The correlation of internal friction angle with SPT N–values from model tests of locally–available sand closely followed the general equation by Dunham (1954) as shown in Figure 6.15.

GROUND IMPROVEMENT BY LIME/CEMENT MIXING METHOD

General

Lime/cement mixing method has been used to improve the properties of soils since Biblical times. The Swedish Geotechnical Institute and Prof. Bengt Broms have done extensive works on the application of lime column technique for foundations and earthworks. Modern applications of deep mixing in–situ soils has been done by Terashi et al. (1979) in Japan. The deep mixing method (DMM) originally was developed to improve the soft ground for port and harbor structures. DMM is now applied to foundations of embankments, buildings and storage tanks. Some applications of lime/cement DMM are illustrated in Figure 6.16.

Method of construction and subsequent soil reactions

Lime/cement piles or columns are constructed by mechanically mixing lime or cement with the soft clay at in–situ conditions. The increase in strength and decrease in compressibility of the soft clay result from the reaction of the clay with lime and/or cement through the processes of ion exchange and flocculation as well as pozzolanic reaction. The divalent calcium ions replace the monovalent sodium ions in the double layer surrounding each clay particle. Thus, fewer number of divalent calcium ions are needed to neutralize the net negative surface charge of each clay mineral, reducing the size of the double

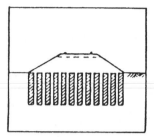

Prevent slope failures and reduce settlement in embankments and structures

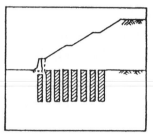

Increase the stability of slopes

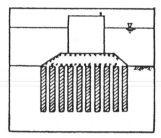

Increase the bearing capacity of soils and used as foundation to structures

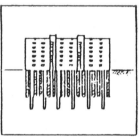

Increase the horizontal resistance of structures

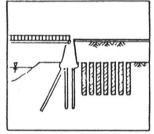

Reduce the settlement and prevent slope failure of the abutments of bridges

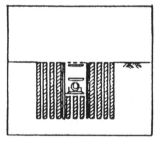

Prevent heaving and reduce the penetration length of sheet piles in excavations

Stabilize cut slopes

Prevent damage to the adjacent structures of construction sites

Prevent settlement of underground embedded structures

Figure 6.16. Some applications of cement columns.

layer, and thereby, increasing the attraction between clay particles leading to flocculated structure. Furthermore, the silica and alumina in the clay mineral react with the calcium silicates and calcium aluminate hydrates to form a cementing gel in a process called pozzolanic reaction.

Effects of natural moisture, salt and organic contents

Miura et al. (1986) reported the influence of natural moisture contents on lime/clay reactions as shown in Figure 6.17. However, Ariizumi (1977) suggested that the salt content of the clay may have influenced the degree of improvement. The addition of salts may have accelerated the improvement as shown in Figure 6.18 up to a certain limit. For clays with high organic contents of more than 8 percent, the use of cement instead of lime is advantageous (Miura et al. 1987).

Dry jet mixing method

The method that uses cement powder or quicklime instead of slurry is called the dry jet mixing method (DJM). In this method, the cement or quicklime powder is injected into the deep ground through a nozzle pipe with the aid of compressed air and the powder is mixed into the clay mechanically by means of rotating wings (Fig. 6.19). In the DJM, the hydration process generates some amount of heat which results in additional drying effect to the surrounding clay and consequently more effective improvement (Yamanouchi et al. 1982). It is necessary to carry out a series of laboratory tests in order to estimate the appropriate percentage of quicklime that satisfies the design strength. Miura et al. (1986) suggested that the laboratory strength should not be less than 4 times the design strength in the field for a certain quicklime content.

Wet jet mixing method

In the wet jet mixing method (WJM), the slurry of lime or cement is jetted into the clay by 20 MPa pressure by a rotating nozzle. In this method, the machine is relatively light as shown in Figure 6.20. The diameter of WJM column tends to vary with depth according to the variations of subsoil shear strength. Miura et al. (1987) found an average field to laboratory strength ratio of 70 percent. In this case, 90 kg of cement was added per cubic meter of clay using water to cement ratio of 2.3. WJM method has been used to stabilize excavations in the soft Bangkok clay in which up to 150 kg of cement per cubic meter of clay was added with a water to cement ratio of 1.11. The average diameter of the resulting cement columns was 1.6 m with center to center spacing of 1.4 m.

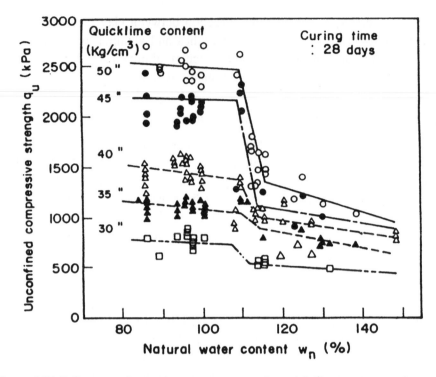

Figure 6.17. Influences of natural water content on the quick lime content on the improvement effect.

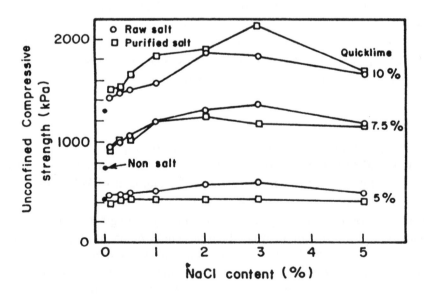

Figure 6.18. Influences of NaCl content on the quick lime improvement effect of Ariake clay (Hasuike Area).

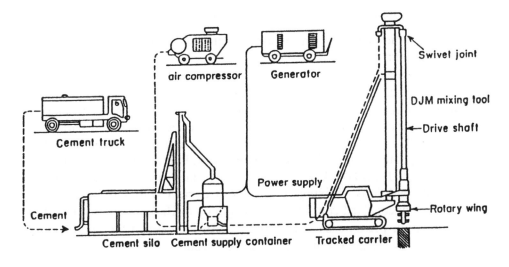

Figure 6.19. Installation of cement column by DJM.

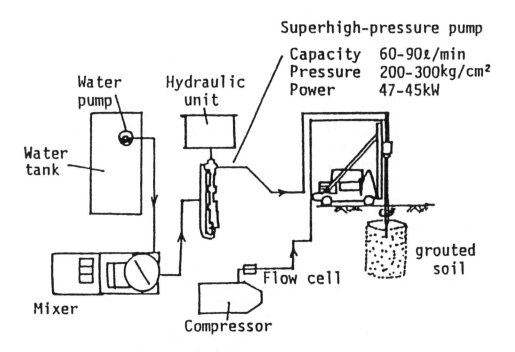

Figure 6.20. Equipment used for jet grouting.

Lime/cement stabilization of soft Bangkok clay

Extensive laboratory studies on the use of lime stabilization have been carried out by Balasubramaniam et al. (1988, 1989). Addition of 5 to 10 percent quicklime is the optimum mix proportion for soft Bangkok clay. The addition of quicklime increased the unconfined compressive strength to about 5 times and increased the preconsolidation pressure by 3 times. The coefficient of consolidation increased by 10 to 40 times. The effective strength parameters also increased, especially the angle of internal friction from 24 to 40 degrees.

Law (1989) conducted studies on the strength and deformation characteristics of cement treated soft Bangkok clay under unconfined compression, oedometer, and triaxial conditions. Mixing 10 percent of cement with soft Bangkok clay increased the unconfined compression strength up to 10 times and increased the preconsolidation pressure by 2 to 4 times. An increase in the coefficient of consolidation by 10 to 40 times was also observed.

Test embankment on DMM improved soft Bangkok clay

A full scale test embankment, 5.0 m high, was constructed on soft Bangkok clay improved with deep mixing method (dry method) using cement powder (Honjo et al. 1991). The plan and section views of the test embankment are shown in Figure 6.21. The cement piles or columns were constructed in two different patterns such as the wall type (south side) and the pile type (north side). The lateral movements and settlements of the wall type are less than that of the pile type as shown in Figures 6.22 and 6.23, respectively. Furthermore, the deformation pattern of the wall and pile types were sliding and tilting, respectively. The wall type was more effective in reducing lateral and vertical deformations. The unconfined compressive strengths in the laboratory were up to 20 times the untreated values for 28 days curing with cement content of 100 kg per cubic meter of clay. The field strength was found to be one-half of the corresponding laboratory test results.

CONCLUSIONS

Vertical drains are used to accelerate the consolidation of soft clay foundations. The state-of-practice of using vertical drains and the case studies regarding the application of vertical drains in the improvement of soft Bangkok clay have been presented. The theory of radial consolidation and its solution have been presented including the effects of well-resistance and smear. The diameter of the smeared zone can be assumed to be twice the equivalent diameter of the mandrel and the horizontal permeability coefficient in the smeared zone is approximately equal to the corresponding values in the vertical direction.

The current state-of-the-art on granular piles scheme has been discussed.

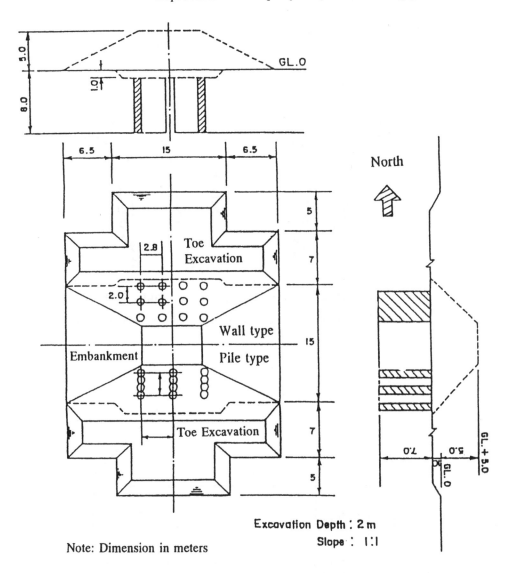

Figure 6.21. Plan and section of the test embankment with soil–cement piles.

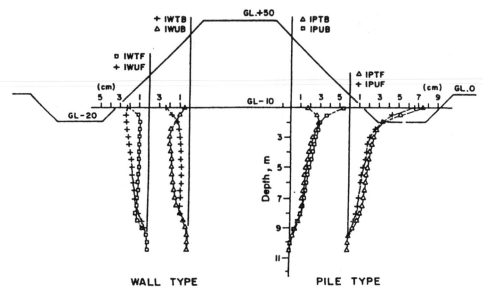

WALL TYPE   PILE TYPE

Figure 6.22. Comparison of lateral movement on improved and unimproved ground (71 days).

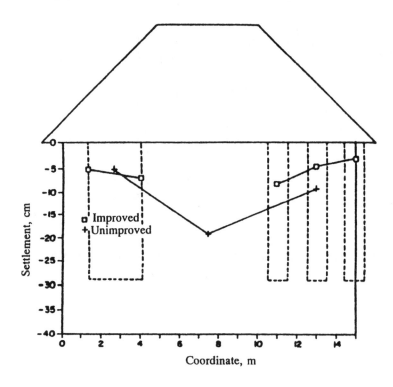

Figure 6.23. Comparison of settlement of improved and unimproved part of 1 m depth (71 days).

There are still loopholes which need to be studied such as the influences of the method of construction, characteristics of pile materials, better estimation of stress concentration factors, stress distribution with depth and time, improvement factors, etc. The performance of a full scale test embankment on soft Bangkok clay improved with granular piles has been described. The improvement factors on bearing capacity, settlement, and gain in strength have been determined and indicated substantial values. Encouraging results have been presented regarding the suitability of locally–available and low–quality silty sands as construction materials of granular piles which are cheaper than clean sands. The utilization of cheaper and locally–available materials is imperative to reduce construction costs and increase the viability regarding the use of granular piles.

The use of lime/cement deep mixing method proved to be an effective method of improving the properties of soft clays including soft Ariake clay in Japan and soft Bangkok clay in Thailand. With the addition of 5 to 10 percent quicklime which is the optimum mix proportion, the unconfined compressive strength increased to about 5 times. Moreover, the preconsolidation pressure improved to as much as 3 times while the coefficient of consolidation increased 10 to 40 times. The effective strength parameters also increased wherein the angle of internal friction increased by about 67 percent. Mixing an optimum 10 percent of cement with soft Bangkok clay improved the unconfined compressive strength and the preconsolidation pressure by up to 10 to 20 times and by 2 to 4 times, respectively. The coefficient of consolidation increased by 10 to 40 times. The in–situ strength of cement stabilized soft Bangkok clay was found to be one–half of the corresponding laboratory test results. Based on full scale test embankment loading on soil–cement piles, the deformation of the improved composite ground depends on the type of construction pattern. The lateral movements and settlements of the wall type construction were lower than the corresponding values of the pile type construction.

REFERENCES

Aboshi, H., E. Ichimoto, M. Enoki & K. Harada 1979. The compozer–a method to improve characteristics of soft clays by inclusion of large diameter sand columns. *Proc. intl. conf. on soil reinforcement: reinforced earth and other techniques*, Paris, 1: 211–216.

Aboshi, H. & N. Suematsu 1985. Sand compaction pile method: State–of–the–art paper. *Proc. 3rd intl. geotechnical seminar on soil improvement methods*, Nanyang Technological Institute, Singapore.

Aoyama, M., N. Nakamura, M. Kuwabara & M. Nuzo 1990. Some examples of field tests for soil improvement methods in Japan. *Proc. symp. dev. of laboratory and field tests in geotech. eng. practice*, Bangkok, Thailand.

Ariizumi, M. 1977. Mechanism of lime stabilization. *J. of Japan Society Soil Mechanics Foundation Engineering*. Tsuchi–to–Kiso, 25: 9–16.

Asaoka, A. 1978. Observational procedure for settlement prediction. *Soils and Foundations*,

18 (4): 87–101.

Balasubramaniam, A.S., N. Phienwej & M. Kuhananda 1988. Coastal development in soft clay deposits. *Proc. Kozai club seminar 1988*, Bangkok, Thailand.

Balasubramaniam, A.S., Y. Honjo, K.H. Law, N. Phienwej & D.T. Bergado 1989. Ground improvement techniques in Bangkok subsoils. *Proc. Kozai club seminar 1989*, Bangkok, Thailand.

Barksdale, R.D. & R.C. Bachus 1983. *Design and construction of stone columns*. Report No. FHWA/RD–83/026, Natl. Tech. Info. Service, Springfield, Virginia, U.S.A.

Barron, R.A. 1948. Consolidation of fine–grained soils by drain wells. *Transactions ASCE* 2346: 718–754.

Baumann, V. & G.E.A. Bauer 1974. The performance of various soils stabilized by vibrocompaction method. *Canadian Geotech. J.*, II: 509–530.

Bell, A.L. 1915. The lateral pressure and the resistance of clay and the supporting power of clay foundations. *Proc. Institute of Civil Engineers*.

Bergado, D.T., G. Rantucci & S. Widodo 1984. Full scale load tests on granular piles and sand drains in the soft Bangkok clay. *Proc. intl. conf. on in–situ soil and rock reinforcement*, Paris, 111–118.

Bergado, D.T. & F.L. Lam 1987. Full scale load test of granular piles with different densities and different proportions of gravel and sand in soft Bangkok clay. *Soils and Foundations*, 27 (1): 86–93.

Bergado, D.T., S.H. Huat & S. Kalvade 1987. Improvement of soft Bangkok clay using granular piles in subsiding environment. *Proc. 5th geotech. seminar on case histories in soft clay*, Singapore, 219–226.

Bergado, D.T., A.S. Enriquez, C.L. Sampaco, M.C. Alfaro & A.S. Balasubramaniam 1990. Inverse analysis of geotechnical parameters on improved soft Bangkok clay. *J. Geotech. Eng. Div., ASCE*, 118 (7): 1012–1030.

Bergado, D.T., H. Asakami, M.C. Alfaro & A.S. Balasubramaniam 1991. Smear effects of vertical drains on soft Bangkok clay. *J. Geotech. Eng. Div., ASCE*, 117 (10): 1509–1530.

Bergado, D.T. & P.V. Long 1992. Numerical analysis of embankments on subsiding ground improved by granular piles. *Proc. intl. conf. on geotech. eng. (NTFE 92)*, Hanoi, Vietnam.

Bergado, D.T., M.C. Alfaro, N.D. Bersabe & K.H. Leong 1993. Model test results of sand compaction pile (SCP) using locally–available and low–quality backfill soil. *Proc. 11th Southeast Asian geotech. conf.*, Singapore, 313–318.

Department of Aviation 1984. *SBIA master plan study and design*. Ministry of Communications, Royal Thai Government, Bangkok, Thailand.

D'Orazio, T.B. & J.M. Duncan 1982. CONSAX: *A computer program for axisymmetric finite element analysis of consolidation*. Report No. UCB/GT/82–01, Univ. of California. Berkeley, U.S.A.

Duncan, J.M., T.B. D' Orazio, C.S. Chang, K.S. Wong & L.I. Namiq 1981. CON2D: *A finite element computer program for analysis of consolidation*. Report No. UCB/GT/81–01, Univ. of California, Berkeley, U.S.A.

Dunham, J.W. 1954. Pile foundations for buildings. *J. of Soil Mech. Found. Eng. Div., ASCE*, 385–1 to 385–21.

Gibson, R.E. & W.F. Anderson 1961. In–situ measurements of soil properties with the pressuremeter. *Civil engineering and public review*, 56 (658).

Greenwood, D.A. 1970. Mechanical improvement of soils below ground surface. *Proc. ground engineering conf.*, Institution of Civil Engineers, 11–12 June, 1970, 9–20.

Hansbo, S. 1979. Consolidation of clay by bandshaped prefabricated drains. *Ground engineering*, 12 (5): 16–25.

Hansbo, S. 1987. Design aspects of vertical drains and lime column installations. *Proc. 9th Southeast Asian geotech. conf.*, Bangkok, Thailand, 2: 8–12.

Honjo, Y., C.H. Chen, D.G. Lin, D.T. Bergado, A.S. Balasubramaniam & R. Okumura 1991. Behavior of the improved ground by the deep mixing method. *Proc. Kozai club seminar 1991*, Bangkok, Thailand.

Hughes, J.M.O. & N.J. Withers 1974. Reinforcing soft cohesive soils with stone columns. *Ground Engineering*, 7 (3): 42–49.

Hughes, J.M.O., N.J. Withers & D.A. Greenwood 1975. A field trial of reinforcing effects of the stone columns in soil. *Geotechnique*, 25 (1): 31–44.

Kamon, M. & D.T. Bergado 1991. Ground improvement techniques. *Proc. 9th Asian regional conf. soil mech. found. eng.*, 2: 521–548.

Law, K.H. 1989. *Strength and deformation characteristics of cement treated clay*, M. eng. thesis, Asian Inst. of Tech., Bangkok, Thailand.

Madhav, M.R., N.G.R. Iyengar, R.P. Vitkar & A. Nandia 1979. Increased bearing capacity and reduced settlements due to inclusions in soils. *Proc. intl. conf. on soil reinforcement: earth reinforcement and other techniques*, 2: 239–333.

Miura, N., Y. Koga & K. Nishida 1986. Application of deep mixing method with quicklime for the Ariake clay ground. *J. of Japan Soc. Soil Mech. Found. Eng.*, 34 (4): 5–11.

Miura, N., D.T. Bergado, A. Sakai & R. Nakamura 1987. Improvements of soft marine clays by special admixtures using dry and wet jet mixing methods. *Proc. 9th Southeast Asian geotech. conf.*, Bangkok, Thailand, 8–35 to 8–46.

Moh, Z.C. & S.M. Woo 1987. Preconsolidation of Bangkok clay by non–displacement sand drains and surcharge. *Proc. 9th Southeast Asian geotech. conf.*, 2: 8–184.

Ranjan, G. 1989. Ground treated with granular piles and its response under load, *Indian Geotech. J.*, 19 (1): 1–85.

Robertson, P.K. & R.G. Campanella 1983. Interpretation of cone penetration test: part I – sand. *Canadian Geotech. J.*, 20 (4): 718–733.

Skempton, A.W. & L. Bjerrum 1957. A contribution to the settlement analysis of foundations on clay. *Geotechnique*, 7: 168–178.

Terashi, M., H. Tanaka & T. Okumura 1979. Engineering properties of lime – treated marine soils and deep mixing method. *Proc. 6th Asian reg. conf. on soil mech. found. eng.*, 1: 191–194.

Vesic, A.S. 1972. Expansion of cavities in infinite soil mass. *J. of Soil Mech. and Found. Eng. Div.*, ASCE, 98 (SM3): 265–290.

Woo, S.M., Z.C. Moh, A.F. Van Weele, R. Chotivittayathanin & T. Transkarahart 1989. Preconsolidation of soft Bangkok clay by vacuum loading combined with non–displacement sand drains. *Proc. 12th intl. conf. soil mech. found. eng.*, Brazil.

Yamanouchi, T., N. Miura, N. Matsubayashi & N. Fukuda 1982. Soil improvement with quicklime and filter fabric. *Proc. ASCE*, 108: 935–965.

7 STABILITY OF SHORELINE PROTECTION STRUCTURES

H.B. POOROOSHASB
Concordia University, Montreal, Quebec, Canada

INTRODUCTION

Reclaimed land must be protected from the action of the sea by a suitable system: the shoreline protection structure. These structures are subjected to "normal loading conditions" and "exceptional loading conditions" as indeed are most other civil engineering structures. The exceptional loading conditions, in the case of shore protection systems, include transient loads which are variable in time as for their magnitude and direction.

Broadly speaking, these transient loads may be divided into two categories. First, those which must be treated as true dynamic cases, such as earthquake loading or impact of objects and waves. Treatment of this subject is outside the scope of the present chapter. Second, those cases which, in view of their rather slow, but nevertheless transient mode of action, may be treated as a quasi-static situation. Thus, when dealing with the latter cases, the equations governing the solution to the problem do not include a term representing the inertia component.

In this chapter, emphasis will be placed on a rather particular, and less understood, aspect of the second (quasi-static) type of the problem. Referring to Figure 7.1, the two shoreline protective structures are subjected to the action of waves which can exert forces of significant magnitude. In addition, the soils surrounding the structure (particularly seabed soils in front of the protective unit) may experience a dramatic change in their strength due, principally, to an increase of the excess water pressure in their pores. These changes may be so severe that they may lead to the instability of the system and its eventual failure. It is this phase of the study that the present chapter will be concerned with.

The sequence of the materials covered here is as follows. First a concise statement of the problem is given. Next, the nature of loads imposed by two types of waves, traveling and standing waves, on the seabed are discussed. This is followed by the presentation of three types of analyses available for the investigation of the stress regime within the seabed soils. These discussions include, of necessity, a section on the non-linear response of soils to cyclic loading. Finally, the chapter

135

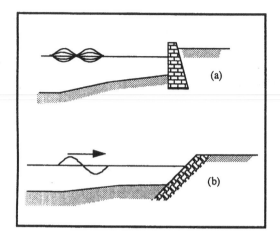

Figure 7.1. Two types of shoreline protection.
(a) Seawall subjected to the action of standing
waves and (b) Armored slope subjected to a train
of traveling waves.

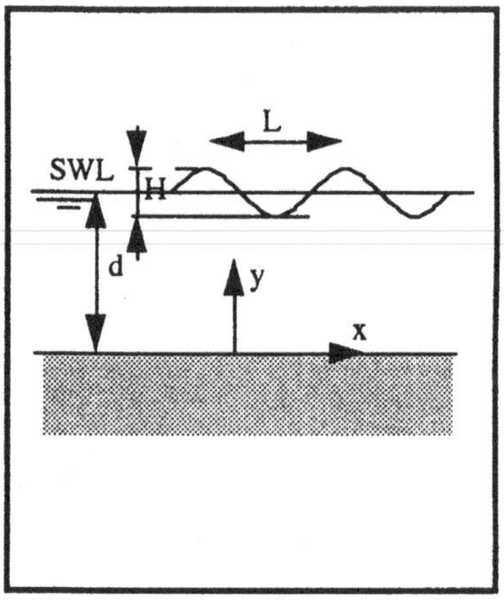

Figure 7.2. Definition sketcn.
(Note: the coordinates refer to
the body of water only).

explores the possibility of the application of a simple model testing technique which
may prove to be of great assistance in the analysis of the problem leading to a more
realistic evaluation of the instability of waterfront structures.

PRELIMINARY CONSIDERATIONS

Statement of the problem

The integrity of the soils in front of the structure subjected to the action of sea
waves is to be analyzed. This is done as a free field analysis. That is, the presence
of the protective unit is ignored and the problem is solved as if no such structure
existed. This simplifying assumption is the most stringent condition imposed but is
necessary in view of the rather complex nature of the problem.

Referring to Figure 7.2, the surface waves of height H and wave length L impose
a time dependent (often oscillatory) water pressure on the seabed, located at a depth
d below the MSL (Mean Sea Level). These pressures are transmitted and redistributed within the seabed changing both the inter-granular forces (the effective
stresses) and the water pressures in the pores of the seabed deposit (the porewater
pressure, p.w.p.). As the number of loading cycles increase, the p.w.p. may accumulate within a typical element of the seabed with a corresponding decrease in the
inter-granular forces. As the "strength" and "deformability" of soils (particularly
cohesionless media) is dependent on the inter-granular forces, a decrease in these
forces can result a great decrease in the "rigidity" of the medium leading to unacceptable displacements of the system (the water front structure, for example) or to a
complete loss of strength (liquefaction) causing total failure.

A concise statement of the problem is as follows. A seabed consisting of a uniform sand deposit is located at a shallow depth d. Surface waves of height H, length
L and period T exert an oscillatory pressure on the surface of the seabed. It is

required to determine the stress and the p.w.p. fields, i.e. to find the distribution of the effective stress and the pore waterpressures within the seabed deposit.

Pressures exerted on the surface of the seabed

The linear wave theory is considered to be sufficiently accurate for the present purpose. According to this theory, also known as the Airy Wave Theory, the pressure at any point within the body of the water may be given by the equation;

$$p = \gamma \frac{H}{2} \frac{\cosh ky}{\cosh kd} \cos(kx - \omega t) + \gamma(d - y) \tag{7.1}$$

where γ is the unit weight of the sea water and ω, the frequency, is given by the equation $\omega = 2\pi/T$. The relation between ω and k is:

$$\omega^2 = gk \tanh(kd) \tag{7.2}$$

In Eq.(7.2) g is the acceleration of gravity and k is the wave number given by $k=2\pi/L$. Often, the wave height and period are known and k (hence L) must be solved for, using Equation 7.2 and a knowledge of the water depth d.

At the seabed elevation $y = 0$. The exerted water pressure (i.e. the pressure exerted due to wave action alone) is given by;

$$p = \gamma \frac{H}{2} \frac{\cos(kx - \omega t)}{\cosh(kd)} \tag{7.3}$$

The above equation relates seafloor pressures to wave characteristics for a wave traveling at speed of $c=L/T$. In the case of standing waves (waves which are the combination of two waves traveling in opposite directions, such a situation is observed in front of sea walls) the corresponding equation is;

$$p = \gamma \frac{H}{2} \frac{\sin(kx)}{\cosh(kd)} \sin(\omega t) \tag{7.4}$$

Equations 7.3 and 7.4 are sufficiently accurate to be used as the loading boundary conditions imposed on the surface of the seabed. The next Section deals with the analysis of the stress and the p.w.p. fields within the seabed sandy deposit, the so called solution domain.

ANALYSIS

Ishihara's analysis

Based on the equations proposed by Yamamoto et al. (1978) and Madsen (1978), Ishihara (1983) proposed an analysis which, while somewhat elementary, it is

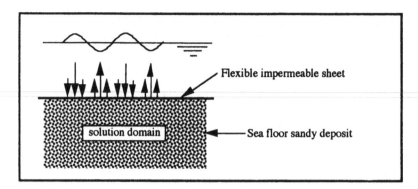

Figure 7.3. Graphical representation of the fundamental assumption in Ishihara's analysis.

possibly the only method that can conveniently be handled by a practicing engineer.

The most fundamental assumption in Ishihara's analysis is that boundary loads are transmitted to the seabed surface in total, i.e. not as a pore water pressure on the surface. Stated otherwise in this analysis it is assumed that an infinitely flexible and completely impermeable sheet covers the seabed surface. Thus the exerted water pressures are treated as ordinary loads acting through an impermeable boundary (Fig. 7.3).

The solution domain itself is treated as a homogeneous isotropic elastic material and the Airy stress function $\phi = A\,r\theta\,\sin\theta$ is used to evaluate the stress components at a given point at a given instant of time.

Denoting by a_0 the expression $\gamma H/2\,\cosh(kd)$ in Equation 7.3 the surface loading may be expressed by;

$$w(x) = a_0 \cos(kx - \omega t) \tag{7.5}$$

where $k = 2\pi/L$ and $\omega = 2\pi/T$ have been substituted in Equation 7.3. Assuming the origin of the coordinate system on the seafloor and the y ordinate pointing downwards into the solution domain, the stress components are evaluated to be;

$$\sigma_{yy} = a_0(1 + ky)\,\exp(-ky).\cos(kx - \omega t) \tag{7.6,i}$$

$$\sigma_{xx} = a_0(1 - ky)\,\exp(-ky).\cos(kx - \omega t) \tag{7.6,ii}$$

$$\sigma_{xy} = a_0(ky)\,\exp(-ky).\sin(kx - \omega t) \tag{7.6,iii}$$

It is emphasized, once again, that these components are total stress components. Also note that these components do not contain terms involving the elastic moduli E, the Young's modulus and ν, the Poisson's ratio, a situation common to plane strain problems of linear elasticity.

In Figure 7.4 the variation of the stress field (variation with respect to time and depth) expressed by its principal components and the orientation of the principal axis with y axis are shown in (a), (b) and (c). Change in orientation of the principal

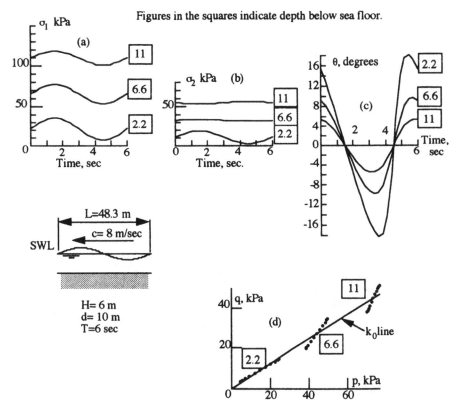

Figure 7.4. Variation of the total major principal stress (a), the total minor principal stress (b) and θ, the orientation of the principal axis (c), with respect to time and depth. The stress paths followed at the depths 2.2, 6.6 and 11 meter below sea floor, (d). Wave characteristics is represented in the lower left section of the figure.

axis with respect to the coordinate system is noteworthy. For example, at a depth of 2.2 meters this orientation may vary from 15 degrees, at a time when the wave front is directly above the element, to -18 degrees after elapse of a time equivalent to about one half of the period of the wave.

The variation of the so called stress path at three depths below the sea floor is depicted in Figure 7.4 (d). The (p,q) axis in this figure are derived from the stress invariants. In the context of the conventional triaxial tests where $\sigma_1 > \sigma_2 = \sigma_3$, these parameters are given by $q=(2/3)^{1/2}.(\sigma_1-\sigma_3)$ and $p=(\sigma_1+2\sigma_3)/3$. The small circles indicate the various stages of total stress that the elements at the three depths shown experience during one passage of the wave. The solid line passing through origin and crossing the three stress paths is the so called k_0-line which represents the state of effective stress under normal (calm sea) conditions.

Having defined the total stress field, one has to proceed with the evaluation of the effective stress field which, after all, is the factor that governs the mode of deformation and failure of the element. To do this Ishihara suggests that a series of undrained tests be performed on samples similar to those in the field and that they be subjected to the same stress paths as experienced in the actual case. Because of

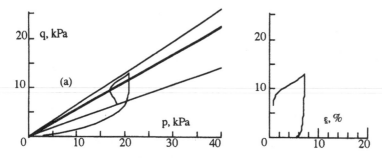

Figure 7.5, (a). The effective stress paths during a "cyclic" loading of a very loose sand element at a shallow depth below the seabed. Undrained loading condition.

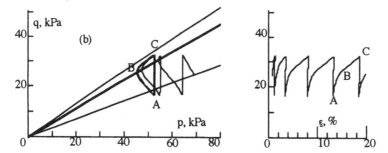

Figure 7.5, (b). The effective stress paths during a "cyclic" loading of a deeper, very loose sand element . Undrained loading condition.

limitations of laboratory testing techniques these tests are often carried out by the triaxial equipment. Figure 7.5 (a) shows a schematic drawing of the effective stress path that might be followed by an element of a very loose sand at a shallow depth in the field. As noted, the element experiences liquefaction (a total loss of strength due to a very large increase in the porewater pressure) at about one half cycle of wave loading.

For deeper elements, the situation is as shown in Figure 7.5 (b). Here the element cannot liquefy in the above sense (i.e. the nature of the exerted stress path precludes the build up of the pore water pressure to values equal to the total mean normal stress). Nevertheless, large distortions are experienced as the element goes through a cyclic loop with every passage of the wave. The portion of the cyclic stress path identified by the letters ABC is the cause of the build up of distortion.

The above analysis represented the behavior of the system under traveling waves. For standing waves the situation is somewhat different as shown in Figure 7.6. Here, for the sake of comparison, the same wave characteristics (height, period and length) and the same depth of water have been used. One point worthy of note is that element C, located at a distance of $L/8$ from the crown of the wave, does not experience a change of stress magnitude (i.e. the stress path is represented by a point). Nevertheless it experiences very high rotation of axis of its principal stresses during the rise and fall of the waves. On the other hand, the stress path for the element A, directly below the crown of the wave, is well defined and it does not experience any

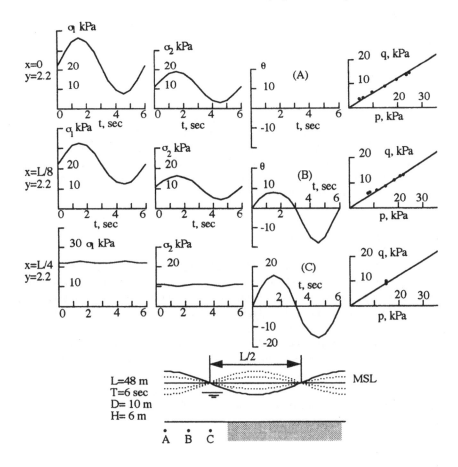

Figure 7.6. State of stress in elements A, B and C. Elements A, B and C are located at a depth of 2.2 meters below the bed. Figures on the extreme right hand side show the stress paths followed by the three elements during the rise and fall of the waves.

principal axis rotation. The results of laboratory cyclic loading may now be used and the response of element A evaluated.

Such evaluations are carried out by computers for which purpose the behavior of the sands under cyclic loading must be coded. This is done through a modeling technique i.e. describing the behavior of the media in a mathematical term that the computer can understand. One such model is given in a later Section entitled "Modeling the non-linear soil behavior" Before leaving this Section, however, a certain point must be made. Ishihara's approach is only statically admissible. That is, the total stress field expressed by Equations 7.6,i to 7.6,iii satisfies only the boundary conditions of the problem and the incremental equations of equilibrium;

$$\frac{\partial \sigma_{xx}}{\partial x} + \frac{\partial \sigma_{xy}}{\partial y} = 0$$

$$\frac{\partial \sigma_{xy}}{\partial x} + \frac{\partial \sigma_{yy}}{\partial y} = 0$$

(7.7)

The displacement field obtained from this stress field (Equations 7.6) is unlikely to satisfy the kinematics constraints. Thus, it is obvious that Ishihara's solution is not an exact solution but an approximate one which, nevertheless, is very attractive in view of its simplicity. Conceptually speaking, this analysis is comparable to the usual technique currently used in geotechnical engineering to evaluate the time dependent and the eventual settlement of structures.

Exact analysis for a poro-elastic seabed

As stated before, Ishihara's analysis is not exact. In this Section the governing equations of the problem are derived for an exact analysis. The seabed soils are assumed to act as a homogeneous linear elastic material with shear modulus G and Poisson's ratio ν. Thus the term exact analysis is perhaps somewhat misleading: soils exhibit a highly non-linear deformation behavior.

According to Terzaghi's principle of effective stress $\sigma_{xx}=\sigma'_{xx}+pwp$, $\sigma_{yy}=\sigma'_{yy}+pwp$ and $\sigma_{xy}=\sigma'_{xy}$ where pwp stands for porewater pressure. The primed parameters are called the effective stress components. Deformation of a soil element is purely the result of a change in the effective stress components. Assuming that the seabed deposit can be modeled as an isotropic elastic solid, for a plane strain problem the corresponding stress-strain relations are;

$$\sigma'_{xx} = 2G\frac{\partial u}{\partial x} + (m-1)Ge$$

$$\sigma'_{yy} = 2G\frac{\partial v}{\partial y} + (m-1)Ge$$

(7.8)

$$\sigma'_{xy} = G[\frac{\partial u}{\partial y} + \frac{\partial v}{\partial x}]$$

where u and v are components of the displacement vector of the solid particles along x and y directions respectively and $m=1/(1-2\nu)$ is a constitutive constant. The parameter e stands for the volumetric strain of the solid phase and is given by;

$$e = [\frac{\partial u}{\partial x} + \frac{\partial v}{\partial y}]$$

(7.9)

Using Terzaghi's principle in conjunction with the last set of equations and substituting the results in the set of Equations 7.7 yields;

$$mG\frac{\partial e}{\partial x} + G\nabla^2 u + \frac{\partial(pwp)}{\partial x} = 0$$

$$mG\frac{\partial e}{\partial y} + G\nabla^2 v + \frac{\partial(pwp)}{\partial y} = 0$$

(7.10)

To account for the consolidation of the element it may be noted that ;

$$\frac{\partial e}{\partial t} + [\frac{k}{\gamma_w}]\nabla^2(pwp) = 0$$

where k is the coefficient of permeability of the soil (not to be confused with $k=2\pi/L$) and γ_w is the unit weight of water . The last equation may be rewritten as;

$$(1+m)G\frac{\partial e}{\partial t} + c\nabla^2(pwp) = 0$$

(7.11)

where $c=(1+m)G(k/\gamma_w)$ is the coefficient of consolidation. Equations 7.9,7.10 and 7.11,the governing equations, are four equations to be solved for the three unknowns, u,v and e, (deformation parameters) and the porewater pressure (one of the stress parameters). Once these unknowns are determined the other stress components may be defined using the set of Equations 7.8.

The above governing equations are a simplified version of the set of equations used by Verruijt (1982) who also includes a term to take into account the relative compressibility of the pore fluid. Verruijt provides solutions where the boundary conditions can be expressed in the form of $\exp(i\omega t)\cos(kx)$, where $\omega=2\pi/T$ and $k=2\pi/L$ as noted before. Details of his exact analysis are contained in the above mentioned paper but an approximation to the governing equations, also provided by Verruijt, is given here.

Let a solution in the form of

$$e=\underline{e} \ \exp(i\omega t)\cos(kx) \ ; \ k=2\pi/L$$

(7.12)

$$pwp=\underline{p} \ \exp(i\omega t)\cos(kx)$$

(7.13)

where $\underline{e}$ and $\underline{p}$ are functions of y only, be sought. Substituting in Equation 7.11 above results in ;

$$i\frac{\omega}{ck^2}(1+m)G\,\underline{e} - (\underline{p} - \frac{d^2\underline{p}}{dy^2}) = 0$$

(7.14)

Denoting the dimensionless parameter ω/ck^2 by ϕ, Verruijt makes the following two observations;

(i) $\phi=\omega/ck^2=L^2/2\pi cT \ll 1$

In this case the imaginary part of Equation 7.14 would be small compared to the real part and $p=A \exp(-ky)+B \exp(ky)$ where A and B are constants. The constant B must obviously be zero (otherwise the *pwp* will increase unboundedly with depth y) and the constant A is determined from the boundary condition of the problem (the *pwp* exerted on the seabed surface). Note that a small ϕ may be caused, amongst other things, by a rather high c value which in turn indicates either a very large permeability coefficient or a very high value of G (an almost rigid material). Also note that in this case, $pwp=p_0\exp(-ky).\cos(kx)\exp(i\omega t)$ where p_0 is the amplitude of the water pressure exerted on the seabed. At any instance of time, the expression for the *pwp* satisfies the Laplace equation.

(ii) $\phi=\omega/ck^2=L^2/2\pi cT \gg 1$

In this case the first term in Equation 7.14 is predominant compared to the rest and it may be concluded that $G(1+m)\underline{e}=0$ leading to the conclusion that $\underline{e}=0$ and hence $e=0$. Differentiating the first equation of the set of Equation 7.10 with respect to x and the second with respect to y and adding them up results in the equation;

$$GV^2 (\frac{\partial u}{\partial x} + \frac{\partial v}{\partial y}) + V^2(pwp) = 0$$

But $(\partial u/\partial x+\partial v/\partial y)=e=0$. Thus the last equation states that the *pwp* must again satisfy the Laplace equation; $pwp=p_0\exp(-ky).\cos(kx)\exp(i\omega t)$.

Note that a large value of ϕ represents either a very large wave length L or a very short period of motion T or a very small value of c which may be caused by either low permeability or high compressibility. Under these circumstances the seabed will behave under what is referred to as undrained conditions.

Verruijt's analysis is exact (for a linear system) and mathematically very attractive. It provides a solution for the parameters involved (in particular the *pwp*) which are periodic in time. For certain special cases ($\phi \ll 1$ or $\phi \gg 1$), the response is without any phase shift. What the analysis does not do, in view of its linearity, is it cannot predict any accumulation of the pore water pressure which is really responsible for the failure and large magnitude movement of the seafloor. It is, of course, possible to use Verruijt's analysis to obtain the total stress field (and thus the stress paths at various points within the domain) and then subject the samples in the laboratory to the same paths (the technique proposed by Ishihara) to observe their behavior. The results are not likely to be too different from those obtained by Ishihara's analysis and in special cases ($\phi \ll 1$ or $\phi \gg 1$), the two approaches would yield identical results.

To truly account for the *pwp* accumulation, it is necessary to use a non-linear constitutive equation for the description of soil behavior subjected to stress reversals. One such non-linear law is presented in the next section followed by the analysis of Yang (1990) which makes use of it.

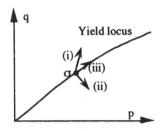

Figure 7.7. Definition of loading, unloading and neutral loading in the (q,p) stress space.

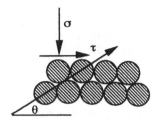

Figure 7.8. An ideal model for an assemblage of soil particles.

Modeling the non-linear soil behavior

Preliminaries

For the sake of simplicity, the discussions presented here will concern the analysis of tests performed in the triaxial equipment only. The model described is by no means the only available model but one which is favored by the present writer. It is called the CANAsand model and its original version can be traced to about a quarter of a century ago (Poorooshasb et al. 1966,67).

The state of stress is given by the two parameters $p=(\sigma_1+2\sigma_3)/\sqrt{3}$ and $q=\sqrt{2/3}(\sigma_1-\sigma_3)$. Note that the definition of q remains as before but parameter p in this section is larger than that defined in an earlier Section by a factor of $\sqrt{3}$. The reason for this choice becomes obvious later on. In what follows, the term stress stands for effective stress unless otherwise noted. An increment of strain is represented by the set of quantities $d\varepsilon_1$ and $d\varepsilon_3$, and it is assumed that each increment has an elastic (reversible) as well as a plastic (non-reversible) component; i.e.;

$$de_1 = de_1^e + de_1^p$$
$$de_3 = de_3^e + de_3^p$$

(7.15)

The elastic components are related to the stress increments $d\sigma_1$ and $d\sigma_3$ by a Young's modulus E (which is a function of p) and a Poisson's ratio. The plastic components are related to the stress increments by a relation of the form;

$$d\varepsilon_i^p = d\lambda \frac{\partial\varphi}{\partial\sigma_i}$$

(7.16)

where i=1,2, φ is the so called plastic potential (a formal definition will be given later in this section) and $d\lambda$ is the loading index. The magnitude of the parameter $d\lambda$ depends on whether the sample is experiencing loading, unloading or a neutral loading. To understand the meaning of these terms reference may be made to Figure 7.7 which shows the stress space for the conventional compression type of loading in the triaxial equipment. The stress path designated (i) shows a loading increment, (ii) indicates an unloading increment and (iii) a neutral loading as it traces an incremental path along the yield locus passing through the stress point σ.

In its simplest form, the yield locus is a straight ray passing through the origin of the stress space. From the form of Equation 7.16, it is obvious that if loading does take place then the ratio of the plastic strain components (i.e. the ratio $d\varepsilon_1{}^p/d\varepsilon_3{}^p$) remains constant and this is indeed observed experimentally. This observation is the permit to use φ in the constitutive relation.

Figure 7.8 shows an idealized model for an assemblage of soil particles. The two parameters τ and σ are "forces" that act on the top row of the particles which is assumed to be in equilibrium. From this simple model three important observation may be made.

(i) If the top row is to move (relative to the bottom row) then an increase in the value of ;

$$\eta = \tau /\sigma \tag{7.17}$$

is required. The magnitude of the individual components τ or σ is of no consequence in this respect.

(ii) If the top row moves, it moves along the direction θ which has no relation to the value of η.

(iii) If the coefficient of friction at points of contact between the particles of the top row and the lower row is designated by μ (=tan ϕ), then for the movement to take place the relation;

$$\tau = \sigma \frac{\tan \phi + \tan \theta}{1 - \tan \phi \tan \theta}$$

must hold. For small values of θ the above relation can be approximated by the expression $\tau = \sigma(\tan \phi + \tan \theta)$.

The term tan θ can be replaced by $-dv/d\varepsilon$ where dv is the vertical movement of the particles (negative when rising) and $d\varepsilon$ is their horizontal displacement. Substituting for tan θ by its equivalent μ results in the equation;

$$\tau + \sigma \frac{dv}{d\varepsilon} = \mu \sigma \tag{7.18}$$

The above three points are of paramount importance in developing the constitutive relation. The first point justifies the use of $\eta = \tau /\sigma$ as the yield surface. Point (ii) justifies the use of a plastic potential and furthermore establishes the possibility of a model following the non-associated flow rule. Finally, point (iii) may be used to evaluate the form of the plastic potential function φ.

Modeling soil behavior for a virgin loading process
When the soil is loaded for the first time (e.g. a sample consolidated to a pressure p_0 then subjected to a compression loading), it is said to be experiencing virgin loading. Modeling the soil behavior during this type of loading is explored below.

The yield function is denoted by f and is given by the simple expression;

$$f = \eta \tag{7.19}$$

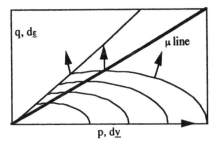 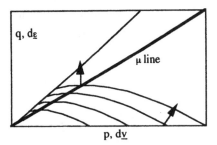

Figure 7.9. The set of curves on the left hand side show the general form of plastic potential curves (i.e., φ=constant curves). The curves on the right hand side satisfy Equation 7.18. The solid line in both graphs represents the μ line otherwise known as the projection in the stress space of the " critical state line", "steady state line", "ultimate state line" and the "phase transformation line".

as a consequence of (i) above. Note that if dη≥0 then the sample is experiencing loading (or is experiencing neutral loading) and the rules outlined in this section apply. If dη<0 then it is experiencing unloading and the behavior must be modeled as stated in the next Section.

Strain increments $d\underline{v}$ and $d\underline{\varepsilon}$ are compatible with stress parameters p and q where;

$$d\underline{v}=(d\varepsilon_1+2d\varepsilon_3)/\sqrt{3} \tag{7.20}$$

and $$d\underline{\varepsilon}=\sqrt{2/3}.(d\varepsilon_1-d\varepsilon_3) \tag{7.21}$$

Note that the work increment associated with an incremental change of stress is simply

$$dw=pd\underline{v}+qd\underline{\varepsilon}=\sigma_1 d\varepsilon_1+2\sigma_3 d\varepsilon_3$$

If the q axis of the stress space is to represent the plastic component of the strain increment parameter $d\varepsilon^p$ and the p axis to represent the plastic component of the strain increment parameter $d\underline{v}^p$ also (in effect superimpose the plastic strain increment space on the stress space), a combined space is obtained in which to every point a unit vector may be assigned to represent the direction of the plastic flow. As this unit vector field is irrotational, it can in turn be represented by the gradient of scalar function; the plastic potential. Thus, for example the component $d\underline{v}^p$ is simply $d\lambda(\partial\varphi/\partial p)$, Equation 7.16.

The general form of the plastic potential function is;

$$\varphi = p\bar{\varphi}(\eta) \tag{7.22}$$

However, if use is made of Equation 7.18 then it may be shown that;

$$\bar{\varphi}(\eta) = \exp(\eta/\mu) \tag{7.23}$$

To show the rationality of Equation 7.18 note that from Equation 7.16 expressed in terms of parameters q (equivalent to τ) and p(equivalent of σ), it may be stated;

$$d\underline{v}^p = d\lambda \frac{\partial \varphi}{\partial p} = d\lambda [1 - \frac{\eta}{\mu}]\exp(\frac{\eta}{\mu}) / N$$

$$d\underline{\varepsilon}^p = d\lambda \frac{\partial \varphi}{\partial q} = d\lambda [\frac{1}{\mu}]\exp(\frac{\eta}{\mu}) / N$$

where $N=[(\partial\varphi/\partial p)^2+(\partial\varphi/\partial q)^2]^{1/2}$, obviously satisfies Equation 7.18. The factor N is used to normalize the gradients of φ.

Having defined the plastic potential, it is required to determine the factor $d\lambda$ for a loading increment. To this end it is postulated that the magnitude of $d\lambda$ is in direct proportion to the increment of yielding df, which is equal to $d\eta$ from Equation 7.19, i.e.;

$$d\lambda=h.d\eta \qquad\qquad (7.24)$$

where h is a proportionality factor with a magnitude depending on the stress level (or rather the η level). To evaluate the magnitude of h it is necessary to establish, experimentally, the form of a plastic base curve relating the parameter ε^p to η. A simple conventional compression test is sufficient for this purpose. For loose sands, this curve can be approximated by the so called hyperbolic relation which has the form $\varepsilon^p = a\eta/(\eta_f-\eta)$ where a and η_f are constants to be determined experimentally. From this equation the increment of plastic distortion is obtained as

$$d\underline{\varepsilon}^p =[a\eta_f/(\eta_f-\eta)^2]d\eta$$

But $d\underline{\varepsilon}^p =d\lambda\ (\partial\varphi\ /\partial q)/N=h.[(\partial\varphi\ /\partial q)/N]d\eta=h.\exp(\eta/\mu)d\eta/\mu N$. Equating the two values of $d\underline{\varepsilon}^p$, factor h is evaluated as;

$$h = \frac{a\mu\eta_f N}{(\eta_f - \eta)^2}\exp(-\frac{\eta}{\mu}) \qquad\qquad (7.25)$$

Note that as η tends toward η_f then h tends to infinity indicating that the sample has reached its failure state.

Equations 7.16, 7.22, 7.24 and 7.25 are the set of equations which describe the mode of deformation of a soil sample during its virgin loading. Thus, a given stress path is divided into a number of increments and the corresponding $d\eta$ evaluated for each increment. Plastic strain components are calculated, using these equations, and added to the total elastic strains to obtain the so called strain path.

As far as the present study is concerned, one particular process plays a central role and that is the so called undrained process. In the previous Sections the role it plays in the analysis was clearly demonstrated, see for example, Figure 7.5. During such a process $d\underline{v}=d\underline{v}^e+d\underline{v}^p=0$.   The elastic component is given by the equation; $d\underline{v}^e= [(1-2v)/E]\ dp$ where E and v are the elastic constants and the plastic component is given by $h.(\partial\varphi/\partial p)d\eta$. Denoting $(1-2v)/E$ by $1/K$ (K here is 3 times the elastic bulk modulus) the above statements may be formulated by the equation;

$$\frac{dp}{K} + \frac{h}{Np}(\frac{\partial\varphi}{\partial p})[dq - \eta dp] = 0 \tag{7.26}$$

having substituted for $d\eta=(dq-\eta dp)/p$.

Let the total stress increment be (dQ,dP). The corresponding effective stress increment will be (dq,dp) where $dq=dQ$ and $dp=dP-\sqrt{3}.du$ where du denotes the change in the pore water pressure. Substituting in the Equation (7.26) results in the value of du as.;

$$du = [dP + \frac{h\frac{\partial\varphi}{\partial p}}{\frac{Np}{K} - h\eta\frac{\partial\varphi}{\partial p}} dq] / \sqrt{3} \tag{7.27}$$

Note that if $dq=0$, then $du=dP/\sqrt{3}=(d\sigma_1+2d\sigma_3)/3$, where now $d\sigma_1$ and $d\sigma_3$ are the total stress increments. This indicates that in an undrained loading process if $(\sigma_1-\sigma_3)$ (and hence q) is kept constant then the effective stress path will stay stationary since $dq=0$ and as a consequence dp $(=dP-\sqrt{3}\ du)$ is also zero.

Modeling of Soil Behavior During Stress Reversals
As noted before, an element of soil in the seabed would be subjected to a number of stress reversals during the passage of waves. It is therefore necessary to code the behavior of the medium under such loading conditions.

The structure of the set of Equations 7.16, 7.22, 7.24 and 7.25 remain essentially the same with the following two modifications. (i) the magnitude of the factor h is judged, by a proper interpolation function, depending on its position with respect to a "datum" and a "conjugate" point in the stress space and (ii) the plastic potential curve is a "reflection" of the virgin curve about the η constant line passing through the current stress point. These concepts (the datum and the conjugate stress point and the reflected plastic potential) will be discussed next.

Referring to Figure 7.10(a) let the highest level of η experienced during a loading process be η_m. Then the conjugate of the stress point S with coordinates (q,p) located on the stress path is a point S_c with coordinates (-q',p) where ;

$$q' = p\eta_m\frac{3 - \sin(\phi)}{3 + \sin(\phi)}$$

ϕ being the angle of friction of the soil. The inclusion of factor $(3-\sin(\phi))/(3+\sin(\phi))$ is to ensure that the conjugate of the stress point S is contained within the envelope prescribed by the Mohr-Coulomb failure criterion. At the instant of failure, on the compression side, the value of q is equal to $\sqrt{2}p[2\sin\phi/(3-\sin\phi)]$ and for the extension side $q=\sqrt{2}p[2\sin\phi/(3+\sin\phi)]$.

Also shown in Figure 7.10(a) is the position of the datum stress point which is located on the η_m ray with coordinates (q",p) where $q"=\eta_m.p$.

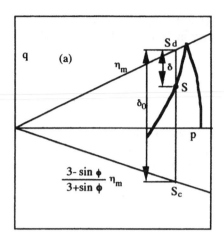
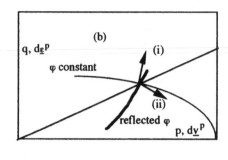

Figure 7.10. (a) defines the datum, S_d and the conjugate, S_c stress points for the state of stress S lying on the effective stress path (shown by the heavy line). Fig. 7.10 (b) shows the plastic potential curve and its reflected form. The unit vector marked (ii) shows the direction of plastic strain increment normal to the reflected potential curve.

The proposed interpolation function is;

$$h = h_c \left(\frac{\delta}{\delta_0}\right)^n \tag{7.28}$$

where n is a positive constant, δ and δ_0 are as shown in Figure 7.10(a) and h_c is to be evaluated at the conjugate stress point. Note that when the stress point is at $\eta = \eta_m$ (i.e. at the beginning of the stress reversal process) $\delta = 0$ and hence h=0. When it approaches its conjugate $h \Rightarrow h_c$ and the material behaves, once again, as a virgin soil.

Figure 7.10(b) shows a φ = constant curve, which is used during the virgin loading and its reflected form, which must be used during the stress reversals.

The components of the unit vector normal to the reflected curve (which are really quantities of interest) are obtained using simple coordinate transformations. Thus let α be the angle between the p axis and the η ray about which the potential curve is reflected, i.e. $\alpha = \tan^{-1}(\eta)$. Then the two components $(\partial\varphi/\partial p)$ and $(\partial\varphi/\partial q)$ would have, about a set of axis parallel and normal to the ray η, transformed components;

$$(\partial\varphi / \partial p)' = (\partial\varphi / \partial p)\cos\alpha + (\partial\varphi / \partial q)\sin\alpha$$

$$(\partial\varphi / \partial q)' = -(\partial\varphi / \partial p)\sin\alpha + (\partial\varphi / \partial q)\cos\alpha$$

For the reflected unit vector;

$$(\partial\varphi / \partial p)'_{ref} = (\partial\varphi / \partial p)'$$

$$(\partial\varphi / \partial q)'_{ref} = -(\partial\varphi / \partial q)'$$

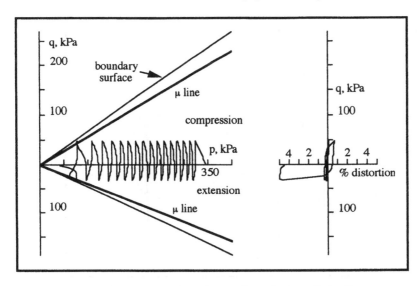

Figure 7.11. Behavior of loose sand subjected to small amplitude cyclic loading.

Note the minus sign in the last equation. A reverse transformation of the axis would yield the components of the reflected unit vector in the original stress space of (q,p):

$$(\partial \varphi / \partial p)_{ref} = (\partial \varphi / \partial p)'_{ref} \cos \alpha - (\partial \varphi / \partial q)'_{ref} \sin \alpha \qquad (7.29,a)$$

$$(\partial \varphi / \partial q)_{ref} = (\partial \varphi / \partial p)'_{ref} \sin \alpha + (\partial \varphi / \partial q)'_{ref} \cos \alpha \qquad (7.29,b)$$

Of course these values must be normalized through dividing them by the factor

$$N = \{[(\partial \varphi / \partial p)_{ref}]^2 + [(\partial \varphi / \partial q)_{ref}]^2\}^{1/2} = \{(\partial \varphi / \partial p)^2 + (\partial \varphi / \partial q)^2\}^{1/2}$$

before they are used.

This completes a brief account of a model capable in describing soil behavior during cyclic loading. Only the bare essentials have been included in view of space limitations. Figures 7.11 and 7.12 show the response of a loose sand sample to a low and a high stress level of cyclic loading respectively. These results may be compared to the published laboratory test results existing in the public domain, for example, the test results reported by Ishihara & Okada (1978).

Yang's analysis

Using a generalized version of the constitutive model described above, Yang (1990) performed the analysis of a seabed deposit subjected to the action of both standing and traveling waves. He formulated the problem within the framework of the mixture theory in conjunction with the finite element and the finite difference techniques. Yang analyzed the system assuming, as an example, a body of water 12 meters deep subjected to a train of waves 73 meters long and 9 meters high with a frequency of 7 seconds. These are the reported characteristics of a storm which caused a great deal of damage through soil liquefaction in Lake Ontario some time

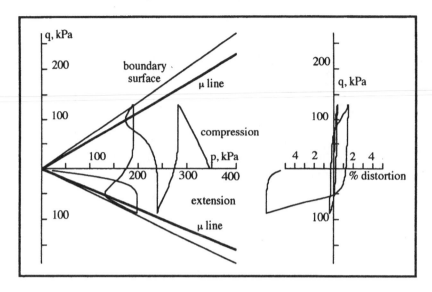

Figure 7.12. Behavior of loose sand subjected to large amplitude cyclic loading.

ago (for details see Christian et al. 1974). Figure 7.13 shows the results of his analysis assuming loose sand condition. Figure 7.13(a) is the computed mean effective stress versus time diagram for an element near the surface of the seabed. At the moment that the mean effective stress component approaches zero liquefaction is imminent (point marked A in the figure). Variation, and in particular accumulation of the porewater pressure with time is shown in Figure 7.13(b), and finally in Figure 7.13(c) is shown the horizontal displacement of the element under study. Large displacements observed as the porewater pressure builds up is a clear indication of failure due to liquefaction.

To conclude this Section, it is emphasized that there are a number of questions still unanswered. For example, as pointed out when dealing with the Ishihara's analysis, the loading caused by the sea waves produces rather large rotation of the direction of the principal axis of stress. The effect of this rotation on the stress deformation behavior of sands is not clearly understood. Thus even the most sophisticated analyses, such as Yang's, are still subject to some criticism. Furthermore, such analyses require main frame computers which are not always at the disposal of professional engineers. In view of the above it is the opinion of this writer that the elementary methods described earlier in this chapter are still the only practical analytical tools available.

Model Testing

The problem may be approached from an entirely different route; that of model testing. Barring centrifuge tests, the approach would require either the use of very large models, such as the case reported by Burger et al. (1988) or alternatively small models in which the "laws of similitude" are observed. These

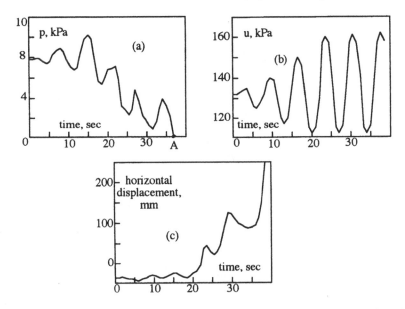

Figure 7.13. Variation of the mean effective stress (a), the porewater pressure (b) and the horizontal displacement (c) with time for a shallow water (depth =12 meters) subjected to Lake Ontario Storm.

laws are of two types; laws that govern the material property and laws that govern the rate of loading of the model. They are stated below without proof.

Material property laws

The first law governs the state of the soils in the model in comparison to that of the prototype. The state of an element of a soil is specified by the effective stress acting on the sample, (p,q) say, and by its void ratio (e). This state may be represented by a point in the state space of (p,q,e).

The "critical state line", as was called originally, or "the steady state line" or the "ultimate state line" trace a curve in the state space the two projections of which are shown in Figures 7.14(a) and 7.14(b). In Figure 7.14(a) the projection is into the (e,p) plane and is labeled the Casagrande line. Figure 7.14(b) shows its projection on to the (p,q) plane and is labeled as the μ line. Also shown in Figure 7.14(b) are two stress paths which are similar in form, i.e. $(q_{model}/q_{prototype}) = (p_{model}/p_{prototype}) = 1:m =$ constant at all stages of loading. Now the first material property law may be stated as: "two elements subjected to similar stress paths would deform similarly if at some stage of their loading (say the initial stage) their effective void ratio is the same". The effective void ratio is the Casagrande void ratio minus the element's void ratio, see Figure 7.14(a). This principle was outlined by Roscoe and Poorooshasb (1963). Recently the writer has provided a rigorous demonstration of this law within the context of plasticity theory (Poorooshasb 1989).

The second law governs the property of the fluid to be used in the model. It states that the viscosity of the fluid used in the model must be so adjusted that the

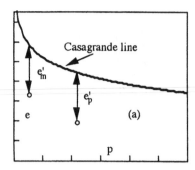
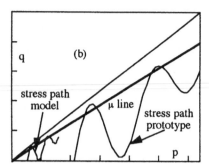

Figure 7.14 (a) *e,p* subspace of the state space. (b) the stress space. The two elements shown in (b) experience similar strains if their effective void ratios, e'_m and e'_p are equal at some corresponding (similar) stress levels.

resulting coefficient of permeability is $1/m$ times that of the actual case, i.e., $k_m=k_p/m$.

Law governing the rate of loading

This law merely states that in order to achieve similarity it is mandatory to load the model at a rate that reflects the scale factor *m*. Thus if the load in the prototype is expressed by $A \sin(\omega t)\sin(kx)$ then the loading of the model should be according to $(A/m) \sin(\omega t)\sin(mkx)$ if similitude is to be achieved.

REFERENCES

Burger,W., H. Oumeraci & H.W. Partensky 1988. Geohydraulic investigation of rubble mound breakwaters. *Proc. 21st coastal engineering conference, Costa del Sol-Malaga, Spain.* ASCE Publication, Billy L. Edge Editor: 2242-2256.

Christian, J.T., P.K. Taylor, J.K.C. Yen & R.E. David 1974. Large diameter underwater pipeline for nuclear power plant designed against soil liquefaction, *Proc. 6th annual offshore tech. conf. Houston* : 597-606.

Ishihara, K. 1983. Soil response in cyclic loading induced by earthquake, traffic and waves. *7th Asian regional conf. on soil mech. & foundation engineering. Haifa.*

Ishihara, K.& S. Okada 1978. Effects of stress history on cyclic behavior of sand. *Soils and Foundations* 18 (4): 31-45.

Madsen, O.S. 1978. Wave-induced pore pressure and effective stresses in porous beds, *Geotechnique,* 28 (4):377-393.

Poorooshasb, H.B., I. Holubec & A.N. Sherbourne 1966, 67. Yielding & flow of sand in triaxial compression, *Canadian Geotechnical Journal,* 3:179-190 4:376-397.

Poorooshasb, H.B. 1989. Description of flow of sand using state parameters. *Computers and Geotechnics.* 8:195-218.

Roscoe, K.H. & H.B. Poorooshasb 1963. A fundamental principle of similarity in model tests for earth pressure problems. *Proc. second Asian regional conf. on soil mech. and foundation engineering, Tokyo, Japan* : 134-140

Verruijt, A. 1982. Approximations of cyclic pore pressures caused by sea waves in poro-elastic half-plane. *Soil mechanics- transient and cyclic loads,* John Wiley & Sons:37-51.

Yamamoto, T., H.L. Koning, H. Sellmeyer & E. Hijum 1978. On the response of a poro-elastic bed to water waves. *J. Fluid Mech.,*87:193-206.

Yang, Q.S. 1990. Wave induced response of sea floor deposits, *Ph.D. Thesis, Concordia University,* 197 p.

8 SEA LEVEL RISE AND ITS CONSEQUENCES

Y. HOSOKAWA* and N. MIMURA**
* *Port & Harbour Research Institute, Yokosuka, Japan*
** *Ibaragi University, Hitachi, Japan*

PRESENT KNOWLEDGE ABOUT CLIMATIC CHANGES AND SEA LEVEL RISE

IPCC establishment

Responding to the growing concern on global climatic changes, the World Meteorological Organization (WMO) and the United Nations Environmental Program (UNEP) jointly established the Intergovernmental Panel on Climate Change (IPCC) in 1988. Prior to its establishment, scientists and meteorologists had already paid attention to the increase in the atmospheric concentration of greenhouse gases such as carbon dioxide. The main reason for their rapid increase, especially in the last thirty years, is attributed to emissions from human activities. The IPCC tried to summarize the existing knowledge on the global climatic changes and associated impacts through intensive scientific discussions. Three working groups were organized by the IPCC in order to deal with the scientific and technical information on specific research fields:

Group 1: on the issues of climatic changes,

Group 2: on the potential environmental and socio–economic consequences of climatic changes, and

Group 3: on the appropriate strategies for mitigating and/or adapting to climatic changes.

In Working Group 3, the Coastal Zone Management Subgroup (CZMS) was responsible for the formulation of adaptive responses along the coastal areas. This subgroup is still in active research on the adaptation to and impact assessment of the potential climatic changes after the reformation of the IPCC organization in the year 1993.

IPCC report on rate and magnitude of climate changes

The First Assessment Report of the IPCC (IPCC 1990) was released in the year

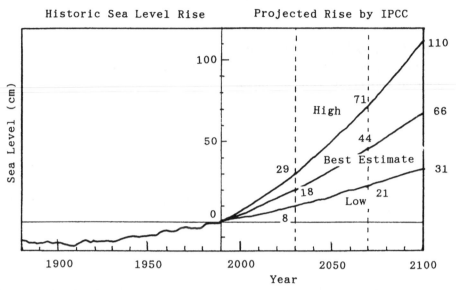

Figure 8.1. Sea level change – historic rise and projected rise from IPCC "Business-as-Usual" Scenario – (based on IPCC 1990).

1990, in which scientists in the IPCC predicted that during the next century the emissions from human activities could increase the concentration of greenhouse gases and could result in an increase of the global mean temperature by 0.2°C to 0.5°C per decade. This prediction was made under the assumptions of Business–as–Usual Scenario for the future economic conditions and with no additional efforts for reducing the emissions. In this First Report, the IPCC also predicted under the same scenario that the global mean sea level could rise by 3 cm to 10 cm per decade. Owing to time lags in the climatic system, some rise in the sea level could be anticipated over the next century even if the strong actions for reducing the emissions were applied right now. As seen in Figure 8.1, the IPCC estimated the magnitude of sea level rise as 66 cm with variation from 31 cm to 110 cm by the end of next century under the Business–as–Usual scenario. The IPCC Supplement Report (IPCC 1992) was published after re-viewing the latest information. The Supplement did not revise the estimation of sea level rise given in the First Assessment Report. Unfortunately, the IPCC could not provide predictions of climatic changes, sea level rise or changes in the frequency and intensity of storms (or other extreme events of climate) for specific region or area.

Coastal impact

Among the widespread effects of the global warming, it is anticipated that the sea level would increase at an accelerated speed in the next century. The

accelerated rise in sea level could cause significant impacts on the world coastal zones, such as exacerbated inundation, flooding, erosion, salt water intrusion, etc. The First Assessment Report indicated that the most vulnerable areas to sea level rise in the world are expected to be the coastal lowlands and the coral islands. This report concluded that coastal nations should begin the process of adapting to sea level rise. The CZMS called upon the coastal nations to carry out case studies to assess the vulnerability to the accelerated sea level rise in 1991 (IPCC CZMS 1991). The following studies are required for each nation or coast:
– identification of the coastal areas at risk to the anticipated sea level rise,
– identification of the effective response measures, and
– implementation of coastal management plans to reduce regional vulnerability.

For the first study, a common methodology has been developed to give a basis for the vulnerability assessment, which consists of practical "seven steps", such as specification of the scenarios for sea level rise and climatic changes, inventory of coastal characteristics, assessment of physical changes and their impacts, formulation of response strategies (IPCC CZMS 1991). In this "Common Methodology", vulnerability was defined as a nation's ability to cope up with the consequences of an acceleration in the sea level rise and other coastal impacts of global climatic changes. The methodology was expected to estimate physical, ecological and socio–economic vulnerability for a country. Population and social assets at risk could be estimated as well. It helped researchers to develop worldwide assessment and to understand the world distribution of vulnerability.

For the second issue, the CZMS classified the response strategies into three categories namely: Retreat, Accommodation and Protection.
– Retreat is an option of abandonment of structures in the currently developed areas, resettlement of the inhabitants and relocation of the developed areas back from the shore.
– Accommodation means continuing to occupy the vulnerable areas with the acceptance of potential floods. This option will include the change of coastal land–use from farming land to fish ponds.
– The last option is Protection which expresses the defense of coastal cities and economical activities by structures, beach nourishment, constructed wetlands and so on.

For the third issue, the CZMS recommended to every coastal country to make up its own comprehensive coastal management plan until 1995 in order to minimize the impacts and damages. As the climatic change tends to aggravate the existing vulnerability to the such extreme events as cyclone attacks, high waves or droughts, present problems threatening the present coasts should also be considered for planning. The CZMS stressed the importance of integration among the various organizations and sectors from community level to the national government.

CASE STUDY ON COASTAL VULNERABILITY

Introduction

This section describes three case studies from different sites namely: Japan (Mimura et al. 1992a), the Tongatapu Island in the Kingdom of Tonga (Fifita et al. 1992), and the Tianjin area in China (Mimura et al. 1992b). These studies were carried out by the Japanese task team under good cooperation with scientists or technical officials of each country. The case study sites have different characteristics, such as developed coastal zones, small coral island, and large low-lying river delta. Sea level rise of 0.3, 0.5, and 1.0 m, and superposed increase due to storm surge were considered as scenarios for future sea level. Impacts of inundation and exacerbated coastal flooding on land areas, population, and assets were evaluated by analyzing geographical and socio-economic data, and satellite remote sensing images. The damages are equally very severe in all the sites, but different in picture depending on the natural, social and historical conditions of the sites.

Vulnerability of Japanese coastal zones

Characteristics of Japanese coastal zones

The Japanese coastal zones are unique in their highly diverse natural conditions and in the intensive concentration of large population, economic activities, and assets in them. A notable feature in the Japanese coasts is the increasing number of man-made coastlines. In addition, the Japanese coasts are being extensively used. There have been severe ocean-related natural disasters. Because of this, about 46% of the entire Japanese coast line of about 34,000 km, is designated as a "protection-needed coast", which should be protected by structures. Coastal protection structures have been built along 27% of the Japanese coastlines.

Since about 72% of the Japanese land area is mountainous, the majority of the population live in the coastal zone, and the economic activities are concentrated in the flat lands close to the coasts. There are about 1100 harbours and ports, as well as about 2950 fishing ports along Japanese coasts. Huge cities like Tokyo, Osaka, and Nagoya are all located in the coastal zones, and facilities for industries, energy production, commerce, fisheries, and recreation are concentrated there. These three areas among themselves account for more than 50% of Japan's industrial production, and about 70% of the national wholesale trade.

Impacts of inundation and flooding

The increasing risk of inundation and flooding would be the most primary impact of the sea level rise. To assess the degree of these impacts on all of

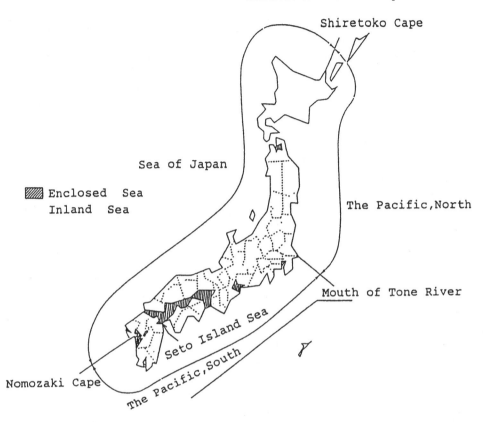

Figure 8.2. Five divisions of Japanese coastal zones (Mimura et al. 1992a).

Japan, affected areas, population, and assets were examined.

To calculate the area that is likely to be inundated or flooded, the data of elevation and land use were taken from "Digital National Land Information". "Population Mesh Data" in the 1985 National Census and "National Accounts" issued by the Economic Planning Agency were used to calculate the population and assets in the flooded area.

An area below the high water level is recognized as an inundated area, because such an area is affected by the sea nearly all the time. An area that will be below the high tide level of a storm surge or a tsunami can be considered as a flood prone area. In order to examine the increase in the extent of these areas due to sea level rise, water levels for both normal and extreme conditions have to be set along the whole coastline.

Tidal range and the water level resulting from a storm surge or a tsunami differs with locations due to their different coastal topographies. In the present analysis, the coastal zones of the entire nation were divided into the following five zones as shown in Figure 8.2: the Sea of Japan, the Pacific North, the

Table 8.1. Scenarios for water levels along Japanese coast (Mimura et al. 1992a).

Zone	Meteorological conditions		
	Mean water level (m)	High water level (m)	High tide level (storm surge or tsunami) (m)
Sea of Japan	0.0	0.6	1.8
	0.3	0.9	2.1
	0.5	1.1	2.3
	1.0	1.6	2.8
North Pacific	0.0	0.6	3.6
	0.3	0.9	3.9
	0.5	1.1	4.1
	1.0	1.6	4.6
South Pacific	0.0	1.0	2.8
	0.3	1.3	3.1
	0.5	1.5	3.3
	1.0	2.0	3.8
Tokyo Bay Osaka Bay Ise Bay	0.0	0.9	4.0
	0.3	1.2	4.3
	0.5	1.4	4.5
	1.0	1.9	5.0
Seto Inland Sea Ariake Bay	0.0	1.6	4.6
	0.3	1.9	4.9
	0.5	2.1	5.1
	1.0	2.6	5.6

Pacific South, and enclosed bays and inland seas. The last zone (enclosed bays and inland seas) was subdivided into two zones according to the tidal range. Based on the estimate of IPCC (1990), three values of 0.3, 0.5 and 1.0 m were taken as the scenarios for future sea level rise, which were superposed over the levels of high water and storm surge. The water levels for normal and storm surge conditions are listed in Table 8.1 for each zone.

Table 8.2 shows the results of the assessment. At present, even under normal conditions, areas of about 861 sq.km. are below the mean high water level. There are about 2 million people living in this zone, together with about 54 trillion Yen (about US$ 415 billion) worth of assets. Once the sea level rises

Table 8.2. Impact of sea level rise on land area, population, and assets at risk for Japanese coasts (Mimura et al. 1992a).

	Area (Km²)	Population (X 10,000)	Assets (trillion Yen)
PRESENT			
Mean water level	364	102	34
High water level	861	200	54
High tide level			
(storm surge/tsunami)	6268	1174	288
0.3 m RISE			
Mean water level	411	114	37
High water level	1192	252	68
High tide level			
(storm surge/tsunami)	6662	1230	302
0.5 m RISE			
Mean water level	521	140	44
High water level	1412	286	77
High tide level			
(storm surge/tsunami)	7583	1358	333
1.0 m RISE			
Mean water level	679	178	53
High water level	2339	410	109
High tide level			
(storm surge/tsunami)	8893	1542	378

by 1 m, this zone will expand to about 2340 sq.km. which is about 2.7 times larger than its present size. The population affected will be about 4.1 million, and the assets worth about 109 trillion Yen (about US$ 838 billion). Moreover, the flood prone zone, which is currently about 6270 sq. km, will expand to about 8900 sq.km. due to 1 m sea level rise. More than 15 million people will possibly be endangered.

Other aspects of sea level rise related impacts
Huge infrastructures such as the disaster prevention systems and transportation facilities protect and maintain socio–economic activity in Japan. If a rise in sea level should cause damage to these infrastructures, the impact will spread to the socio–economic systems as a whole.

As stated earlier, the coastlines of Japan are becoming increasingly "man–made," with the number of sandy beaches and tidal flats decreasing. The recognition and the need to protect natural beaches and coastal ecosystems is becoming rapidly popular. In Japan, since many embankments and other protective facilities have been built along coastlines, in many places, there is not much room left for the natural geographical features and ecosystems to retreat inland. Therefore, the impact of the sea level rise on natural beaches would also be strong.

Vulnerability of Tongatapu Island, The Kingdom of Tonga

Characteristics of coastal zones

The Kingdom of Tonga was chosen for a case study site as a typical coral island nation. The Japanese task team has collaborated with Tongan researchers to carry out the vulnerability assessment.

The Kingdom of Tonga is located in the South Pacific. The total land area is about 747 sq.km. There are about 171 named islands, but only 36 are permanently inhabited. The largest is Tongatapu, and the capital, Nuku'alofa, is on it. The present assessment focused on the Tongatapu Island, since it occupies about one third of the land area and contains about 70% of the Kingdom's population of about 97,000.

The islands of Tonga are fresh volcanic and uplifted limestone islands. The limestone islands, while founded on older volcanic substrates, are essentially biological constructs closely interlocked with the living fauna and flora. This is the basic natural condition of the islands of Tonga. Accelerated sea level rise could bring about some particular impacts.

Tongatapu has broad tidal flats, coastal swamps, and mangrove forests along the northern edge. A narrow fringing reef surrounds the east, south and west coasts, while coral reefs extend northward from the north coast of the island. A shallow lagoon exists in the central part of the island, and mangrove forests fringe the lagoon and the north coast of Tongatapu.

One of the most distinctive features of the region's meteorology is the occurrence of tropical cyclones. Although tropical cyclones may occur all year round, they are principally confined to the wet season from December to the following April, which is also called the cyclone season.

The distribution of population in Tonga is very uneven. Over the past 50 years there has been a pattern of urbanization with people migrating from the outer islands to the Greater Nuku'alofa area, and from rural areas of Tongatapu to the capital.

Tonga has a unique land tenure system. Families generally have a small piece of land in the 'api kolo (place to reside in town) and an 'api 'uta (place to garden). This pattern of land use is still in effect today and is one of the major facets of Tongan life and environment. The Act of 1882 established the right of every Tongan male above 16 years of age over both of these lands, i.e. a town 'api and a garden (or bush or tax) 'api.

Land allocation is one of the major issues in Tonga today. Increasing population and urbanization have used up the available agricultural and town land.

Population increase and migration have forced the Government to supply new lands and, since all of the useful agricultural land is already allocated, there has been an unfortunate tendency to subdivide and register environmentally sensitive lowlands and lands with low productivity or hazardous potential.

Impacts of inundation and flooding
In order to evaluate the area likely to be affected by sea level rise and flooding, a topographic map with at least 1–m contour was necessary. Although such a map was not available, a set of airborne photographs was found for the whole region of Tongatapu. A new map was made using these photographs to obtain contours at 1 m interval up to an elevation of 5 m.

As for the water levels, several scenarios can be drawn. In order to take various situations into account, the following water levels were used in this study.
  a) Present mean sea level : 0.75 m above chart datum
  b) High water level : 1.0 m above chart datum
  c) Extreme event (storm surge): 2.8 m above chart datum
  d) Sea level rise : 0.3 m and 1.0 m
According to the scenarios of sea level rise (SLR), four water levels were set to evaluate the inundated area (area at loss) and the flooded area (area at risk), as shown in Table 8.3.

Figure 8.3 shows the areas below the 2 to 5 m contours. Since the position of the present coastline almost corresponds to the high water level, i.e. 1 m above the chart datum, these contours indicate 1 to 4 m elevations. It can be seen that lowlands extend along the north shore, and that the land elevation is particularly low at Nuku'alofa. The areas and populations which would be affected by the increase in the water levels are listed in Table 8.4.

Table 8.3. Scenarios for water levels for Tongatapu (Fifita et al. 1992).

	Present condition (m)	SLR 1 (m)	SLR 2 (m)
Ordinary case (High water level)	1.0	1.0 + 0.3 = 1.3	1.0 + 1.0 = 2.0
Extreme case (Storm surge)	2.8	2.8 + 0.3 = 3.1	2.8 + 1.0 = 3.8

Elevations are based on the chart datum. The contour line of 1 m nearly corresponds to the high water level i.e. the present coast line.

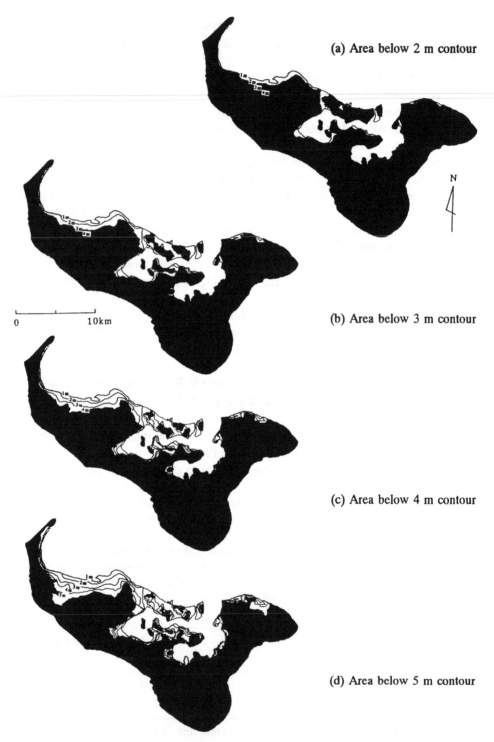

Figure 8.3. Areas in Tongatapu Island at risk due to sea level rise (Fifita et al. 1992).

Table 8.4. Impact of sea level rise on land area and population at risk for Tongatapu Island (Fifita et al. 1992).

	Present condition	SLR 1 (+0.3m)		SLR 2 (+1.0m)	
INUNDATED AREA (AREA AT LOSS)					
Agricultural land area (Km^2)	0.0	3.1	(1.3%)	10.3	(3.9%)
Residential area (Km^2)	0.0	0.7		2.2	
Population affected	0	2700	(4.3%)	9000	(14.2%)
FLOODED AREA (AREA AT RISK)					
Agricultural land area (Km^2)	23.3 (8.8%)	27.9	(10.6%)	37.3	(14.1%)
Residential area (Km^2)	4.9	5.9		7.6	
Population affected	19,880 (31.3%)	23,470 (37.0%)		29,560 (46.6%)	

Percentages of affected areas and population are based on the total values of the Tongatapu Island.

The 0.3 and 1.0 m sea level rises cause land loss of 3.1 and 10.3 sq.km. respectively, which occupies about 1.3% and 3.9% of the total area. The population likely to be affected by such inundation will be 2700 and 9000, i.e. 4.3% and 14.2% of the total population of Tongatapu respectively. For the situation of extreme event, about 20,000 people who currently live in the low-lying areas are likely to be affected by the storm surge of 2.8 m. If a storm surge occurs under the 0.3 m sea level rise, an area of 27.9 sq.km. (10.6% of the Tongatapu Island) and 23,470 people (37% of the Tongatapu population) will be at risk. These will increase to 37.3 sq.km. (14.1%) and 29,560 (46.6%) for 1 m sea level rise. These results show that the impacts of sea level rise are not limited to inundation; the danger of the cyclone induced–storm surge would increase significantly, in particular for a small island.

Other impacts
Other than the danger of tropical cyclone, various impacts were examined through the present study. They include impacts on water resources, agriculture, natural coastal ecosystem, culture, and sewage disposal.

Vulnerability of Tianjin area, China

Background
It has been pointed out that low deltaic areas of large rivers are particularly vulnerable to the sea level rise. China has huge low–lying depositional plains,

such as the Lower Liao River deltaic plain, the North China coastal plain, the East China plain, and the Pearl River deltaic plain. In order to examine the possible impacts on a low–lying and highly populated area, the Japanese task team and researchers of the Peking University jointly started a study for Tianjin area located in the North China plain.

The North China plain was formed jointly by several rivers, such as Yellow, Heihe, and Luanghe Rivers, and faces the Bohai Sea. The Tianjin area is located in the middle of this plain, and its large part is below +5 m. The economic and industrial activities are high in this area, including heavy industries such as iron industry, oil fields, and salt manufacturing. The Tianjin Port is an important international trade port and plays the role of a gateway to Beijing. Because of such economic and political importance, the city of Tianjin administrates the surrounding eleven prefectures. The case study site of the present study is this greater Tianjin area.

Impacts of inundation and flooding
As for the scenarios for future sea level rise, 0.3 m and 1.0 m were taken. A historical value of 2.9 m recorded during a typhoon in 1985 was used as the storm surge level to calculate the flooded area, though a higher record of 4.45 m was reported for a storm surge that occurred in 1895. On the basis of the known mean high water level, the storm surge level, and sea level rise scenarios, areas likely to be inundated and flooded were determined on a topographic map.

In order to assess the socio–economic impacts related to the danger of inundation and flooding, wide variety of data were collected, including population, land use, industrial products, roads and railways etc. Satellite remote sensing data of LANDSAT was also used to obtain a detailed pattern of land use. These data were combined with a topographic map by means of a computer–aided geographical information system (GIS) to analyze the degree and spatial distribution of the impacts.

Figure 8.4 shows the affected areas by inundation and flooding under the sea level rise scenarios. The inundated areas amount to about 12% and 44% of the greater Tianjin area, respectively, for 0.3 m and 1.0 m sea level rise. The flooded areas would be about 44% and 52% respectively. In the prefectures located along the coast, the affected areas would be more than 50% or 90% for each scenario.

Figure 8.5 shows the potential impacts on population and industries. The high water levels for the 0.3 m and 1.0 m sea level rise scenarios are set at 2.5 m and 3.2 m respectively, and the storm surge levels for both sea level rise scenarios are 3.2 m and 3.9 m respectively.

It can be seen that even a "static" rise in mean sea level would cause significant impacts on the population and industrial activities. If a storm surge is superposed on the sea level rise, the potential damages would be devastating.

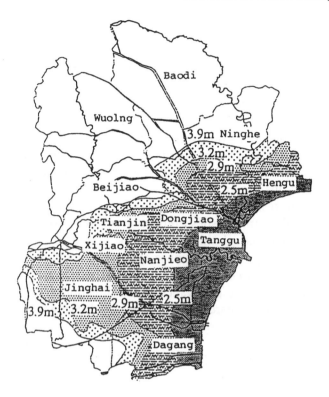

Figure 8.4. Area at risk due to sea level rise – Tianjin area (Mimura et al. 1992b).

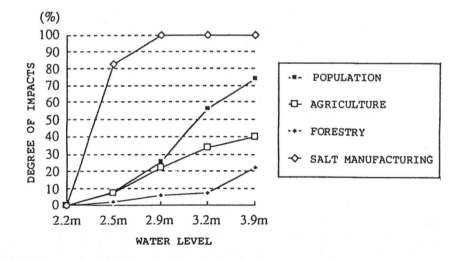

Figure 8.5. Impact of sea level rise on population and industries at risk – Tianjin area (Mimura et al. 1992b).

Conclusions

In this section, three case studies carried out by the Japanese task team were introduced. The case study sites, i.e. Japan, the Tongatapu island, and Tianjin area, are extremely different in natural and socio–economic conditions. However, it was found that the anticipated range of sea level rise would increase potential impacts of inundation at all the three sites. Moreover, danger of coastal flooding due to a storm surge would be intensified to a level of fatal significance. For example, 47% of population in the Tongatapu Island and 52% of land in the Tianjin area could be threatened by storm surge if the sea level rises by 1 m above the present level.

Another important result of the case studies was that the vulnerability to sea level rise has local characteristics, reflecting natural, social, and historical conditions in each site, though this could not be described in detail.

We are now at the stage where response strategies should be established. The CZMS of IPCC proposes to develop a coastal zone management plan as a basic responsive measure. We have to advance in this direction to preserve good and safe environment and to prepare for future sea level changes.

STUDY ON COASTAL FACILITIES AND PROTECTION SYSTEMS

Introduction

As our life and activities in coastal areas depend deeply upon the climatic conditions, it is natural to consider that the impacts spread towards wider range. Various facilities, which are supporting our social and economic life there, are also expected to receive some impacts. In this chapter, Japanese coast is taken as a typical example of highly populated and intensively utilized coast. First, the methodology for the analysis of the complicated impacts is discussed. Then, we start the evaluation of the changes of physical conditions by the primary impacts directly due to the sea level rise. Next, we discuss the possibility of damages or disasters of some facilities. Finally, an example to integrate individual effects into the regional impact is reviewed as a case study.

Methodology for impact analysis

Diagram tree for impact propagation analysis
Changes in climatic conditions or sea levels affect the magnitudes of typhoons, salt intrusion to rivers and aquifers, and wave shoaling along the coast. Change of wave shoaling alters the wave height attacking the coastal structures and beaches. Then, beach erosion is also likely to occur. These affecting chains are summarized in a tree–diagram as shown in Figures 8.6 and 8.7. Figure 8.6 is

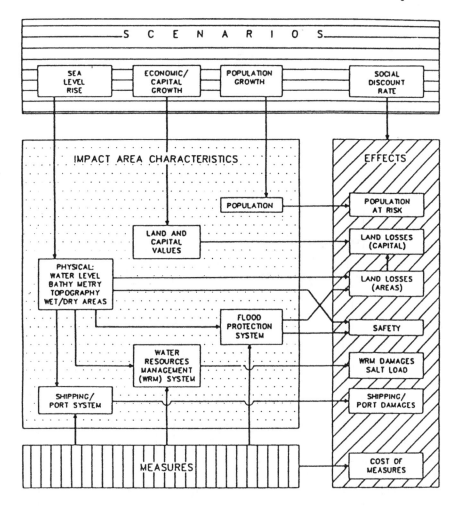

Figure 8.6. ISOS diagram (Wind & Vreugdenhil 1987).

a schematic diagram of ISOS study at the Netherlands (Wind & Vreugdenhil 1987). In this figure, final results are also shown in the right column from the several intermediate indices, opposite to the impact propagation from the initial scenarios to the intermediate indices. For the port facilities, Japanese Ministry of Transport proposed Figure 8.7 (Takeshita et al. 1991). This figure shows the details of the impact paths from the climatic changes to the society in hinterland through physical changes or natural effects. Tree branches show impact–transfer paths. The rectangular boxes express physical phenomena or influenced facilities. We can easily understand the whole process of impact propagation by such an arrangement as these. These trees will be modified from country to country, or site to site, because the main coastal disasters to be protected from

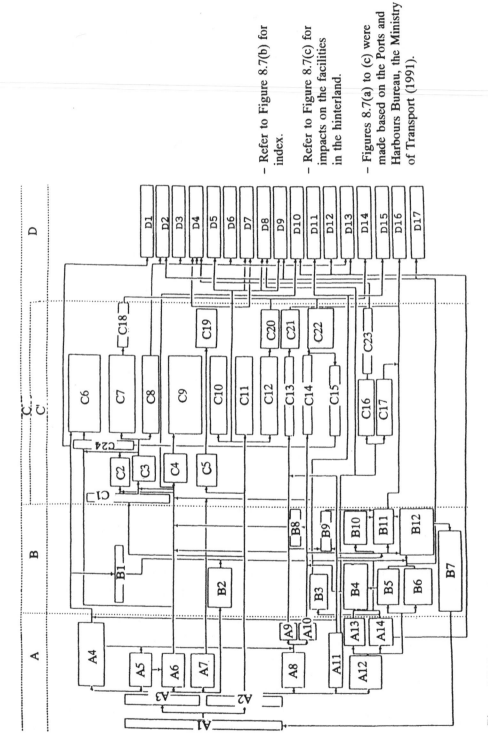

Figure 8.7(a). Impact–propagation–tree along coast (Takeshita et al. 1991)

– Refer to Figure 8.7(b) for index.

– Refer to Figure 8.7(c) for impacts on the facilities in the hinterland.

– Figures 8.7(a) to (c) were made based on the Ports and Harbours Bureau, the Ministry of Transport (1991).

A: PHYSICAL CHANGES

A1:   Sea level rise
A2:   Rise in relative sea level
A3:   Increase in water depth
A4:   Changes in direction and velocity of local current
A5:   Changes in wave–breaking position
A6:   Increase in wave height and wave force
A7:   Increase in buoyancy
A8:   Changes in standard surface for erosion
A9:   Erosion
A10:   Deposition
A11:   Retreat of shoreline
A12:   Intrusion of salt water into river and groundwater
A13:   Increase in river stage
A14:   Increase in groundwater table

B: ENVIRONMENTAL IMPACTS

B1:   Changes in water quality
B2:   Decrease in light penetrating into sea bottom
B3:   Liquefaction of ground
B4:   Changes in deposition due to flocculation
B5:   Increase in salinity in rivers
B6:   Increase in salinity in underground aquifers
B7: CO_2 increase due to decreasing carbon dioxide assimilation
B8:   Debris disaster
B9:   Loss of national land
B10:   River mouth clogging
B11:   Change in landscape
B12:   Impact on ecosystems. Loss of tidal flat, impact on marine animals and plants

C: IMPACTS ON THE NATIONAL LAND
C':   IMPACTS ON HARBOURS, PORTS, RIVERS, AND COASTS

C1:   Decreased crown height
C2:   Increase in transmitted waves
C3:   Increase in wave overtopping
C4:   Decrease in stability
C5:   Reduced drainage slope

Figure 8.7(b). Index to Figure 8.7(a).

C6:   Loss of operative days for surface storage facilities. Increased difficulty in ship operation in navigation channels and harbour basin

C7:   Flooding of facilities for mooring, cargo handling, storage, passenger, transport and environmental protection

C8:   Functional losses to shore protection facilities

C9:   Safety losses to outlying facilities, moorings, and harbour transportation facilities, safety loss in the shore protection facilities

C10:   Mismatch of a moored ship relative to the fenders

C11:   Changes of the grade of ramp used for passengers and vehicles

C12:   Decreased clearance below bridges

C13:   Decreased width of artificial beaches

C14:   Shoaling of harbour

C15:   Increase in reflected waves

C16:   Decreased crown height at river banks

C17:   Loss of river terrace

C18:   Loss of operative days

C19:   Loss of drainage capacity

C20:   Interference with ship navigation

C21:   Restricted beach use

C22:   Slowing of ship navigation and cargo handling

C23:   Increased overflow

C24:   Reduced calmness

D: TYPES OF IMPACTS

D1:   Changes in dispersion of heated water discharge and polluted water

D2:   Damage to plant growth

D3:   Poor visibility

D4:   Inundation and flooding

D5:   Inflow of detritus and polluted water

D6:   Seepage and adsorption into landfill materials and coal

D7:   Decreased water intake capacity

D8:   Danger of ship collisions

D9:   Corrosion and concrete deterioration

D10:   Lack of strength of structures

D11:   Subsidence and building up–lift

D12:   Increased water spillage

D13:   Damage to road beds

D14:   Reduced cargo handling capacity

D15:   Damage to cable piping equipment

D16:   Damaged landscape

D17:   Loss of water purifying function

Figure 8.7(b) (contd.). Index to Figure 8.7(a)

Types of impacts	Roads	Airports	Buildings	Underground shopping malls	Power stations	Railways	Subways	Telephone, electric power and city gas facilities	Recreational facilities	Water works	Sewage	Waste treatment facilities	Storage base
Changes in dispersion of heated water discharge and polluted water					*						*	*	
Damage to plant growth	*	*			*				*				
Poor visibility	*	*				*			*				
Inundation and flooding	*	*	*	*	*	*	*	*	*	*	*	*	*
Inflow of detritus and polluted water					*					*	*		
Seepage and adsorption into landfill materials and coal					*							*	
Decreased water intake capacity					*					*			
Danger of ship collisions	*	*				*				*			
Corrosion and concrete deterioration	*	*	*	*	*	*	*	*		*	*		*
Lack of strength of structures	*	*	*	*	*	*	*	*	*	*	*	*	*
Subsidence and building up–lift	*	*	*	*	*	*	*	*	*	*	*	*	*
Increased water spillage	*		*	*		*	*	*	*	*	*	*	*
Damage to road beds	*	*				*		*		*			
Reduced cargo handling capacity					*							*	*
Damage to cable–piping equipment	*	*	*	*	*		*	*		*	*		*
Damaged landscape			*						*				
Loss of water purifying function										*	*		

Figure 8.7(c). Impact on the facilities in the hinterland.

may be different owing to the uniqueness of the natural and the social conditions of each area.

Damages of whole system and individual facilities
As seen in the figures of the tree analysis, damages to individual facilities can be caused by various elementary impacts which may occur directly due to the sea level rise. If we can find an easily–occurring path among the tree branches and trace to the individual damages, then we can design the effective countermeasures. For comparing the importance among these paths, we have to estimate the magnitudes of the impacts. So, sensitivity of the primary physical changes to the sea level rise should be first compared, and then the vulnerability of each coastal facility to these physical indices should be estimated.

Primary changes due to sea level rise

Direct effects of sea level rise
Global warming is reported to change tidal amplitude; magnitude and intensity of typhoon; precipitation pattern; ocean currents; strength of seasonal winds and other regional climatic factors besides the sea level rise. Changes of storm intensity and precipitation patterns are not direct impacts due to the sea level rise but expressions of the climatic changes itself. These changes bring large effects from off–shore to the coastal area. Magnitude of the response of each coastal physical phenomenon to above changes can be estimated by hydraulic calculations under the assigned scenario.

Tidal amplitude
Nakatsuji et al. (1991) reported after numerical analysis for Osaka Bay that high water level for semidiurnal tide will raise at almost the same height as the sea level rise itself. The dynamic effects of the sea level rise on the tidal amplitudes are small enough under the same boundary conditions at the bay mouth. It was pointed out that the effect of the sea level rise on the tidal amplitudes is of comparable magnitude with that of several reclamation works proposed along the coast line. Hosokawa & Sekine (1992) also showed that the sea level rise increases the progressive speed of tide in Tokyo Bay, though changes in its amplitude are negligibly small. Kioka et al. (1992) calculated the change of tidal amplitude at the inner bottom end of Ise Bay as –10 cm to –20 cm after 0.55 m to 1.0 m of the sea level rise. They attribute this decrease mainly to the peak–cut effect of the tsunami breakwater in the bay. We can easily guess that for the estimation of high water level after the sea level rise, the magnitude of sea level rise itself is important, on the assumption that the boundary condition of water level remains the same and the effect of the ocean current stays similar. In some cases, attention must be paid to the artificial morphological changes owing to the coastal development besides the sea level change.

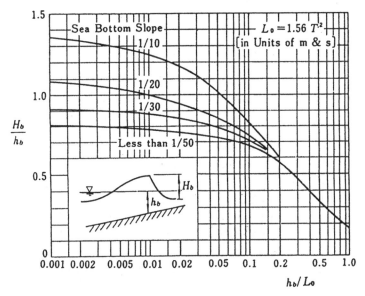

Figure 8.8. Chart of breaker height (Goda 1985).

Shoaling and breaking of offshore waves

When approaching the shallower areas near the coasts, progressive waves from the off–shore start to feel the bottom friction. Group velocity and the wave height changes with water depth due to the shoaling effect. Breaking point of the waves also moves landward after the sea level rise. If the bottom slope is stable, change of wave height at the breaking point is easily estimated by the shoaling charts (Goda 1985 and OCDI 1991). Figure 8.8 is a typical shoaling chart of limiting breaker height of regular waves. In this chart, h_b is water depth at breaking, L_0 is wave length in deep water, T is wave period, and H_b is wave height at breaking. Then, the changes of incident wave height and breaking point alter the exerted forces on coastal structures and natural beaches.

Storms and storm surges

Emanuel (1987) discusses the dependency of hurricane intensity on climate. Based on his chart of the world distribution of the minimum central pressure, around Japan global warming will increase the typhoon intensity by about 15% to 20%. But, we have to note the resolution and accuracy of his GCM model as well. After an intensive review of the historical records and diaries over 300 years, Kawata & Hohana (1989) and Kawata (1991) pointed out that at the beginning of cold period the number of typhoons passing through old Kyoto was larger than that in warmer period as shown in Figure 8.9. Number of writings on surge disasters was also larger at early and end stages of cold periods. This might be the result of the unstable climatic conditions during and after the transition periods.

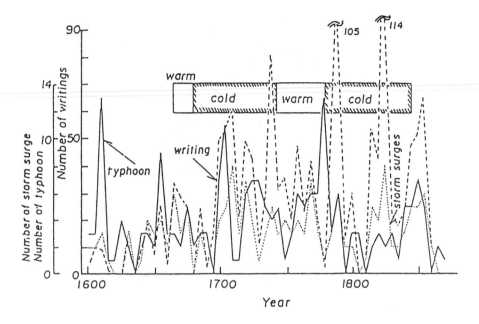

Figure 8.9. Number of storm surges, typhoons passing through Kyoto and writings on surge disasters for three centuries (Kawata 1991).

For storm surges, increase of typhoon intensity makes the surge height larger by both the stronger suction effect of low pressure and the stronger set-up effect of wind blowing. But, increase of water depth due to the sea level rise makes the set-up height for the wind blowing effect smaller. Detailed discussion can be found in a later section on case study for Tokyo Bay.

Increase of wind velocity and blowing fetch may effect the height and force of wind waves.

Rainfall
The changes of precipitation pattern predict that it is possible to have more frequent draught and inundation in Japan. Yoshino (1990) summarized the hydrological data of about a hundred years and found that frequency of precipitation "more than 300 mm in two consecutive days" during the most warm decade (1958–67) is significantly larger than that during the most cold decade (1901–10). It was also predicted from the water resources point of view that annual precipitation in Japan will increase due to the global warming and there will easily occur, at the same time, very small precipitation for as long a period as 60–90 days.

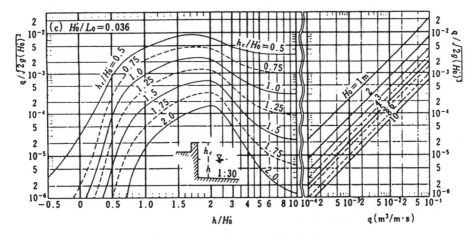

Figure 8.10. Design diagram of wave overtopping rate for vertical revetments on a bottom
slope of 1/30 (Goda 1985).

Impacts to individual coastal facilities

Overtopping

For protection from high waves or surges, coastal lowlands are often surrounded
by the high seawall. Crown height of the wall is designed from the view point
of wave overtopping (OCDI 1991). Water mass running over the crown top
increases non-linearly when the crown height is below the critical value. Sea
level rise means decrease of the relative crown height from the mean water
surface. Accordingly, effect of the sea level rise will amplify the overtopping
largely, even if the attacking wave height remains the same. Figure 8.10 is a
chart for the estimation of overtopping ratio (q) per unit length per unit period
against the relative crown height (Goda 1985). In this figure, h_c is crown height
from the mean sea level, and H_0' is equivalent deepwater wave height
(corresponding to the significant wave height). Overtopping ratio (q) can be
seen increasing non-linearly with water level in front of the wall. Assigning the
magnitude of the sea level rise, we can estimate new wave height in front of the
concerned seawall by the shoaling model, and then new overtopping ratio by
the overtopping–chart. Bank raising is expected as one of the direct
countermeasures to decrease overtopping. Before raising, checks are necessary
to avoid circular slip and excess toe pressure (failure of the foundation) for the
new weight.

Stability of armour stones and blocks

On the slope surface of a structure, larger rubble or concrete blocks are usually
placed as armoring materials against wave attacks. The increase of wave force

attacking the rubble can be evaluated by the Hudson's formula (OCDI 1991). The necessary weight of rubble is proportional to the third power of significant wave height-- $(H_{1/3})^3$. If the wave height increases much due to the sea level rise, some of the surface rubles should be replaced by heavier ones in order to prevent drifting or destruction damages.

Sliding of breakwaters
One of the most important facilities for large ports in Japan is breakwaters which protect waves from outer sea and maintain calm water surface inside the port area for the safe navigation and cargo handling. Breakwaters in Japan usually set at the depth of 5 m to 20 m for design wave–height of about 5–10 m. Normally, caisson–type breakwater is adopted, which is designed as a combination of rubble mound and a heavy concrete–caisson box with sand inside. Total length of 900 km has already been arranged as a main port facility against severe wave attacks. Takayama (1990) and Takayama & Ikeda (1992) conducted statistical calculations for sliding damage of breakwater due to the sea level rise. Assuming simplified beach with uniform slope and off shore wave, the dimensions of each caisson at several depths were first designed based on the official Design Standard (OCDI 1991). This design includes estimations of shoaling, wave height and exerted wave force on the caisson. Then, under the same assumption except for the sea level, again the exerted force with new water depth and shoaling effect was estimated. Increase of water

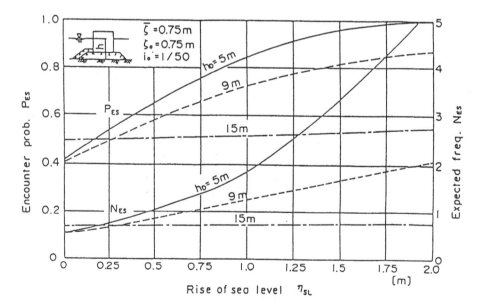

Figure 8.11. Encountered probability and expected frequency of sliding after sea level rise (h_d is water depth) [Takayama 1990].

depth increases the floating effect on the caisson also. Probability and encounter frequency during life time (50 years) for the sliding failure were simulated for each breakwater at different depths. One of the results for 1/50 slope beach is shown as Figure 8.11. It is clear that the caisson at shallow depth (–5 m) is affected significantly by the sea level rise and encounters frequency increases sensitive to sea level rise. But, the caisson at deep point (–15 m) is relatively stable against sliding even after the sea level rise. Care has to be exercised regarding the stability of structures at shallow area when the sea level rises.

Bearing capacity and liquefaction
Sea level rise may promote salt water intrusion into aquifers and rise of groundwater level at the shore line. Bearing capacity of shallow foundations in lowlands may be changed due to the rising level of groundwater. Excess porewater pressures will cause liquefaction during earthquakes. The possibility of liquefaction and effects are studied by Yasuhara & Mimura (1992).

Beach erosion
Bruun rule is proposed for the estimation of beach erosion due to the sea level rise (Committee on Engineering Implications of Changes in Relative Mean Sea Level 1987). This rule is based on the assumption that the same equivalence between wave force and beach slope is satisfied as before. Figure 8.12 is a schematic figure for this rule. Beach slope at the shoreline (point A) is decided by the force balance between the wave and the particle weight which is a

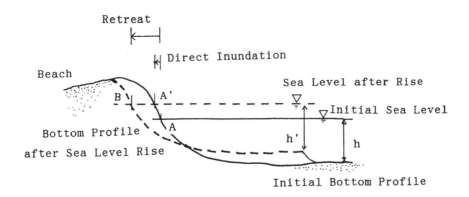

Figure 8.12. Schematic figure of Bruun rule (New shoreline is at point B, not at A') – Based on Committee on Engineering Implications of Changes in Relative Mean Sea Level 1987.

function of the sand diameter. After the sea level rises, point A' is exposed to the wave as new shoreline. So, beach slope at point A' changes similar to that at point A. As the total mass of beach sand is maintained constant, sand must be supported from the upper part of beach to make a new equivalent slope. Hence, beach is eroded to the point B. Finally, after this rule, new shore is placed at point B, not A'. At present, we cannot verify the Bruun rule yet by the field data at subsidence beaches.

Lowland drainage system

For providing protection from the storm surges at the river mouth delta or canal areas, the required length of the seawall along waterways becomes very large. From the safety and economic point of view, some of the lowland areas need a different type of protection system, a combination of a surge–barrier wall line

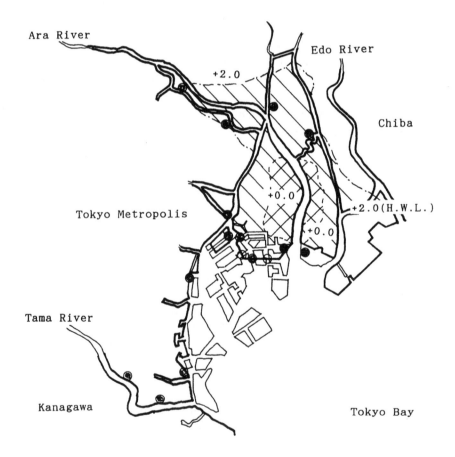

Figure 8.13. Arrangement of barrier embankment and pumping stations (O) in Tokyo Metropolitan district.

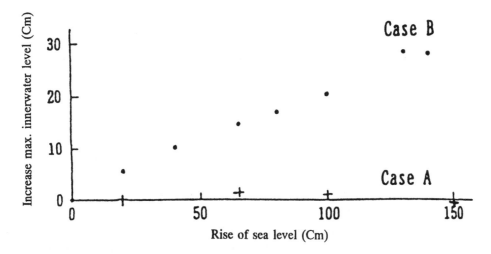

Figure 8.14. Increase of maximum inner water level at a storm with sea level rise {Ratio of (inner water surface/catchment area) is 0.11 for Case A and 0.009 for Case B}.

with water gates and drainage pumping stations. A barrier wall with high crown is set along the front line of the lowland or at the bay mouth. Water gates are placed at the river and canal mouths to assure navigation in normal weather. During storm weather, gates are closed and pumps are operated for the drainage of associated strong rainfall. Figure 8.13 shows a typical arrangement of barrier wall, gates and pumping stations along the Tokyo Port. Inside the barrier line, canal network is developed for navigation in the lowland areas.

After sea level rise, a pumping station may become smaller in its drainage capacity due to the increase of the water–head difference between outer and inner water levels. Using the relationship between drainage capacity vs. head difference of the pumps, Hosokawa & Sekine (1992) calculate the maximum inner water level for a given storm. Sea level rise affects very little the inner water levels if the barrier system has: a) large area of inner water surface compared to its catchment area, b) enough capacity for pumping or extra number of pumps, and c) earlier operation of gates at the approach of a storm. In Figure 8.14, increase of maximum inner water level is plotted against the magnitude of the sea level rise. In this figure, the effect of the water surface ratio to the catchment can be compared as two lines. The contribution of urbanization in the catchment area to the change of inner water level is also calculated. In some cases, the change of urban run–off pattern after precipitation is comparatively large as that of the sea level rise. Maintaining large water surface of inner area is said to be also useful to minimize: a) the effects of peak flush runoff from tributaries, and b) the water quality impact by the flush to the coastal sea.

Case study for increase of storm surge height along Tokyo Bay

Following the impact propagation path, Tsutsui & Isobe (1992) simulated the possible risk of inundation in Tokyo area by a storm after the global warming. Their numerical calculations included: a) model typhoon setting, b) estimation of the storm surge due to the model typhoon's passing across, via several routes, c) evaluation of wind waves, and d) prediction of the maximum water level distribution along the coast. Figure 8.15 shows selected coastal points for the prediction of maximum water levels during a typhoon.

The results of calculations for a certain route are shown in Figure 8.16. This figure shows that: a) water level along the coastal line increases gradually toward the inner bottom end, and b) it becomes the highest near the bottom end of the bay, due to the storm surge and wind waves. Through the numerical study, Urayasu, Narashino and Chiba were found as vulnerable areas for the storm surge under the sea level rise condition. Comparing the present crown height of the barrier wall, Matsui et al. (1992) found the distribution of overtopping ratio along these vulnerable cities.

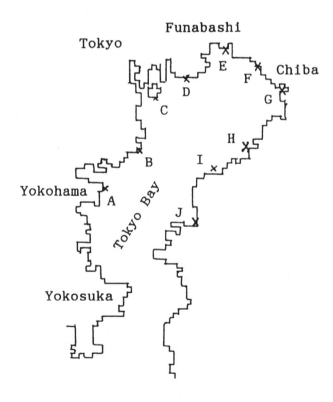

Figure 8.15. Calculation points (A–J) of water levels during a typhoon attack (Tsutsui & Isobe 1992).

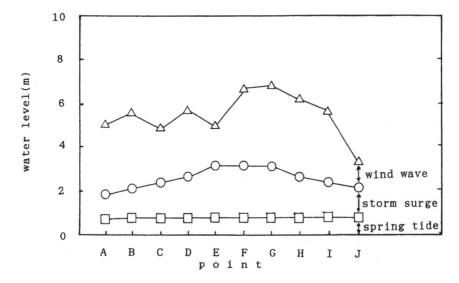

Figure 8.16(a). Calculated water levels for designed typhoon (940 hPa) on route A without sea level rise (Tsutsui & Isobe 1992).

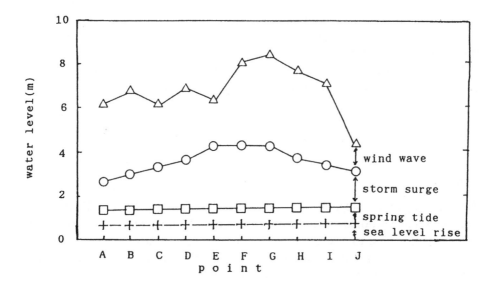

Figure 8.16(b). Calculated water levels for larger–intensity typhoon (925 hPa) on route A with 65 cm sea level rise (Tsutsui & Isobe 1992).

Conclusions

Reviewing the studies on the coastal impacts to the sea level rise, we can conclude from the engineering point of view as follows:

a) Regional scenarios for climatic changes such as sea level rise, precipitation patterns and magnitude of storms, are not yet clear. But, under appropriate assumptions, the impact estimation for the coastal lowlands is not quite impossible based on our present knowledge.

b) Qualitative estimation of primary physical impacts such as changes of wave shoaling and tidal amplitude is possible based on our knowledge of the coastal hydraulics, if the magnitude of sea level rise and other related information are assigned.

c) Some of the secondary impacts can also be estimated, if the process is expressed clearly by equations or response charts. Overtopping and sliding are categorized in this group.

d) For the analysis of the regional vulnerability, we must understand the whole system of the disaster prevention facilities, water supply and transportation. A "Fault Tree" for the coastal protection system is presented as an example. Using this tree, a case study for Tokyo Bay was tried. Numerical simulation is useful for the estimation of the damage response for the various stages of the sea level rise.

e) Simulation will give us important information about the spatial and temporal distribution of the vulnerability along the coast. This will help us to consider the effective countermeasures against the sea level rise.

As we cannot detect or observe the actual sea level rise immediately, it is essential for us to understand what will happen before it occurs.

COUNTER MEASURE PLANNING

Bank raising

Among the three categories of measures classified by IPCC CZMS (1991), engineers have studied in depth the protection option than others, so far. Sea level is anticipated to rise gradually with years. Bank raising with large height at early phase might be unnecessary and uneconomical. Especially for those countries which have a long coastal line to protect, financial resources may be limited for bank raising at once. Accordingly, protection policy should be planned strategically. Strategies comprise combinations of specific measures. Even for the single "bank raising" measure, the design method for which is relatively clear and well–known, it is possible for us to choose various combinations of raising height and raising timing in order to maintain a certain safety level against flooding.

A simulation model called ISOS–Model was developed as a decision–supporting tool in the Netherlands (Peerbolte et al. 1991). This model covers wide range of areas with different protection systems, and bank raising is studied as an example. Three parameters such as bank raise height, free board height and implementation years can be selected and altered as policy options by model users. Comparing to the sea–level–rise scenario assigned, the model estimates safety level (or expected probability for flood). Principle of the bank raising program is shown in Figure 8.17. For maintaining a safety level (or maintaining the required crown height), necessary measures with preferable combinations of the three parameters can be estimated for each of the different scenarios. With the help of Geographic Information System (GIS) data, the model can also calculate total cost for this selected combination for a given period. National cost per next century was estimated quantitatively as substantial for the Netherlands. For the severe scenario, immediate action was proved to be necessary.

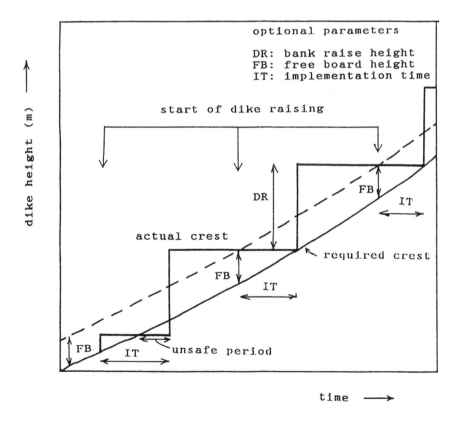

Figure 8.17. Principle of bank raising program in ISOS Model (Peerbolte et al. 1991).

Coastal zone management

The vulnerability of the coastal zone is exacerbated by the accelerated sea level rise and climatic changes. Potential severity and extent of propagation paths of the impacts should be taken into account for long–term coastal planning. Ill–combined and isolated responses by different coastal sectors could reduce the effectiveness of mitigating measures. Moreover, many of the response measures to the anticipated sea level rise are similar to those required to address existing coastal management problems under population pressures to these areas. Thus, planning for the future sea level rise should be integrated with coastal zone management practices at present.

IPCC CZMS (1992) recognized the necessity of Integrated Coastal Zone Management (ICZM) and decided to support the development and implementation of Coastal Zone Management Plans by the year 2000. ICZM is defined as a dynamic process in which a coordinated strategy is developed and implemented for the allocation of environmental, socio–cultured, and institutional resources to achieve the conservation and sustainable multiple uses of the coastal zone. It was found through discussions among the eastern hemisphere countries (Organizing Committee of IPCC Eastern Hemisphere Workshop 1993), that several countries such as Thailand, Malaysia, Philippines and Sri Lanka have already undertaken pilot projects to introduce ICZM at selected coastal locations. Some of them include the establishment of building setbacks codes along the vulnerable coasts (McLean & Mimura 1993).

Future directions of study

IPCC CZMS (1992) directs its future activities towards technical cooperation, and systematic observation and monitoring as well as data and information exchange. Monitoring of the sea level change and climatic changes is necessary to reduce the uncertainty of the phenomena and detect the regional changes. Vulnerability studies are also important for each specific coast, especially for small reef islands and coastal lowlands. New scientific tools and systems are requested to support such new concepts as coastal vulnerability and integrated coastal management.

REFERENCES

Committee on Engineering Implications of Changes in Relative Mean Sea Level 1987.
   Responding to changes in sea level. National Academy Press, 148p.
Emanuel, K. A. 1987. The dependency of hurricane intensity on climate. *Nature* 326(2):
   483–485.
Fifita, N. P., N. Mimura & N. Hori 1992. Assessment of the vulnerability to sea level rise
   for the Kingdom of Tonga. *Proc. IPCC CZMS Workshop "the rising challenge of the*

sea" (in press).

Goda, Y. 1985. *Random seas and design of maritime structures.* University of Tokyo Press, 323p.

Hosokawa, Y. & Y. Sekine 1992. *Effect of global warming on drainage function of pump stations in ports.* Report of Port and Harbour Research Institute. 31(1): 51–115 (in Japanese).

IPCC CZMS 1991. *A common methodology for assessing vulnerability to sea level rise.* Ministry of Transport, Public Works and Water Management, Tidal Water Division, Rijkswaterstaat, the Netherlands.

IPCC CZMS, WG3 1992. *Global climate change and the rising challenge of the sea.* Report of the Coastal Zone Management Subgroup. Ministry of Transport, Public Works and Water Management, Tidal Water Division, Rijkswaterstaat, the Netherlands, 35p.+Appendix.

IPCC 1990. *Climate change – the IPCC scientific assessment–.* Report prepared for IPCC by Working Group I, Cambridge University Press, Cambridge, 364p.

IPCC 1992. *Climate change 1992*: The supplementary report to the IPCC scientific assessment, Cambridge University Press, Cambridge, 200p.

Kawata, Y. 1991. Effects of long term temperature changes on typhoon characteristics. *Proc. coastal engineering,* JSCE, 38(2): 931–935 (in Japanese).

Kawata, Y. & S. Hohana 1989. *Characteristics of occurrence frequency of natural disasters in Japan and China.* Annuals, Disas. Prev. Res. Inst., Kyoto Univ.,32 B–2: 891–908 (in Japanese).

Kioka, W., N. Nagashima & Y. Ookura 1992. Prediction of change of tidal current in Ise Bay due to sea level rise. *Proc. Chubu region annual conf. JSCE.* 255–256 (in Japanese).

McLean, R. & N. Mimura (Eds.) 1993. Vulnerability assessment to sea level rise and coastal zone management. *Proceedings of the IPCC eastern hemisphere workshop. Governments of Japan & Australia, Tsukuba, Japan:* 429+39p.

Matsui, S., H. Tateishi, M. Isobe, A. Watanabe, N. Mimura & R. Shibazaki 1992. Prediction of inundation area at risk along Japanese coast due to sea level rise. *Proc. coastal engineering,* JSCE. 39(2): 1031–1035 (in Japanese).

Mimura, N., M. Isobe & Y. Hosokawa 1992a. Impacts of sea level rise on Japanese coastal zones and response strategies. *Proc. IPCC CZMS workshop "the rising challenge of the sea"* (in press).

Mimura, N., Y. Hosokawa, M. Han, S. Machida & K. Yamada 1992b. Vulnerability assessment of coastal zones to sea level rise –case study of Bohai Bay area in China–, *Environmental System Research,* JSCE, 20, 176–183 (in Japanese).

Mimura, N., K. Yasuhara, M. Isobe & Y. Hosokawa 1992c. Vulnerability assessment to sea level rise. *Proc. ILT 92 on POLD, Saga Univ., Saga, Japan:*99–106.

Nakatsuji, K., H. Kurita, S. Karino & K. Muraoka 1991. Prediction of changes in water environment due to bay area developing plan for Osaka Bay. *Proc. coastal engineering,* JSCE, 38(2), 206–210 (in Japanese).

OCDI 1991. *Technical standard for port and harbour facilities in Japan.* New edition. The Overseas Coastal Area Development Institute of Japan, 438p.

Organizing Committee of IPCC Eastern Hemisphere Workshop 1993. Report of IPCC eastern hemisphere workshop on the vulnerability of sea–level rise and coastal zone management, *submitted to World Coast Conference '93, on vulnerability assessment to sea level rise and coastal zone management. Edited by R. McLean & N. Mimura.* Governments of Japan & Australia, Tsukuba, Japan:429+39p.

Peerbolte, E.B. et al. 1991. *Impact of sea level rise on society – a case study for the*

Netherlands–, Final report, UNEP and Government of the Netherlands, the Hague, the Netherlands, 104p.+Annex.

Takayama, T. 1990. Effect of sea level rise in failure on stability of breakwater. *Proc. coastal engineering, JSCE*, 37: 873–877 (in Japanese).

Takayama, T. & N. Ikeda 1992. *Estimation of sliding failure probability of present break waters for probabilistic design*. Report of Port and Harbor Research Institute, 31(5):1–30.

Takeshita, M., S. Takatsu, T. Oogama, K. Izumi & S. Miyazaki 1991. Study on coastal impacts due to sea level rise. *Proc. coastal engineering, JSCE*. 38(2):941–945 (in Japanese).

Tsutsui, J. & M. Isobe 1992. Prediction of storm surge in Tokyo Bay under global warming. *Proc. Japan Assoc. Coastal Zone Studies*. 4:9–19 (in Japanese).

Wind, H.G. & C. B. Vreugdenhil (eds.) 1987. *Impact of sea level rise on society –* Report of a project–planning session, Delft, the Netherlands. 27–29 August.,1986 (Published for the International Association for Hydraulic Research), 230 pp., Hfl.155, – A.A. Balkema, P.O. Box 1675, Rotterdam, Netherlands.

Yasuhara, K. & N. Mimura 1992. Impacts of global warming on soil structures and ground foundations. *Proc. 26th Japan Soc. Soil Engineering* (in Japanese).

Yoshino, F. 1990. On the change of hydrological cycle by green–house effect. *Jour.Japan Society of Civil Engineers*, 75(4), 18–21 (in Japanese).

9 COMPREHENSIVE FLOOD CONTROL AND MITIGATION SYSTEMS

T. TINGSANCHALI
Asian Institute of Technology, Bangkok, Thailand

INTRODUCTION

Many large cities are located in flood plains and lowland areas such as Bangkok, Taipei, Jakarta, Shanghai, and Osaka,etc. These cities are highly urbanized and populated. They are flooded often due to heavy rainfall, overbank flow, insufficient flood drainage, land subsidence and lack of effective flood control measures. Flooding periods in lowland areas normally prolong over a long period of time and in many cases have serious consequences on socio-economic and environmental conditions. Flood control and protection methods can be categorized into two major components namely: structural measures and non-structural measures. Structural flood control measures may be over-designed or not fully effective if their design or operation do not consider non-structural flood control measures. Therefore, a comprehensive flood loss prevention and mitigation system should be developed incorporating both structural and non-structural flood control measures. However, it should be realized that in general, structural flood control measures mainly contribute to the overall effectiveness of the comprehensive flood loss prevention and mitigation system. Structural flood control measures include dams, reservoirs, flood bypass channels, river dikes, tidal barriers, flood control gates and polders, etc. Figures 9.1, 9.2 and 9.3 show the structural flood control measures of the Bicol river basin, the Philippines; of Bangkok, Thailand and of Jakarta, Indonesia (Tingsanchali 1988b). Non-structural flood control measures include land use planning and control, flood proofing, flood forecasting and warning, etc. Given a design hydrological condition, the geometrical dimensions of selected flood control measures can be determined by using mathematical models or physical hydraulic models. Both types of models have to be calibrated using measured field data. After model calibration, the calibrated model is applied to determine the hydraulic effectiveness of the flood control measures on flood flow conditions. With the advancement of computers, mathematical models are now being widely used with more flexibility, higher

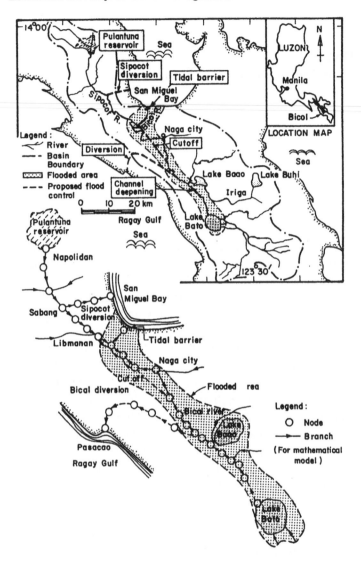

Figure 9.1.   Proposed flood control scheme and mathematical model
for flood flow simulation, Bicol river basin, Philippines.

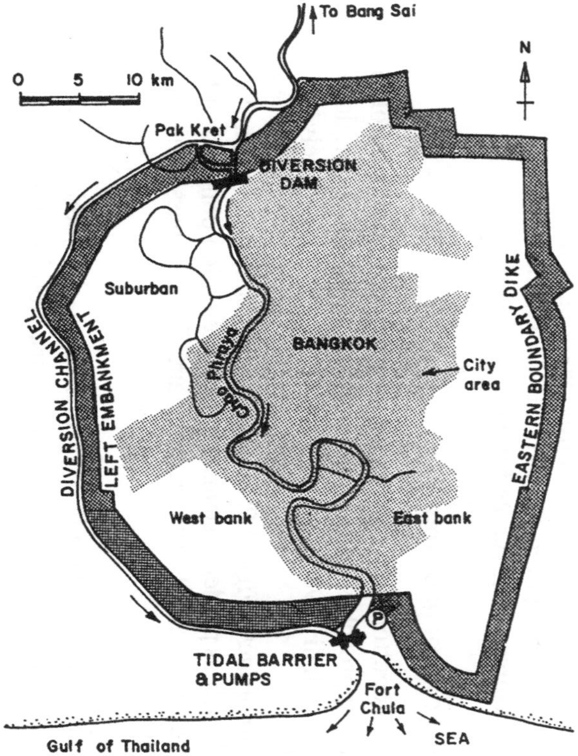

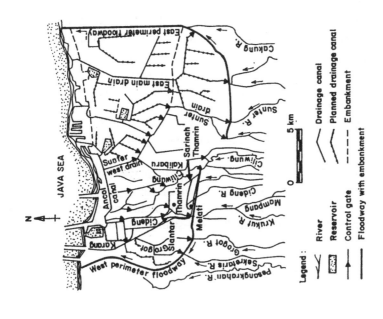

Figure 9.3. Existing and proposed flood control scheme for Inner Jakarta, Indonesia.

Legend:

——	River
▨	Reservoir
↑	Control gate
══	Floodway with embankment
⌐	Drainage canal
⌐	Planned drainage canal
----	Embankment

Figure 9.2. Proposed flood control scheme of Bangkok (Chao Phraya-2 Project), Thailand.

accuracy and less economical implications.

Additionally, the planning and operation of the comprehensive flood loss prevention and mitigation system can be carried out effectively by the use of flood risk analysis and mapping. In general, publicizing of flood risk map would help the residents and the agencies concerned to understand flooding situations and minimize flood damages. The flood risk map can be developed by three different approaches depending on the purpose of the applications, the required accuracy and the availability of data. These approaches are geomorphological approach, past flood approach and hydrological/hydraulic approach.

The planning and development of a comprehensive flood loss prevention and mitigation system in a watershed and in an urban area are described. Both structural and non–structural measures are cited, including the development of the flood risk map which is useful for the non–structural flood control measures.

FLOOD CONTROL MEASURES

As previously described, a comprehensive flood loss and mitigation system should consist of both structural flood control measures and non–structural flood control measures. In most Asian and Pacific countries, the provision of structural and non–structural flood control measures are not adequate. Moreover, the non–structural flood control measures which are relatively a new concept are not fully recognized (ESCAP 1990b). The details of both types of flood control measures are given as follows:

Structural flood control measures

Most of the flood control works are based on structural flood control measures. There are three approaches in the implementation of structural flood control measures (Tingsanchali 1988a and 1990, ESCAP 1990a) namely:

Inland control approach (Figure 9.4). In this approach, flood control structures are built inland such as polder systems and flood retention ponds for urban flood control, check dams in head water regions and flood bypass channels in river basin flood plains.

Water control approach (Figure 9.5). In this approach, flood control structures are built in waterways such as reservoirs, dams, flood control gates and tidal barriers, etc.

Combined inland and water control approach (Figure 9.6). In this approach, the combination of the above two approaches is utilized such as the construction of a polder surrounding a river portion with an upstream diversion dam and a

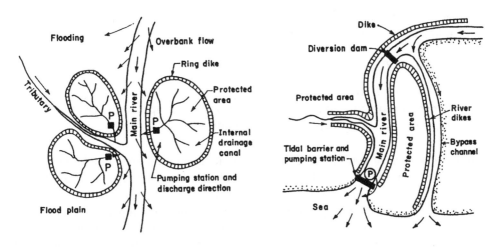

Figure 9.4. Inland control approach (polder system).

Figure 9.5. Water control approach.

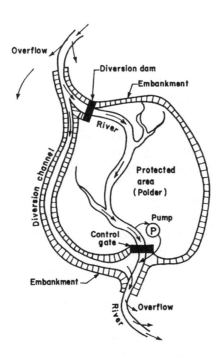

Figure 9.6. Combined inland and water control approach.

diversion channel along the outer perimeter of the polder. The construction of river dikes is also another example of this combined approach.

Non-structural flood control measures

The non-structural flood control measures involve less socio-economic and environmental problems and also involve less costs compared to structural flood control measures. They can be done in many ways as follows:

Land use planning and control. The encroachment of upper catchment areas such as the deforestation and urbanization should be restricted. Encroachment of forest area for agriculture or human residences should also be carefully controlled. Retention storages or natural ponds should be maintained and should not be converted for other purposes. Dense vegetation results in high interception losses and evaporation and acts as a retarding basin for flood control.

Flood proofing. This involves the implementation of flood protection scheme which prevents flood water to enter housing residences, private and public buildings by enclosing such buildings within surrounding walls or by providing shutters at ground floor entrances.

Restriction of development. This would minimize potential damage in flood prone areas. Flood risk map would provide additional support in implementing the restriction of development.

Flood forecasting and warning. Flood forecasting and warning made sufficiently accurate and in advance would significantly reduce flood damages and provide enough time for evacuation and emergency activities . Also, the indication of the most susceptible area to flooding to the public would ensure the effectiveness of flood forecasting and warning.

Public information. Publication of flood risk maps of flood control areas and its dissemination may be helpful in easing flood problems. The understanding of flooding situation and the method of solving flooding problems would support the activity in flood control, prevention and mitigation. Public information would help the public to understand the significance of flood control and cooperation in implementing the required measures. Effective preparedness for flood control and evacuation would be achieved. Public information would discourage people to encroach into flood prone areas.

Emergency activities. The organizations responsible for flood control and mitigation must actively involve in emergency rescuing people from flooded areas and in providing relief including food supply, clothing and medical treatment to affected people.

SELECTION OF FLOOD CONTROL SCHEMES

In order to have a comprehensive flood loss prevention and mitigation system, both structural and non–structural measures have to be considered. Structural flood control measures are costly but effective. However, structural flood control measures can create significant socio–economic and environmental problems. On the other hand, non–structural flood control measures are less costly and have less socio–economic and environmental problems. The effectiveness of the non–structural flood control measures depends largely on the cooperation of the people. Various types of structural flood control measures are, for example: river dikes, flood bypass and flood drainage channels, reservoirs and retention storages, channel training structures, tidal barriers, flood control gates and polders, etc. The selection criteria on the types of flood control measures depends on 1) land topography, 2) extent of flooding and its location, 3) type of land use, 4) land cost and area to be protected, 5) effectiveness of flood control measures , 6) socio–economic and environmental conditions and 7) benefit and cost of flood control measures.

As there are various possible alternatives of flood control schemes, the determination of the hydraulic effectiveness of each flood control scheme can be done by using the mathematical models or physical hydraulic models given design floods. Their hydraulic effectiveness can be compared based on the selection criteria described above. In many cases, a careful environmental impact assessment has to be carried out to assure public on safeguarding the environment.

Bangkok, the capital of Thailand, is located in the large flood plain of the Chao Phraya river. The city suffers from frequent floodings due to overbank flow from the Chao Phraya river over a distance of 100 km along the river banks. To protect the low lying areas of the city from flooding, polders were built around these low lying areas. Drainage facilities including drainage channels, pumping stations and control gates were provided to protect the areas from inundation due to heavy rainfall within the city. These flood control schemes saved tremendous amount of losses due to flood damages in the past. However, a larger polder scheme is still required to save the outer area of Bangkok. This is proposed as shown in Figure 9.2.

DESIGN OF FLOOD CONTROL WATER LEVELS USING FLOOD FLOW SIMULATION MODELS

The design of flood control levels can be done by using either mathematical models or physical hydraulic models or the combination of the two types of models. The application of both types of models in the design can be explained as follows:

Model calibration

Before applying the models for the design purpose, one has to be sure that the model can simulate actual flood flow conditions in rivers and flood plains. The models can be calibrated using actually measured flood flow data and the model parameters such as Manning's n which can be adjusted to match the computed water levels and/or discharges with the measured data. After model calibration, the model is considered to represent the prototype condition.

Model application in the design of flood control level

After model calibration, the model can be applied for the design purpose. By changing the geometry of the river channel or flood plains and Manning's n according to the design condition, the changes in the flow conditions are then determined by the model. The dimensions of the flood control work can be modified and the model is rerun until the changes in the flow conditions meet the requirements.

Flood flow modelling

As previously described, the modelling of flood flow can be done by mathematical models or physical hydraulic models. With the advancement of computer technology and numerical methods, the use of mathematical models become more widely accepted. With the increasing flexibility of mathematical models and its reduction in computational cost, the mathematical models become more widely used. However in many circumstances, the use of the physical hydraulic model is still needed in many applications especially in complicated localized hydraulic problems such as in the close vicinity of the flood control structures.

Mathematical models

The basic components of mathematical models used in calculating the hydraulic effects of flood control structures are the watershed model, river flow model and flood plain flow model (Tingsanchali 1979). The calculated runoff from the watershed model is routed along river channels by using the river flow model. Where overbank flow occurs, the flood plain flow model is used to calculate flood depths and velocities in flood plains. Well known watershed models are available such as the Tank model, STANFORD model and SSARR model, etc. For the river flow model, there are two main types namely: 1) 1–D channel flow model and 2) channel network model (or quasi 2–D model). The channel network model may be applied for a network of river and flood plain channels.

For flood plains, the flow can be calculated by using the channel network model or a 2–D flood plain flow model. The advantage of the channel network model is that it requires much less computation time and its numerical computation is more stable than the 2–D model. The accuracy of the channel network model is nearly the same as the 2–D model.

One dimensional river flow model. This model consists of two governing equations namely: continuity and momentum equations (Tingsanchali 1979), i.e.,

$$\frac{\partial Q}{\partial x} + B\frac{\partial h}{\partial t} = 0 \tag{9.1}$$

$$\frac{1}{gA}\frac{\partial Q}{\partial t} + \frac{2Q}{gA^2}\frac{\partial Q}{\partial x} - \frac{Q^2}{gA^3}\frac{\partial A}{\partial x} + \frac{\partial h}{\partial x} - S_0 + \frac{n^2 Q \mid Q \mid}{A^2 R^{4/3}} = 0 \tag{9.2}$$

where Q = discharge, h = depth, B = topwidth, x = distance, t = time, g = gravitational acceleration, A = cross–sectional area, S_0 = channel bed slope, n = Manning's roughness coefficient and R = hydraulic radius.

Channel network flow model. This model is applicable to both river channel network or river and flood plain channel network (Tingsanchali 1979). There are three governing equations namely: 1) storage equation for a channel junction 2) continuity equation for the channel connecting the junctions and 3) momentum equation for the channel. The storage equation for a junction can be written as follows:

$$F\frac{dH}{dt} = \Sigma Q_{in} - \Sigma Q_{out} \tag{9.3}$$

where F = surface area of the junction, H = water level at the junction, ΣQ_{in} = summation of all inflows to the junction, ΣQ_{out} = summation of all outflows from the junction and t = time. The 1–D continuity and momentum equations for the channels interconnecting the junctions are the same as Equations (9.1) and (9.2). The channel network model is found to be very practical and gives accurate results in simulating flow in river and flood plain channel network.

Two–dimensional flood flow model. The governing equations of the 2–D flow model are the depth–averaged equations of continuity, x– and y– momentum conservation (Tingsanchali & Sriamporn 1991), i.e.,

$$\frac{\partial h}{\partial t} + \frac{\partial Uh}{\partial x} + \frac{\partial Vh}{\partial y} = 0 \tag{9.4}$$

$$\frac{\partial Uh}{\partial t} + \frac{\partial U^2h}{\partial x} + \frac{\partial UVh}{\partial y} = -gh\frac{\partial H}{\partial x} - \frac{\tau_{xb}}{\rho} \tag{9.5}$$

$$\frac{\partial Vh}{\partial t} + \frac{\partial UVh}{\partial x} + \frac{\partial V^2h}{\partial y} = -gh\frac{\partial H}{\partial y} - \frac{\tau_{yb}}{\rho} \tag{9.6}$$

where U, V = depth – averaged velocities in x and y directions, h = water depth, H = water level, τ_{xb} and τ_{yb} = bed shear stresses in x and y directions and they are expressed by

$$\tau_{xb} = \frac{gn^2U\sqrt{U^2 + V^2}}{h^{1/3}} \tag{9.7}$$

$$\tau_{yb} = \frac{gn^2V\sqrt{U^2 + V^2}}{h^{1/3}} \tag{9.8}$$

The flow domain in the flood plain can be discretized into rectangular grids in which the 2–D depth averaged continuity and momentum equations are applied in a finite difference form and the finite differences equations are solved numerically by a computer.

Physical hydraulic models

In open channel flow, the effect of gravity force is more important than the effect of viscous force. The effect of viscous force in the model could be diminished by keeping the Reynolds number in the model in the range of fully rough turbulent flow. Allen (1947) suggested that the model's Reynolds number should be greater than 1400 to assure that the viscous force is insignificant. Under such a condition, the drag coefficient in the model would approach the value in the prototype. Based on the Froude similarity, the modelling criteria for flow investigation are as follows:

The velocity ratio V_r has to satisfy the relationship:

$$V_r = Z_r^{1/2} \tag{9.9}$$

where V_r and Z_r are the velocity ratio and the vertical scale ratio respectively. These ratios are defined by the prototype/model values. The flood travel time ratio T_r is therefore

$$T_r = \frac{X_r}{V_r} = \frac{X_r}{Z_r^{1/2}} \tag{9.10}$$

where X_r is the horizontal scale ratio.

The ratio of prototype/model Manning's roughness coefficients n_r is given by

$$n_r = \frac{n_p}{n_m} = \frac{R_r^{2/3}}{X_r^{1/2}} \tag{9.11}$$

where R_r is the ratio of hydraulic radius or hydraulic mean depth, n_m and n_p are the Manning's roughness coefficients of the model and prototype respectively. The discharge in the model Q_m has to satisfy the following relationship:

$$Q_m = \frac{Q_p}{X_r Z_r^{1/2}} \tag{9.12}$$

where Q_p is the discharge in the prototype.

The selection of the scales of a physical hydraulic model should be such that the model Reynolds number based on the hydraulic mean depth is greater than 1400. That is

$$M_m \geq 1400 Z_r^{3/2} \tag{9.13}$$

where $M_m = V_m R_m / \upsilon_m$ is the model Reynolds number, V_m, R_m, and υ_m are the velocity, hydraulic mean depth and kinematic viscosity of model fluid respectively.

FLOOD RISK MAPPING

Flood risk mapping is essential for the implementation of non–structural flood control measures especially in the aspect of flood plain management. The objectives of flood risk mapping are to provide information concerning flood risk maps to residents in the flood prone areas, to establish a flood protection system, warning and evacuation. Flood risk maps should be disseminated to agencies concerned and to the public, so that they have a better understanding about flooding in the study area and to protect floodings and minimizing the damages. The identification of flood risk areas may hurt land owners financially in terms of land price. But on the other hand, it is very useful for flood control planning and implementation. There are three approaches in preparing flood risk maps, namely: 1) geomorphological approach, 2) past flood approach and 3) hydrological/ hydraulic approach. The selection of each approach depends on the purpose of the application, the required accuracy as well as the available data and the condition of land use, etc.

Flood plains are normally populated due to its favorable characteristics for residence, transportation, agriculture and industries, etc. However, intensive urbanization in flood plain areas may create flooding due to acceleration of runoff caused by rooftop, pavement and insufficient drainage. To solve these flooding problems usually structural flood control measures are constructed despite their high costs and difficulty in land acquisition. Without proper control of urbanization, the existing flood control facilities become less effective. Moreover, upstream urbanization, irrigation project and deforestation cause more floodings and damage in the downstream area. Therefore, it is important to introduce the non–structural flood control measures to support the design, operation and effectiveness of the structural flood control measures. For the non–structural flood control measures, there are two main approaches:

1) Flood plain management such as control of land use, location of activities in the flood plain, flood proofing for buildings and residence.
2) Flood forecasting, warning and evacuation

ESCAP (1988) described the three methods of preparing the flood risk maps as follows:

Geomorphological approach

The flood risk map is developed based on land forms or geomorphological units such as natural levees, old river courses, flood plains, sediment deposits which indicate flood susceptibility. The study of topography and sediment can reveal much of the history of past floods. This approach is useful when streamflow records are lacking and it is suitable for large and wide flood plains. However, the extent of flooding and flood prone areas can be estimated on a qualitative basis. The effect of flood control measures and urbanization cannot be estimated by this approach. The cost of preparing the flood risk map is less than the other two approaches.

Past flood approach

This approach is a simple way to outline the inundation area and depth of inundation based on the past flood data of known magnitude and frequency. However, this method cannot indicate the effects of future flood control activities or any human interferences that may cause more future floodings. The flood risk map, based on this approach, is very convincing to the community since the residents in that area have experienced the past inundations.

Hydrological and hydraulic approach

In this approach, the flood risk map is prepared by rainfall–runoff computation and flood routing. The inundation area, depth and duration can be calculated for given flood magnitudes in which corresponding flood frequencies can be determined. In this approach, the effects of flood control measures, urbanization, river and flood plain modification can be determined by routing computation. This approach is the most expensive and time consuming and it requires detailed information of rainfall data, water levels and/or discharges of past floods, geometry of rivers, flood plains and flood control structures.

The rainfall–runoff model is used to convert basin rainfall to runoff and the runoff is routed along rivers and flood plains using a 1–D river flow model, or a river and flood plain channel network model or a 2–D flood plain flow model. The computed maximum flood levels at each cross–section of rivers and flood plain channels (in the case of 1–D or channel network model) or along the boundary of the flooded area (in the 2–D model) are used to delineate the flood risk map.

CONCLUSIONS

In order to have an effective and efficient flood loss prevention and mitigation system, flood control measures both structural and non–structural should be implemented. The structural flood control measures are expensive and their feasibility depends on many factors such as land appropriation, budget availability, engineering technology, socio–economic and environmental conditions. Normally, government authorities provide financial support for planning and implementation of comprehensive flood control measures. In general, structural flood control measures are more effective than non–structural flood control measures and they may be classified into 3 types of approach namely:inland control approach, water control approach and combined inland

and water control approach. The selection of structural flood control measures depends on land topography, extent of flooding, types of land use, land cost and land availability, benefit and cost of the project, socio–economic and environmental conditions. The design of flood control levels can be done by using mathematical models, i.e. a 1–D river flow model, a river–flood plain channel network model and/or a 2–D flood plain flow model. The design of flood control levels can also be done by using a physical hydraulic model. With the advancement of computers and numerical methods, the use of the mathematical models become more widely accepted and the results become more accurate. From practical experiences, it is found that the river and flood plain channel network model is a very powerful tool in calculating flooding conditions in rivers and flood plains and hence the design of the flood control structures. The use of the 2–D flood plain flow model requires more computation time and large computer memory. This restricts the applicability of the 2–D flood plain flow model in some practical applications.

Though non–structural flood control measures are less effective than structural flood control measures, non–structural measures have much less effects on socio–economic and environmental conditions. Moreover, non–structural flood control measures, if properly implemented, can effectively reduce the sizes of structural flood control measures and flood magnitudes.

A flood risk map should be developed for the area considered for implementation of flood loss prevention and mitigation system. The flood risk map would provide better understanding to the residents and organizations concerned. The flood risk map can be developed by using three approaches namely: geomorphological approach, past flood approach and hydrological/hydraulic approach. The selection of each approach depends on the purpose of application, the data availability and the accuracy required.

REFERENCES

Allen, J. 1947. *Scale models in hydraulic engineering*. Longmans. Green and Co., London.

ESCAP 1988. Improvement of flood loss prevention system based on risk analysis and mapping. *Proc. expert group meeting*. ESCAP, United Nations, Bangkok, Thailand: 59–114.

ESCAP 1990a. *Urban flood loss prevention and mitigation*. Water Resources Series No.68, ESCAP, United Nations, Bangkok, Thailand.

ESCAP 1990b. Integrated approach to flood disaster management and rural development. *Proc. workshop on integrated approach to disaster management and regional development planning with people's participation*, ESCAP, United Nations, Bangkok, Thailand: 1–9.

Tingsanchali, T. 1979. A composite flood routing model for an entire watershed. *Proc. 18th IAHR congress, Cagliari, Italy*. 5: 211–218.

Tingsanchali, T. 1988a. Chao Phraya–2 project : A multipurpose flood control project for Bangkok. Proc. conf. urban flood loss prevention and mitigation, ESCAP, United Nations, Bangkok, Thailand: 81–86.

Tingsanchali, T. 1988b. Flood control investigation in low land areas in Southeast Asia. *Proc. intl. symp. on shallow sea and low land.* Saga University, Saga, Japan: 55–66.

Tingsanchali, T. 1990. Investigation on flood control and drainage of Chao Phraya river for Bangkok, *Proc. intl. seminar on geotechnical and water problems in lowland.* Saga University, Saga, Japan: 11–15.

Tingsanchali, T. & W. Sriamporn 1991. Prediction of floods due to an assumed failure of Bhumibol Dam, Northern Thailand. *Proc. intl. conf. computer applications in water resources.* Taipei, R.O.C. 1: 235–242.

10 MANMADE COASTAL ISLANDS

T. TSUCHIDA

Port and Harbor Research Institute, Yokosuka, Japan

INTRODUCTION

Japan is a mountainous country consisting of 4 main islands and about 4000 small islands. The total land area is no more than 378,000 km², ranking 51st in the world in extent. Compared to its small land area, Japan has a long coast line of nearly 34,000 km. It has a large economic sea zone area of 200 nautical miles, which amounts to almost about 12 times the land area of the country. Traditionally, the sea-front areas have played an important role in the various developmental activities of Japan. In order to further expand and utilize these important areas, reclamation has been carried out almost since the beginning of the Japanese history.

The speed of reclamation has been accelerated in the last 40 years or so, because of the rapid economic growth. Figure 10.1 shows the cumulative area of reclamation in Japanese ports since 1955. The dotted line shows a prediction made in 1982. As seen in Figure 10.1, the reclamation has been carried out at an almost constant rate. Figure 10.2 shows the changes in the use or the purpose of the reclaimed lands over the years. As seen therein, most of the reclaimed land in the 1960s was utilized for industrial uses such as steel production, oil refinery, power plants, automobile and ship building industries,

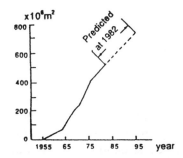

Figure 10.1 Cumulative area of reclamation in Japanese ports since 1955.

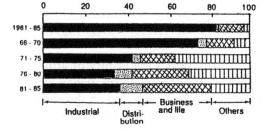

Figure 10.2 Change of use of reclaimed lands.

Table 10.1. History of artificial islands in Japan.

Date	Name	Main purposes	Area (ha)
1880–1920	Sea Fort 1st, 2nd and 3rd	Defense	
1950–1970	Mi–ike 1st, 2nd and 3rd island	Coal mining	
1953–1979	Osaka South Port	Distribution base, Housing, Park, Business, Recreation	937
1965–1980	Kobe Port Island 1st Stage	Port facilities, Housing, Urban development,	436
1971–1974	Nagasaki Airport	Airport	163
1971–1975	Ogishima	Industrial use	515
1972–1989	Higashi Ogisima	Port, Urban development	443
1972–1991	Rokko Island	Port, Urban development	580
1972–1993	Osaka North Port	Disposal of waste, Port, Industrial use, Sports, Recreation	615

etc. However, using the reclaimed land for urban developmental activities such as housing and recreation has started to increase of late. In accordance with the increasing need and demand for reclaimed lands, there has also been a change in the type of reclamation. Instead of the traditional on–shore type reclamation, the artificial island type reclamation is getting more and more popular. This is because most of the shorelines are already highly developed, and also that the artificial island type of reclamation is becoming more acceptable from the environmental point of view.

HISTORY OF ARTIFICIAL ISLANDS IN JAPAN

Table 10.1 shows the history of artificial islands in Japan. The construction of artificial islands in Japan began with an artificial island built for defense purposes at the Tokyo bay in the early 19th century. In the 1950s, the artificial islands were constructed for purposes of coal mining from the seabed. During the period of high economic growth of the 1960s, construction of large scale islands for industrial use and urban development began in 3 major bays. In the 1970s, construction of airports became a major purpose for the construction of artificial islands in accordance with the increased demand of air transportation, and with it the problems of increased noise pollution. Since the 1970s, some islands have been reclaimed with waste materials, and recently the safe disposal of urban waste is becoming an important purpose of artificial island projects. Of late, since the distance of the artificial island projects from the coast line is getting larger, more severe problems are being faced during their construction such as large water depths, severe wave conditions and soft seabeds.

The Ministry of Transport and local governments have been studying twelve proposed projects to develop artificial islands, whose locations are indicated in Figure 10.3. The outline of these projects is listed in Table 10.2, wherein it can be seen that the proposed artificial islands have various purposes, such as international trade, distribution base, marina, convention facilities, recreation and sports, fishery etc. Construction activities in one of these projects "Wakayama Marina City" has already begun. Figure 10.4 shows the plan of

Table 10.2. Outline of artificial island projects in Japan.

Location	Water Depth (m)	Area (ha)	Main purposes
Kisarazu	1–23	240	Container terminal, Distribution base, Convention, Hotel, Shopping mall, Marina, Sports
Kasai–Urayasu	5–10	395	Marina, Marine sports, Park, Artificial beach, Training center, Housing
Yokosuka (Tokyo Bay)	5–45	163	Commerce, Business, Park, Distribution base
Yokosuka (Kanada Bay)	15–25	116	Recreation, Marina, Artificial beach, Hotel, Research on fishery
Simizu	8–25	240	Container terminal, Marina, Hotel, Convention, Research on marine development
Wakayama	3–9	49	Marina, Cruising base, Sports, Housing, Park
Tamano–Kurashiki	3–5	330	Marina, Artificial beach, Fishery, Research and Development base, Housing
Simonoseki	9–20	706	Ferry terminal, Container terminal, Marine stock farm, Research, Recreation, Airport
Kitakyusyu	7–10	265	Container terminal, Artificial beach, Park, Fishing
Beppu	9–25	16	Marina, Convention, Park, Hotel, Sports, Shopping mall, Commerce

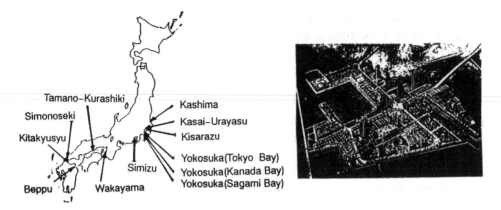

Figure 10.3. Proposed offshore artificial islands.　　　Figure 10.4. Plan of Wakayama Marina City.

the Wakayama Marina City, which is an international marine recreation base composed of marina, various facilities for sports, leisure and residential accomodations.

GEOTECHNICAL ENGINEERING FOR CONSTRUCTION OF ARTIFICIAL ISLANDS

Large scale reclamation needs to be carried out for the construction of artificial islands. In Japan, since most of the seabed strata are composed of soft alluvial layers, geotechnical engineering design often becomes a dominating factor for the feasibility of these projects. Geotechnical engineering problems related to these projects can be listed as follows:
(i) Investigation and testing of seabed;
(ii) Stability of seawall;
(iii) Settlements of seawall structure and reclaimed land;
(iv) Bearing capacity and settlement of foundations of facilities constructed on the reclaimed land;
(v) Modification of seabed and fills; and
(vi) Seismic stability of the reclaimed ground.

A typical artificial island project wherein the geotechnical engineering played a significant role is the Kansai International Airport (KIA) project. KIA, which is located on the largest artificial island in the world, is being constructed in the Osaka Bay, since 1987. Artificial islands for the purpose of airport facilities often need large scale reclamation and short construction periods. For the safe operation of the aircrafts, the gradient of the ground surface in the airfield must be kept within the restricted values given in Table 10.3. However, in major Japanese ports, the time available for the routine maintenance works once the airport is in operation, is very limited. For these reasons, one of the important

Table 10.3. Maximum and minimum values of gradient of pavement surfaces in airfields.

Facility	Max. (%)	Min. (%)
Runway (longitudinal)	0.8	0.5
Runway (crossing)	1.5	0.5
Apron	1.0	0.5
Taxiway	1.5	0.5

problems in the construction of airports on soft grounds, is the control of the consolidation settlements. The consolidation process will have to be accelerated by some means and the residual settlements will have to be kept at its lowest level possible.

KANSAI INTERNATIONAL AIRPORT

Kansai International Airport (KIA) is being constructed as the first airport in the world to be built on a "perfectly artificial" island. The KIA island, whose location and layout are shown in Figure 10.5, is located 5 km offshore from Sensyu area in the southeastern part of the Osaka Bay to avoid any undesirable impact due to the noise of the air traffic. The total area of the island is 5,110,000 m^2. The total construction costs are estimated at 1170 billion yen, of which a 510 billion yen are for the island itself; another 510 billion yen for

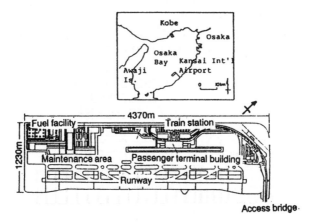

Figure 10.5. Location and plan view of the Kansai International Airport.

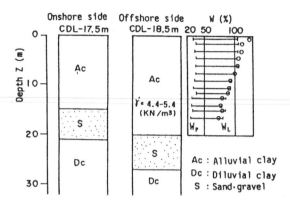

Figure 10.6. Soil properties at the site.

the airport facilities such as buildings, pavements, roads and utilities; 150 billion yen for the access bridge. The reclamation work was started in 1987 and completed in 1991.

The water depths at the construction site varied from 16.5 m to 19.0 m, and the soil properties are shown in Figure 10.6. The seabed consists of extremely soft alluvial clay of about 16.0 m to 20.0 m thickness. The consolidation yield stress, p_c, of the alluvial layer is in a slightly overconsolidated state. It was predicted that the construction of the island and the maintenance of the airport facilities will be difficult without the ground improvement. This was because of the low bearing capacity available for the seawall structure, and the large settlements that are likely to occur due to the reclamation. To improve vast areas of soft clay deposits, the vertical drain method, including sand drain method, pack drain method and prefabricated drain method, was thought of to be the most economical method. In the case of the KIA project, large scale sand drain works were carried out.

Figure 10.7 shows the profile of the reclamation works of the airport island. Sand drains were installed first at the location of the seawall, and further in the

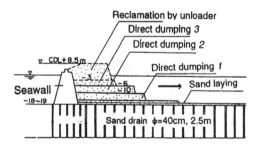

Figure 10.7. Profile of the reclamation works of the airport island (Arai et al. 1991).

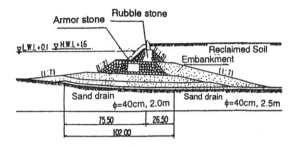

Figure 10.8. Typical cross-section of the seawall (Arai 1991).

entire reclamation area. For three years since 1987, more than one million sand drains were installed. This was the first case in which the sand drain method was applied in such a large scale to the soft deep-water seabed within a short period.

The seawall structure surrounding the 5,110,000 m² airport island is approximately 11 km long. In Figure 10.8 is shown the typical cross section of the seawall structure, which consists mainly of rubble mound and is at its lowest cost. The reclamation was carried out in the following sequence: (i) direct dumping (1, 2 and 3 in Figure 10.7) by bottom hopper barge and (ii) reclamation by the soil unloader. Stability analysis with consideration of the settlement and the strength increase was made for each execution step. The minimum safety factor against circular slip was set at 1.2. In view of that, a

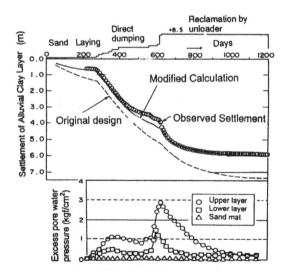

Figure 10.9. Observed and calculated time-settlement relationships (Arai et al. 1991).

six month consolidation period had to be allowed between dumping 2 and 3.

Figure 10.9 shows the comparison between the predicted settlements of the island based on soil investigation results and the observed settlements. As shown therein, the observed settlements increased in proportion to the reclamation load and almost ceased after about nine months after the reclamation. The effect of the sand drain was clearly felt. Figure 10.9 also shows the changes in the porewater pressures in the sand mat and the clay stratum. In the sand mat, no excess porewater pressures were observed, indicating that the sand mat was functioning as a drainage layer as expected. The excess porewater pressures in the clay stratum were dissipating rapidly, indicating again that the sand drain worked effectively as expected. In Figure 10.9 the observed settlements are smaller than the predictions, right from an early stage. This discrepancy seems to widen with the progress of the reclamation. This was explained by considering that the clay stratum was slightly overconsolidated and that the measured unit weight of the sand fill 2.0 tf/m^3 was larger than the predicted value of 1.8 tf/m^3.

For the airport construction projects on reclaimed lands over soft clay deposits, the prediction of the consolidation settlements is fundamental to the design of the various facilities. In the KIA project, it is predicted that more than 1 m of consolidation settlement of the deep diluvial clay layer will take place after all the construction activities have ceased, in spite of the full improvement of the alluvial clay layer. Therefore, the prediction of both the total and the differential settlements, due to nonuniform ground and loading conditions, in the reclaimed land, were the critical engineering problems faced.

A new method for evaluating the differential settlements was developed and

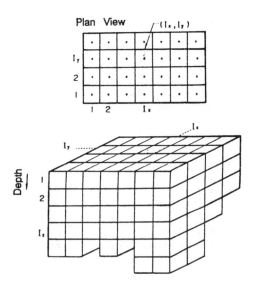

Figure 10.10. Three–dimensional modeling of the ground.

used for the design of the various airport facilities (Tsuchida & Nishida 1990). According to this method, at first, the ground is modelled as a group of small soil blocks in three dimensions, as shown in Figure 10.10. The settlements at the center of the surface blocks are obtained by calculating the consolidation settlements from the Terzaghi's one–dimensional consolidation theory, and then by adding them vertically. The differential settlement is simply defined as the difference in the settlements between the center of the neighboring surface blocks. To consider the variability of the ground conditions, the Monte–Carlo simulation was used in this study. In each trial calculation, the soil parameters such as the compression index C_c, the initial void ratio e_0, the consolidation yield stress p_c and the thickness of the layer H, were assigned randomly according to their probabilistic distribution models. The spatial auto–correlation characteristics of the soil properties and the three–dimensional distribution of the settlements among the soil blocks were taken into consideration. The number of trial calculations were 50. At each trial, the following parameters of the settlements at the ground surface were calculated as the indices of the statistical properties of the settlements;

m_s, d_s, V_s : mean, standard deviation and coefficient of variation of settlements;

t : differential settlement between the center of neighboring surface soil blocks;

t_{mean} : mean of t

t_{max} : maximum value of t

r_{mean} : rate of the mean differential settlement defined as (t_{mean}/m_s)

r_{max} : rate of the maximum differential settlement defined as (t_{max}/m_s)

Case study was conducted with the measured settlement data at the Tokyo

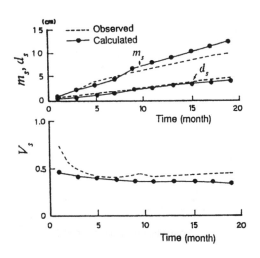

Figure 10.11. Mean, standard deviation and coefficient of variation of settlements.

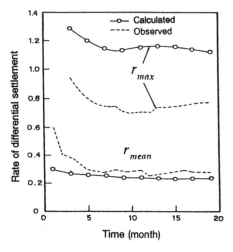

Figure 10.12. Rate of differential settlements.

Table 10.4. Curvature radius of pavement surface critical to the structural damage.

Pavement	Critical curvature (m)
Asphalt Concrete	1070
Plain Concrete	2500
Prestressed Concrete	1500

International Airport, where a runway was constructed on reclaimed land over soft clay layer in the Tokyo Bay (Tsuchida & Ono 1989). Figure 10.11 shows the comparison between the measured and the calculated settlements. As shown in Figure 10.11, the calculated mean and standard deviation of the settlements agreed fairly well with the measurements. Comparison of the rates of the differential settlements, r_{mean} and r_{max}, are shown in Figure 10.12. Although there is some discrepancy in the value of r_{max}, the prediction gave good estimation of the value of r_{mean}.

Since the Kansai International Airport is going to be the first international airport that will be open all day, the runway must be free from serious maintenance works once its starts to function. The structural damage to the pavement such as cracks due to the differential settlements was studied by finite element analysis, and the critical curvature of the pavement surface was given as shown in Table 10.4. In order to evaluate the effect of the differential settlements on the pavements of the runway, a numerical simulation was carried

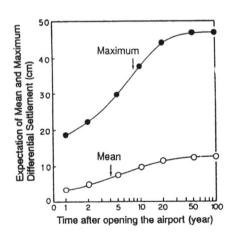

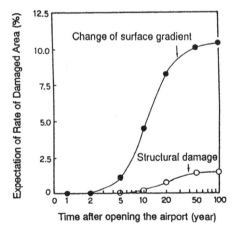

Figure 10.13. Expected average and maximum differential settlements in the runway.

Figure 10.14. Rate of the damaged area with time after opening of the airport.

out. Figure 10.13 shows the expected mean and maximum values of the differential settlements in the runway. These values are 10 cm and 38 cm, respectively, after 10 years from the opening of the airport. Using the criteria shown in Tables 10.3 and 10.4, the rates of the damaged area in the runway were also calculated and are shown in Figure 10.14. As shown therein, damages due to changes of the surface gradient are more likely to occur than due to structural damage. Estimating the frequency of the repair work of the runway based on Figure 10.13, it is predicted that the necessary repair work will not be so much as to affect the normal operation of the airport. The conclusion of the simulation was that the future differential settlements will not cause serious damages to the runway of KIA. In the same manner, the expected values of the differential settlements calculated by this method were used for the design of other airport facilities, such as terminal buildings, pavements, roads and utilities.

CONCLUSIONS

One of the major approaches to utilize the coastal ocean space is the construction of artificial islands. For a small and mountainous country like Japan, the development of the sea–front areas through reclamation has been vital. Recently the purpose of reclaimed lands is changing from industrial use to multi–purpose uses such as urban development, housing, airport and recreation. The type of reclamation is also changing from the traditional on–shore type reclamation to the artificial island type reclamation. This is because most of the shorelines are already highly developed and that the artificial island type reclamation is becoming more acceptable from the environmental point of view. Twelve artificial island projects in Japan have been outlined.

Geotechnical engineering for soft ground conditions is one of the most important technologies required for the construction of artificial islands. Recent developments in this field are reviewed with regard to a case history of the Kansai International Airport project, which is being constructed 5 km offshore in Osaka bay since 1988. Airport projects on soft ground conditions have the following characteristics: (i) the schedule of the projects is usually tight and the extention of the completion time is extremely difficult; (ii) the gradient of the surface of the airport pavement must be kept within some specified values from the view point of the safe operation of the aircrafts. For these reasons, one of the most important problems faced is the control of the consolidation settlements, i.e. acceleration of the consolidation and reduction of the residual settlements. In Kansai International Airport project, large scale sand drain works were carried out to improve the soft alluvial clay. The effectivness of the sand drains were ascertained by field measurements. A new technique to predict the differential settlements in reclaimed lands was developed and used for the practical design of the runway and other airport facilities.

REFERENCES

Arai, N. 1991. Construction of an artificial offshore island for the Kansai International Airport. *Proc. intl. conf. on geotechnical engineering for coastal development, Yokohama* 2:927–943.

Arai, N., N. Oikawa & N. Yamagata 1991. Large scale sand drain works for the Kansai International Airport island. *Proc. intl. conf. on geotechnical engineering for coastal development, Yokohama* 1:281–286.

Ministry of Transport 1991. *Offshore artificial islands in Japan.*

Tsuchida, T. & N. Nishida 1990. Prediction of differential settlements in airport constructed on soft ground. *Proc. 10 Southeast Asian geotechnical conference, Taipei* 279–284.

Tsuchida, T. & K. Ono 1989. Evaluation of differential settlements. *12th intl conf. on SMFE, Rio de Janeiro* 873–876.

11 GROUNDWATER HYDROLOGY IN LOWLAND

J. C. van DAM
Delft University of Technology, Delft, the Netherlands

INTRODUCTION

Groundwater hydrology in lowlands is an important subject for two reasons. First, as lowlands are densely populated, there is a need for exploitation and control of groundwater. Second, groundwater in lowlands originates from recharge in the uplands and locally, and, due to its geologic history, the groundwater at some depth in lowlands is in many cases still saline and brackish. Moreover, human activities, as groundwater abstraction, land reclamation and drainage, make groundwater systems in lowlands even more complex than under natural conditions.

Therefore, this chapter is subdivided into three sections which deal with successively:
- occurrence, origin and behaviour of groundwater in lowlands;
- the effects of human activities on groundwater in lowlands;
- groundwater management in lowland areas.

OCCURRENCE, ORIGIN AND BEHAVIOUR OF GROUNDWATER IN LOWLANDS

Geologic and geomorphologic conditions

Most of the lowlands on earth occur in coastal and deltaic areas. The subsoils in such areas consist mainly of sedimentary deposits, carried by rivers and deposited in fluvial or marine environment. The present courses of the rivers in these areas are, in many cases, fixed by dikes. In the geologic past these rivers have changed their courses almost continuously. These changes were due to a number of processes, such as meandering and cut–offs, glacier action in ice ages and withdrawal of glaciers during interglacials, upheaval or downwarping by tectonic forces and changes of sea level. The latter two causes have alternately led to regressions and transgressions of the sea. As a result of all these processes, the composition of the subsoil is generally rather capricious, with great variations in material composition (gravel, sand and clay and also shells and peat) both in lateral and in vertical directions.

During transgressions marine sediments have been deposited. Marine sediments occur generally over larger areas and are less heterogeneous than fluvial deposits, which have been deposited more locally in the inner curves of rivers and in settling basins behind the natural levees. Moreover, the fluvial deposits have many times been removed by changing watercourses and deposit-ed elsewhere in the coastal or deltaic zones or even on the sea bottom up to a certain distance off the shoreline.

From the geohydrologic point of view, acquaintance with the degree of interconnectedness of aquifers and semi–permeable layers is of great importance for the understanding of the groundwater flow pattern and the magnitudes of groundwater flow (IAHS 1988). As the structure of the subsoil can never be fully known, it is worth noting that at present, in mathematical geology, methods are being developed to generate geologic profiles which satisfy the stochastics of the behaviour of, for instance, avulsions, as observed in borelogs, outcrops and excavations. After generating such profiles it is possible to come to areal average values of the permeability, together with its standard errors (Bridge & Leader 1979, Stam et al. 1989, Tetzlaff & Harbaugh 1989, Koltermann & Gorelick 1992).

So far this description of the geologic and geomorphologic conditions was in terms of unconsolidated sediments. Depending on the age of the formations and the chemical composition of the clastic deposits and the water, some degree of consolidation and cementation may have occurred, transforming clays into shales, sand into sandstone and gravel into conglomerates or breccia, rendering the primary porosity and permeability of each of them smaller than those of the corresponding original clastic deposits. In this connection carbonate formations should also be mentioned, being the consolidated version of shell deposits.

Apart from the primary porosity and permeability of the consolidated formations, there may be a secondary porosity and permeability due to fracturing as for instance due to tectonic forces or due to dissolution of the material as in carbonate rocks, thus forming karstic formations with interconnected hollows (cracks, dolines, chambers) where the groundwater flow can concentrate.

Another formation which belongs in this overview is peat. It can occur in layers of a few centimeters up to many meters thickness. Generally, it occurs in rather vast areas and offers a high resistance against vertical groundwater flow. Peat layers below the groundwater table are subject to consolidation under the pressure of the overburden which leads to land subsidence and to an increase in the hydraulic resistance against vertical flow of groundwater. As long as peat occurs at the land surface, the upper part of it is subject to oxidation, which is another cause for lowering of the landsurface.

Geohydrologic conditions, flow regimes, recharge and seepage

The geohydrologic conditions are largely determined by the geologic and geomorphologic conditions. The flow regimes are determined by the geohydrologic conditions and the boundary conditions like topography, natural and artificial recharge, groundwater abstraction and artificially controlled groundwater tables and piezometric levels.

Coastal lowlands are mostly situated at the foot of the uplands from where the sediments come and have come, after erosion and transport by rivers. Depending on the supplies of sediments and water and on the sea levels, now and in the geologic past, the coastal areas can have a steep gradient (see Figure 11.1), as for instance the Cho Shui Chi alluvial fan at the west coast of Taiwan, or be very flat (see Figure 11.2) as for instance the coastal plains and the deltas of the rivers Rhine and Meuse in the Netherlands.

In the case of flat coastal plains, such plains may be intersected by tidal creeks, either still natural or dammed off by men. Under natural conditions these creeks may or may not be subject to tides and the water in these creeks can be saline, brackish or fresh. Depending on their water levels with respect to the adjacent land levels, the natural or artificially controlled groundwater tables therein, these creeks drain or recharge the groundwater.

Many natural creeks in flat coastal plains have first developed and intersected the original deposits of clay and peat and silted up later with coarser sediments. Due to land subsidence of the adjacent areas with clay and peat soils, the land elevation of such former creeks is now relatively high, as shown in Figure 11.3. This is indicated by the term "inverse landscape". Such a geomorphologic and topographic situation provides, locally, ideal conditions for natural recharge, sometimes in an otherwise saline environment. After recognition of this mechanism, the groundwater in many such former creeks is now being abstracted for various purposes. In other cases, this is impossible because villages have been built there, making use of the better subsoil conditions for the foundation of houses, buildings and roads. It is obvious that their presence reduces the rate of natural recharge.

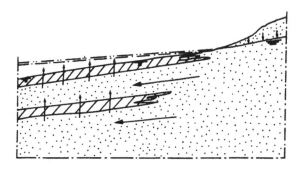

Figure 11.1. Geohydrologic profile of an alluvial fan.

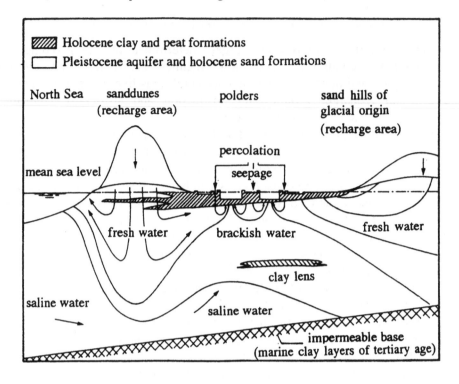

Figure 11.2. Geohydrologic profile of a flat coastal zone with sanddunes, deep polders and a recharge area.

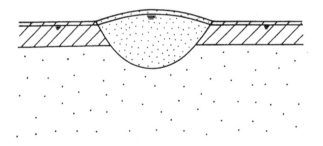

Figure 11.3. Inverse landscape.

The phenomenon of creeks and silted up creeks did not only occur in recent times. It has occurred all along the geologic history. So, in the subsoil of the present lowlands there are many buried silted up former creeks. As more permeable elements in otherwise less permeable clay and peat layers, they interconnect the aquifers above and below it, in particular in those cases where they intersect the clay and peat layers completely.

Besides the inverse landscapes, there are other natural features in coastal zones, such as local hills which can be recharge areas or outcrops of older formations, effecting the transmissibility of the aquifers.

The seaside of the coastal plains can, under natural conditions, be flat or very gently sloping or have sandbars or even have sanddunes of several kilometers width and several tens of meters in height, as shown in Figure 11.2. This is the case along the greater part of the coast of North and South Holland, the two western provinces of the Netherlands.

At the inland side of sandbars or sanddunes there are, in many coastal plains, lagoons as for instance in the south of France or the Hachiro–Gata lagoon in Japan. The sandbars or sanddunes are locally interrupted by rivers or even estuaries or just as an outlet for the excess water of the lagoons. Depending on the streamflow and the tidal volumes (thus by the cross–sectional area and the tidal range) the salinity of the water in the lagoons ranges from fresh via brackish to saline.

With this picture in mind, the present natural groundwater flow regimes in coastal zones and deltaic areas can be characterized as follows.

Apart from the local natural recharge, which depends very much on the climate, the local topography and the present drainage density, the coastal zones can have groundwater inflow from recharge in the uplands or from recharge in the sanddunes, where present. Depending on the climate and the nature of the formations, the uplands may have natural recharge which finds its way through the subsoil into the adjacent lowlands. Another form of inflow from the uplands is natural infiltration of water in the river bed in and above the alluvial fan. This form of recharge can under certain topographic and geologic conditions be increased artificially by rising the river level by the construction of a dam just upstream of the apex of the alluvial fan.

The inflow from natural recharge in the sanddunes, where present, depends on the recharge area, thus on the width of these sanddunes. This inflow is only a fraction of that natural recharge as the other fraction is, under natural conditions, lost by groundwater outflow towards the sea.

In modern geohydrology, the concept of groundwater flow systems has been introduced (Engelen & Jones 1986). In this concept, an area is subdivided in separate closed systems, bounded by water divides, which may have changed or change over time. In each groundwater flow system in steady state there is recharge and discharge at an equal rate (both expressed in $L^3.T^{-1}$). The discharge of a groundwater flow system occurs as seepage or as outflow into surface waters. In this concept it is possible to study flow paths and the changes in chemical and isotopic composition of the water along the flow paths (Stuyfzand 1993). The present chemical and isotopic composition of the groundwater is, in turn, an indication for the origin of the groundwater and for possible changes in the past of the groundwater flow systems which react very slowly on such changes because groundwater flows slowly and immense volumes of groundwater in the subsoil are involved. Groundwater flow systems are sometimes nested. The groundwater flow systems approach has proven, in a relatively short period of time, to be a useful tool.

Salinity distribution in groundwater

The salinity distribution in groundwater in coastal areas, both natural and due to human activities, is dealt with intensively in the biennial so–called Salt Water Intrusion Meetings (SWIM), held since 1967 in western Europe (Jinno et al. 1990).

From the foregoing description of the geologic and geomorphologic conditions, and more in particular the description of the genesis of the subsoil of coastal and deltaic areas, it will be clear that the groundwater present in the sedimentary deposits originates not only from natural recharge in present and past times but also from rivers, estuaries and the sea in the past (Stuyfzand 1993). Consequently, the composition of the groundwater varies from fresh water to saline water. Salts dissolved in water are mainly, in terms of cations, Na^+, K^+, Ca^{++} and Mg^{++}, and in terms of anions, Cl^-, HCO_3^- and SO_4^{--}. The predominant combination in seawater and groundwater originating from former transgressions is NaCl. Its salinity is usually expressed in $mgCl^-/l$. More inland, the relative fractions of HCO_3^- and SO_4^{---} ions in groundwater may be greater due to the fact that the water from natural recharge has flown through e.g carbonate rocks, where Ca^{++} and HCO_3^{--} ions were dissolved.

Fresh water can have a salinity as low as that of precipitation, total dissolved salts roughly a few tens of mg/l. Saline groundwater has a salinity which can even exceed that of the present seawater which is roughly 19,000 $mgCl^-/l$. Such high salinities are due to evaporation in the past, leaving the salt ions in less water, thus giving rise to a higher concentration in the remaining water. Between these two extremes there is a continuous range of concentrations from fresh via brackish to saline groundwater and even hypersaline groundwater occurs. The salinity gradients, in lateral and in vertical directions, vary from rather sharp to very smooth, depending on the geologic history, the composition of the subsoil and human activities.

In general, the salinity increases with depth, either gradual or with a relatively sharp interface, which is in the latter case with a transition zone from fresh to saline groundwater of only 5 to 10 m thickness. In so–called inverse profiles, saline or brackish groundwater occurs above fresh groundwater. As saline water is heavier than fresh groundwater this situation is either unstable or the underlying fresh groundwater – which is either relict groundwater or currently recharged from the uplands – is covered and protected by a vast clay or peat layer of high resistance against vertical flow. So the inverse salinity distribution – which can easily be determined by geo–electrical prospecting – can be an indirect indication for the presence and extent of such a layer (van Dam 1976b).

Apart from the effects of human interference, discussions about which follow later, the present natural spatial distribution of the salinity of the groundwater is generally not yet in a state of dynamic equilibrium, because some of the relevant physical and chemical processes have not yet fully worked out. The processes are Flow, Dispersion, Diffusion, Evaporation, Ion exchange and dissolution.

Flow. Flow leads to replacement of groundwater of a certain composition by water of another composition. As groundwater flows very slowly, and immense volumes of groundwater may be involved it will take considerable time, in the order of centuries, before a new state of dynamic equilibrium is reached. Groundwater flow is strongly affected by the density distribution, which is a function of salinity and, to a lesser extent, of temperature and, to a far lesser extent, of pressure.

Dispersion. Dispersion occurs only in combination with flow. As the subsoil is a porous medium, flowlines on a microscale can never be such smooth curves as can be calculated for continuous media. Due to the tortuosity of the pores, water particles and the salts dissolved therein will deviate to either sides of the theoretical flowlines. The effect is that an instantaneous point injection of some tracer (or salt) spreads on its way as a cloud of increasing dimensions, and at the same time, decreasing concentrations. Similarly, a continuous injection in one point spreads in a zone of which the width is gradually increasing with distance. The same happens along interfaces between fresh and saline groundwater. If they would ever have been sharp, a transition zone will develop gradually.

Dispersion occurs moreover on different levels of scale.

a. In the pores, the velocity profile of the laminar flow has low velocities along the sides formed by the sediment particles and higher velocities in the centre of the pores.

b. The aquifers are generally anisotropic. So the velocity in the coarser layers is greater than that in the finer layers of the aquifer and, thus, a sharp front will always become more and more diffuse.

c. On the macroscale, the local or regional presence or absence of less permeable layers forces the flow paths to pursue their ways partly above and partly below the less permeable layers and to unite further on. Thus, the travel times of the water particles above and below such obstacles will generally be different.

The effect of dispersion, on all scale levels, is that of mixing.

Diffusion. Diffusion is the physical process of migration of ions from high concentration to low concentration, even in stagnant groundwater. The rate of transport is, according to Fick's law, proportional to the gradient of the concentration. This process takes place very slowly. It can be neglected in relation to dispersion, but on a geological time scale it may have had considerable effect in stagnant groundwater.

Evaporation. By evaporation, pure water returns to the atmosphere and the dissolved salts stay in the topsoil in increasing concentrations. Also in lakes without outlets, in the so-called internal drainage basins, the salt concentration increases. Examples are the Dead Sea in Israel and the Great Salt Lake in the USA. Salt domes (big bodies of solid salt) in the deeper subsoil are the remainders of former inland seas.

Ion exchange and dissolution. Ion exchange, a physico-chemical process between water and soil particles, changes the chemical composition of groundwater during its passage through the subsoil. Dissolution of salts in the subsoil (for instance $CaCO_3$ from carbonate formations) leads to increasing concentrations of such salts in the groundwater during its passage through the subsoil.

State of natural equilibrium

Apart from the disturbing effects of human activities over the past millennia, which will be dealt with in the next section, even the natural processes have generally not yet fully worked out (Meisler et al. 1984). First of all, the geologic and geomorphologic processes, described in a previous subsection, are still going on. The same holds for the effects of sea level changes. So, even the natural systems in which the groundwater flows are subject to slow changes. The input into such systems, the natural recharge, as determined by the continuously, but slowly, changing climate, is also changing very slowly. As a result of these slow changes, the groundwater flow pattern and magnitude will change only very slowly. As has been set out before, the changes in salinity distribution lag far behind the changes in the flow. This holds equally for the related density distribution of the groundwater, which in turn affects the groundwater flow considerably. So, in conclusion, even the natural groundwater flow systems are most likely not in a state of natural equilibrium. This holds in particular for the salinity distribution of the groundwater.

Effects of sea level rise

Sea level rise, in connection with climate change, has attracted much attention of the scientific community over the past decade (IPCC 1992). In this chapter the impacts of sea level rise on groundwater (van Dam 1993a, German IHP/OHP National Committee 1993, Jelgersma 1993) deserve serious consideration because of their social and economic consequences.

For proper understanding of the impact(s) of sea-level rise on salt water intrusion in aquifers, distinction should be made between various geohydrologic profiles.

Sedimentary deposits in coastal plains and deltaic areas mostly have a small slope downward in seaward direction; sometimes very small or even zero. In such cases the rising sea will invade the land at a rate $l = \eta/i$, where η is the rate of sea–level rise and i the slope of the land surface (Figure 11.4). Depending on the values of η and i, l may be in the order of hundreds or thousands of meters per century. It is obvious that in those cases where the groundwater was originally fresh, it will gradually be replaced by the heavier saline water, which penetrates vertically into the subsoil. The fresh groundwater is lost together with the land.

Many coastal areas built up of sedimentary deposits are protected, in a natural way, by a belt of sanddunes along the coast. Such a belt, as shown in Figure 11.5, is locally interrupted by estuaries, rivers or outlets of lagoons. In such cases, the rising sea may enter the low–lying area at the inland side of the sanddunes, as an attack in the back. The low–lying areas will be flooded. Due to erosive forces at both sides the width of the sanddunes will decrease, as shown in Figure 11.6. The consequences thereof are twofold. The total natural recharge of the groundwater will decrease accordingly and the thickness of the fresh water lens will also decrease. As, in the state of dynamic equilibrium, this thickness is proportional to the width L of the sanddunes, the volume of the lens is proportional to L^2, which can imply in the long run a big loss of fresh water in storage.

In many situations the openings in the sanddunes have been closed by human activities such as the construction of dikes and dams with outlet structures where required. See Figure 11.7. In such cases there is no attack from the backside. The rising sea level will make the coastline to retreat by erosion, unless it is fixed by coastal defence works. If the coastline moves inland some land will be lost and together with it the fresh groundwater in the disappearing sanddunes.

There is another very important aspect to be described here, the inflow of saline groundwater under the fresh water lens. In many coastal areas the groundwater table is controlled by artificial control of the surface water in an intensive network of canals and ditches. The artificially controlled level of the surface water is, under present conditions, in many coastal areas or parts thereof, already below mean sea level; the controlled levels ranging from a few decimeters up to several meters below mean sea level. The latter is the case where lagoons or lakes have been reclaimed. The magnitude of inflow q_s of saline groundwater underneath the fresh water lens, as shown in Figure 11.8, can roughly be described by:

$$q_s = kH. \; h \; / \; L \qquad\qquad (11.1)$$

where kH is an average value for the transmissibility of the aquifer between the

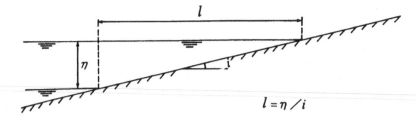

$$l = \eta / i$$

Figure 11.4. Rates of sea–level rise η and land invasion l at slope i.

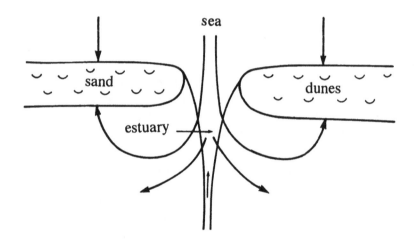

Figure 11.5. Sanddunes interrupted by an open estuary.

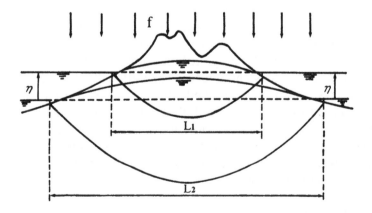

Figure 11.6. Fresh water lens in sanddunes before and after sea–level rise.

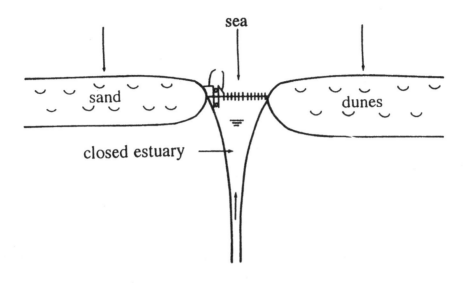

Figure 11.7. Sanddunes interrupted by an estuary closed with a dam and structures.

fresh water / salt water interface and the impermeable base; h is the difference between mean sea level and the controlled water table at the inland side of the sanddunes; L is the distance between the coastline and the area where the water table is controlled at h below mean sea level.

In case the aquifer is covered by semi–permeable top layers at one or both sides with hydraulic resistances against vertical groundwater flow c_s and c_l [T], as shown in Figure 11.9, Equation 11.1 has to be adapted as follows:

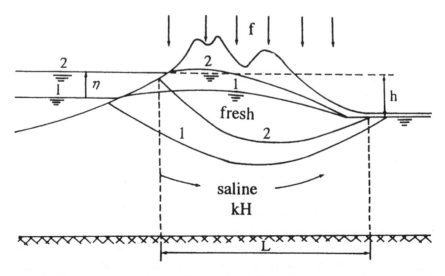

Figure 11.8. Fresh water lens in sanddunes with inflow of saline water between the fresh water lens and the impermeable base.

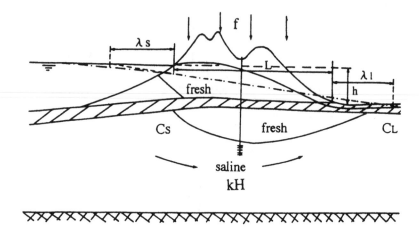

Figure 11.9. Fresh water lens in sanddunes with inflow of saline water between the fresh
        water lens and the impermeable base in case the lower acquifer is covered by
        semi-permeable layers.

$$q_s = kH. \, h \, / \, (\lambda_s + L + \lambda_l) \qquad\qquad (11.2)$$

where λ_s and λ_l are the so-called characteristic or spreading lengths (which are
better terms than the often used term leakage factor):

$$\lambda_s = \sqrt{(kHc)_s} \; [L] \qquad\qquad (11.3)$$

and

$$\lambda_l = \sqrt{(kHc)_l} \; [L] \qquad\qquad (11.4)$$

    Depending on the values of c_s and c_l, which can be as much as several or even
many thousands of days, and in accordance therewith, λ_s and λ_l which can be
up to several thousands of meters, the presence and the hydraulic resistance of
semi-permeable top layers can have a considerable and favourable influence on
reduction of the inflow q_s of saline water.
    The same formulas hold for the increase in inflow Δq_s due to a rise Δh of the
sea level. The inflow, q_s or $q_s + \Delta q_s$, occurs as seepage in the areas with
controlled low water tables, and must be pumped out together with the excess
precipitation. This is generally not a big problem, as it is only a fraction of the
excess precipitation. More serious is the fact that in many cases the seepage
water is or becomes saline as will be further elaborated in the next section.
    More favourable are those situations where the impermeable base is at shallow
depth, such that the fresh water lens in the sanddunes reaches naturally to the
impermeable base, thus making inflow of saline water impossible. See Figure

11.10. This is for instance the case in the Belgian coastal zone and in the delta of the Vistula river in Poland.

In other cases, the geohydrologic profile perpendicular to the coastline is as drawn in Figure 11.11. The groundwater in the aquifer is confined between two impermeable layers. The expression for the length L of the salt water wedge is:

$$L = \alpha k H_0^2 \,/\, 2q_f \tag{11.5}$$

where α is the relative density difference $(\rho_s - \rho_f) \,/\, \rho_f$ [–]; k is hydraulic conductivity of the aquifer for fresh water [L.T^{-1}]; H_0 is full thickness of the aquifer [L]; q_f is natural outflow of fresh groundwater from recharge in the uplands eventually diminished by pumping groundwater [L^2.T^{-1}].

From this formula it appears that L is independent of the depth A of the top of the aquifer below mean sea level. This implies that the length of the salt water wedge does not increase as a consequence of a sea–level rise.

As the sea level is a boundary condition for the water levels in open estuaries (Savenije 1992), any sea level rise entails an immediate rise of the water levels in estuaries, regardless of (any change in) the elevation of the bottom of the estuary. Where the adjacent land areas are protected by dikes, and the water tables in these areas are artificially controlled at a constant level, the differences between the water levels in the open estuaries and those controlled water levels increase accordingly. This implies more seepage from the estuaries into the adjacent land areas. See Figure 11.12.

Depending on the salinity distributions in the estuary (Savenije 1992) and in the subsoil of the land areas, the salinity of the seepage water will change over time until a new state of dynamic equilibrium has been reached. The mechanisms are the same as described before for the impacts of sea–level rise along the coastline. As groundwater flows slowly and large volumes of groundwater are involved, there will be a considerable time lag between the cause – the sea–level rise – and the effect, the ultimate salinity distribution in the subsoil and that of the seepage water.

EFFECTS OF HUMAN ACTIVITIES ON GROUNDWATER IN LOWLANDS

Human activities

In the previous section, the description of the geohydrologic conditions was mainly related to natural conditions. Human activities are equally important. Therefore in this subsection, a brief review of the relevant activities is given as a basis for understanding their effects, which will be described in the next three subsections.

Human activities are first of all the various forms of land use and misuse. Coastal and deltaic areas are among the most densely populated areas of the

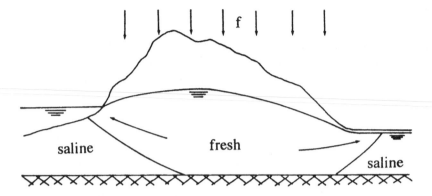

Figure 11.10. Fresh water lens on a shallow impermeable base; inflow of saline groundwater is impossible.

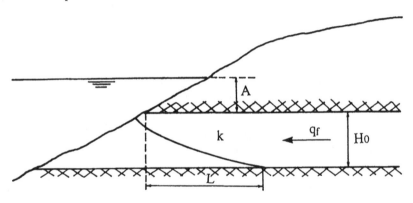

Figure 11.11. Geohydrologic profile perpendicular to the coastline with confined groundwater.

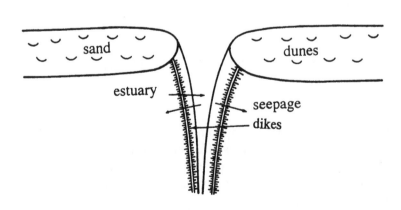

Figure 11.12. Estuary with dikes and seepage.

world because they are flat and fertile and offer good possibilities for agriculture and fisheries. Moreover, harbours at the coastline form the connection between sea–transport and inland transport, by road, rail and inland navigation. So, apart from the rural areas, in many cases, urban, industrial and harbour areas have been developed. Some of them are very large. These areas in particular must be protected against flooding by the sea and by rivers. So, either the land levels have been raised artificially or dikes have been built and even estuaries have been dammed off with provisions for discharge of streamflow and, in many cases, with shiplocks. These activities have led to artificial control of the surface water levels and therewith, indirectly, of the groundwater table.

Another important human activity is land reclamation, either at the coastline, stepwise by diking, or inland where lagoons and other shallow lakes or parts thereof are or have been transformed into land areas. In the latter cases, the surface water table is artificially controlled at some depth below the former lake bottom level, which can be up to several meters below the land surface or water table in the surrounding areas. It is obvious that such large scale works, both in areal extent and in drawdown of the water levels, can have considerable effects on the groundwater flow system.

For most of the human activities, water is required, either surface water or groundwater. After use, part of it is discharged. Depending on the degree of treatment its quality has at least been changed (Hooghart 1985, van de Ven & Hooghart 1986, Hooghart 1988, Tjallingi et al. 1993). Apart from waste water, the human activities produce also increasing volumes of solid waste. The fraction of it, which is not reused or incinerated, is or has been dumped, either on or below the original land or water surface.

For the above–mentioned human activities – construction of dikes, roads and railways, local raising the land level and for the construction of houses, industrial and utility buildings – there is a need for clay, sand and gravel. This is present in the subsoil in big quantities and in a great variety of properties. When borrowing these minerals from the subsoil, by excavation and by dredging, to depths of up to some tens of meters, lakes are left which can have other useful functions such as water storage reservoirs, nature reserves, areas for recreation in, on and along the lake. In other cases, such lakes have been filled in by dumping waste products from coal–mining. It is obvious that such changes of the original geohydrologic profiles together with the newly established water levels, whether constant or variable over the year, natural or artificially controlled, can have considerable effects on the groundwater flow systems.

Effects of human activities on groundwater tables, piezometric levels, flow regimes, recharge and seepage

The human activities as described in the previous subsection have their effects on recharge, groundwater tables and piezometric levels, and thus on the regimes of groundwater flow, the groundwater flow systems (Custodio 1987). Due to most of these activities, the original natural groundwater flow systems change considerably; the water divides, as boundaries change. New groundwater flow systems can develop, others may disappear or join together. The pattern of recharge (infiltration) and seepage changes accordingly. The effects of the activities, as listed in the previous subsection, will be indicated here in the same order of sequence.

For optimal crop yield in agriculture, the groundwater table should be controlled at a certain depth below land surface, depending on the soil type and the crop. In wet periods, the groundwater table should be controlled by an adequate drainage system. The required depth and spacing of the open or closed drains depends on the soil type and the drainage specifications (i.e. tolerance limits for the groundwater tables). In dry periods there should be an adequate supply of water by capillary ascent from the groundwater table to the plant roots in the unsaturated zone. In case of deep groundwater tables, the water supply by capillary ascent is insufficient. Then irrigation should be applied by spreading of water at the land surface, in furrows or by sprinkling.

These practices have their effects on the recharge of the groundwater. Intensive drainage makes any recharge to be quickly drained off and discharged, to the effect that it can hardly be called recharge. In cases where the groundwater table is deep and irrigation is applied, there will always be some irrigation losses in the form of percolation. Such losses reappear as return flow. These losses cause the groundwater tables to rise as happened, for instance, in the Nile delta in Egypt. In order to prevent too high groundwater tables, drainage should be applied, either by a horizontal drainage system or a vertical drainage system, i.e. by means of wells.

Depending on the climatic conditions, and more in particular on whether the annual depth of evapotranspiration exceeds the annual depth of precipitation, the water requirement for crop growth must be augmented with an additional supply for leaching the topsoil in order to prevent accumulation of salts therein. This additional supply finds its way as return flow, with an increased content of dissolved salts, ultimately back to the surface waters.

The water required for irrigation is either withdrawn from rivers, with or without regulation by reservoirs upstream, with intakes upstream of the salt water intrusion zone in the estuaries, or from groundwater. In the latter case, the groundwater is pumped by the individual farmers in a great many of wells of small capacity and at small depths in their own fields. The consequences of pumping groundwater for irrigation will be discussed further on in this subsection, in combination and in comparison with the consequences of groundwater abstraction for domestic and industrial use.

Urbanization, including industrial areas, harbour terrains, roads, airports and glasshouses for horticulture, etc. reduce the area for infiltration of precipitation. A greater part of the precipitation is discharged quickly as surface water and the natural recharge of groundwater is reduced accordingly (Hooghart 1985, 1988; Tjallingii et al. 1993). Both these effects are detrimental.

Dikes and dams are constructed first of all to protect the adjacent land against flooding. The effect of these works and also that of weirs in rivers is a change of the river or estuary water level with respect to the groundwater tables in the adjacent land which follow the controlled water levels in the networks of watercourses in that land. This change can be even greater in cases where the controlled water level in those areas is lowered, for instance in order to keep pace with land subsidence. So, the interaction between the water in rivers and estuaries and the groundwater in the adjacent areas is affected. In many cases this has resulted in an increase of seepage in the adjacent areas.

In the previous subsection it was already mentioned that land reclamation in shallow lakes and lagoons can have considerable effects on the groundwater flow system. The effect is (increased) seepage in the reclaimed polder area itself together with (increased) infiltration in the areas around the polder. The average rate of seepage over the polder may vary from less than 1 mm/day to several tens of mm/day. It is proportional to the difference in water level between the controlled water tables in the polder and in its surroundings, and depends very much on the presence or absence of semi–permeable top layers in and around the polder and on the hydraulic resistance c of such layers against vertical groundwater flow (c is expressed in days and ranges from a few tens of days to a hundred thousand of days, which corresponds to almost impermeable) and on the transmissibility kH (expressed in m^2/day) of the underlying aquifer. The geohydrologic constants c and kH are combined in the so–called characteristic length or spreading length λ ($\lambda = \sqrt{kHc}$). The greater the λ, the larger will be the area which is influenced outside the polder. In case of low values of c, the rate of seepage in the polder is high and the costs of pumping to evacuate that much of water may be economically prohibitive and the polder should not be reclaimed. Existing polders could even be abandoned as such and become water recreational areas or water nature reserves. The effects of such actions on groundwater flow systems are the reverse of those of reclamation of polders.

Abstraction of groundwater, by means of wells, drains or galleries leads to drawdown of groundwater tables and/or piezometric levels around the abstraction works. The drawdowns are greatest in the aquifer where the water is abstracted and near the abstraction works. The corresponding drawdowns in deeper and shallower aquifers, where present, are less, not cone–shaped but smoother.

A comparison of groundwater abstractions for irrigation and for domestic and industrial use and their effects leads to the following observations.

1. Groundwater abstraction for irrigation is generally distributed over a great number of small wells spread over a large area with filters at shallow depths, whereas abstraction for domestic and industrial use is preferably concentrated in well fields, with great capacity per well, and filters in an aquifer of great transmissibility, covered by one or more semi–permeable layers. In such

situations, the drawdowns of the phreatic groundwater table is reduced with respect to that in the pumped aquifer, be it still over a large area. Moreover, from the hygienic point of view, the semipermeable layer(s) above the pumped aquifer protect and prevent the aquifer against pollution by accidents in the terrain in the vicinity of the well field.

2. Groundwater abstraction for irrigation shows clearly a seasonal variation, whereas groundwater for domestic and industrial use is generally much more uniform over the year. Exceptions are for the use as cooling water and in tourist resorts where the number of consumers is much greater during the holiday seasons.

After usage, the water, or rather the fraction which is not lost by evaporation, is discharged, treated or not, mostly into surface waters. Most of the rivers are effluent, but there where they are influent, this water can, mixed with other water, re-enter into the subsoil. Depending on the conditions, this may be a point of attention. This holds even more where the waste water is infiltrated in ponds. The groundwater quality may be affected in a negative way, making it unfit for use or only after intensive treatment. Other sources of groundwater pollution will be dealt with later in a special subsection. Groundwater used for cooling is sometimes re-injected after use. In many cases this is even prescribed by the local or regional water authorities with the aim to conserve groundwater for other uses, like domestic water supply.

Sand and gravel borrow-pits bring about great changes in the original geohydrologic profiles. Together with the newly established water levels, whether constant or variable over the year, natural or artificially controlled, this can have considerable effects on the groundwater flow systems.

Whatever the cause may be, drawdown of groundwater tables and/or piezometric levels entails a decrease of the groundwater pressures and an equal increase of the intergranular stresses in the subsoil. The latter effect leads subsequently to compaction of the subsoil and, thus, to land subsidence and, depending on their foundation, also damage to buildings and structures. This damage can be irreparable, as for instance damage to historic buildings.

Drawdown of groundwater tables can also be harmful for the flora and fauna in nature reserves. This environmental impact cannot be expressed in terms of money, neither can it be compensated by just money.

Effects of human activities on the salinity distribution in groundwater and seepage water

The effects of human activities on the salinity distribution in groundwater and seepage water have been recognized since long. In recent literature, attention is paid to the description and the quantification of these effects (CHO–TNO 1980, Custodio 1987, Jinno et al. 1990, van Dam 1976a, 1983, 1986, 1992).

In the foregoing subsection, it has already been set out that human activities can bring about big changes in the groundwater flow systems. This holds also for the distribution of groundwater quality. Salinity is one of these water quality aspects, but a very particular one for two reasons. First of all, salt water can not be used unless after costly desalting or it can be used just for cooling. In both cases, the problem of disposal of the brine after desalting or the saline or brackish water after cooling remains. The second reason is its behaviour due to its higher density.

A general rule is, based on the principle of Badon Ghijben (Badon Ghijben en Drabbe 1889) and Herzberg (Herzberg 1901), that any drawdown of the groundwater table or piezometric level works out in a rise of the interface in the same aquifer, whether sharp or diffuse, between the fresh groundwater and the underlying saline groundwater. According to this principle, the rise is a multiple (40 to 50 times) of the drawdown of the groundwater table or piezometric level.

This phenomenon occurs both on the local or small scale and on the regional or large scale. The small scale refers, in this context, to the very small area around and below an abstraction well. This phenomenon at this scale is called upconing as the interface deforms locally in the shape of a cone with its top in the lower end of the well screen. It is obvious that this makes the well unfit for further abstraction of fresh groundwater.

The regional or large scale effects occur in much larger areas where the interface moves slowly and smoothly in upward and/or in inland direction. The large scale displacement is caused by groundwater abstraction and/or by large scale lowering of the controlled groundwater table as in reclamation projects, new polders or lowering of the polder water level in existing polders.

Most of the groundwater flow systems in coastal areas are not yet in a state of dynamic equilibrium as both the natural changes and the effects of increasing human activities during the past centuries have not yet fully worked out (van Dam 1976a, CHO–TNO 1980). In the fictive case of no further sea level change and no human activities other than maintaining the present situation, an ultimate situation of dynamic equilibrium will be reached. Such a situation is dictated by, among others, the groundwater abstractions and the water tables maintained in, sometimes large and deep, polders. During the period of transition, big volumes of the present saline and brackish groundwater will be displaced. As in reality, the sea level will change continuously and new human activities will take place and the ultimate situation will also change with time.

The total volume of saline groundwater can decrease by outflow of it as seepage in deep polders. On the other hand, there may be inflow of saline water from the seaside. The ultimate situation, which will be reached without sea level rise and without further human interventions, may either be one with a permanent inflow of saline water from the seaside and an equal outflow as saline seepage in deep polders at short distance from the coastline or one with stagnant or almost stagnant saline groundwater below fresh water in the state of dynamic equilibrium between recharge, seepage and abstraction.

As groundwater flows slowly and big volumes of groundwater are involved (van Dam 1986), the period of transition will last long, in the order of magnitude of centuries (van Dam 1992). This is illustrated by the Figures 11.13a to 11.13d (van Dam 1993b) which depict, for four different situations, very roughly, the volume fraction of saline seepage in deep polders since their reclamation as a function of time; the time scale should be read in centuries.

Figures 11.13a and 11.13b hold for deep polders at short distance inland of the coastline, which is formed either of natural sanddunes, with inflow of saline groundwater underneath the fresh water lens, or of man–made dikes. These polders are located in the first zone. Figures 11.13c and 11.13d hold for deep polders further inland, located in the second zone, separated from the first one by a groundwater divide. The second zone is another groundwater flow system.

In Figure 11.13a, the groundwater in the subsoil of the deep polder at the moment of reclamation was supposed to be completely saline and so the seepage in that polder. Ultimately, a fraction of the seepage will be fresh due to infiltration of fresh water in the near surroundings of the polder in the newly created groundwater flow system. Then, the fresh seepage occurs, permanently, in a circular belt at the inner side of the polder embankment and the saline seepage occurs, also permanently, in the central part of the polder.

In Figure 11.13b, the groundwater in the subsoil of the deep polder at the moment of reclamation was supposed to be fresh to a depth of several tens of meters, followed by saline groundwater at greater depths. So, the seepage in that polder was initially completely fresh and ultimately partly fresh, partly saline. The origin of both fractions of the seepage in the ultimate situation is the same as described in the case of Figure 11.13a, with different conditions at the moment of reclamation.

In Figure 11.13c, the groundwater in the subsoil of the deep polder at the moment of reclamation was supposed to be completely saline and so the seepage in that polder, as is for instance the case in polders reclaimed on the bottom of former tidal bays. After some time, a fraction of the seepage will be fresh due to infiltration of fresh water in the near surroundings of the polder in the newly created groundwater flow system. In the ultimate situation, the seepage will be completely or almost completely fresh. This does not mean that there will be no saline groundwater left; big volumes of stagnant or almost stagnant groundwater will stay behind, which can even reach to the controlled water level in the polder or slightly below it.

Figure 11.13d refers to the situation of a deep polder reclaimed in the second zone where the groundwater in the subsoil at the moment of reclamation was supposed to be fresh to a depth of several tens of meters, followed by saline groundwater at greater depths. The seepage in such a polder is initially fresh with a break–through of saline water after a period of decades or centuries. After reaching a maximum, the fraction of saline seepage will decrease gradu–ally to reach the ultimate situation where the seepage will be completely or almost completely fresh.

Another human activity, described in the previous subsection, is the construction of dams in estuaries which keep the saline sea water outside. In a

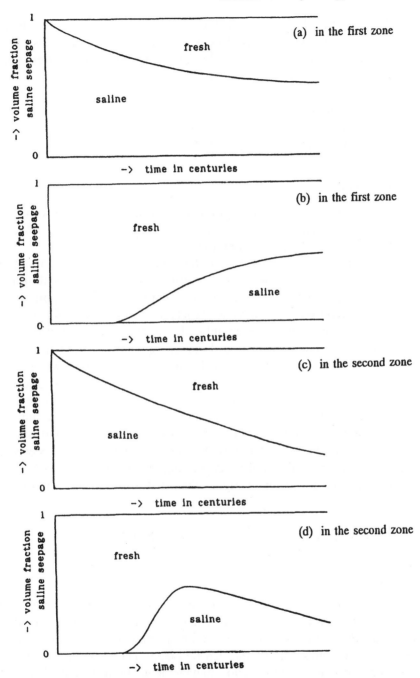

Figure 11.13. Volume fraction of saline seepage as a function of time in a deep polder in the first zone (a) and (b) and the second zone (c) and (d). At the moment of reclamation, the groundwater was completely saline in (a) and (c); in (b) and (d) the aquifer contained a thick layer of fresh groundwater above the saline groundwater at greater depth.

relatively short period of time, months or years, the water in the estuaries becomes fresh. This is due to the replacement of the saline or brackish water in the estuary at the moment of closure by inflowing fresh river water. So, in case of seepage in the adjacent areas, the salinity of the groundwater in those areas can gradually become more fresh due to replacement of the original groundwater by fresh water newly infiltrated in the bed of the estuary. So, depending on the present salinity of the groundwater in such areas, the salinity of the seepage may, in the long run, change accordingly.

Pollution of groundwater

Pollution of groundwater is a serious problem worldwide, also in lowlands. In recent literature (van Duijvenbooden & van Waegeningh 1987, D'Itri & Wolfson 1987, IAHS 1990) much attention is paid to it.

In addition to what was mentioned in the previous subsection about the disposal of waste water and its possible effects in terms of pollution of groundwater, there is another point of serious attention. This is the pollution of the soil and the groundwater by a great variety of diffuse and point sources. Diffuse sources of pollution are mainly the application of fertilizers in agriculture, and the abundant quantities of manure spread over the fields. These practices can make the groundwater unfit for the production of drinking water or only at very high costs. Point sources are solid waste disposal sites, certain industrial activities, petrol stations, leakage from sewers and pipelines, etc. After percolation through the unsaturated zone and reaching the groundwater table, the pollutants may either follow the flowlines of the groundwater or spread out in a thin layer just below the groundwater table or percolate downward to the bottom of the aquifer, depending on whether the density of the polluted water is equal to, lower or higher than that of the local groundwater.

GROUNDWATER MANAGEMENT IN LOWLAND AREAS

Types of measures

In order to overcome adverse effects of the above–described natural conditions and human activities on groundwater, whether in terms of available quantities, groundwater quality, groundwater tables and or piezometric levels, occurrence of seepage, etc., measures should and can be taken. There exists even a great variety of measures for control and management, both technical and by legislation.

The choice of measures or combinations thereof depends on the objectives and the possibilities. The objectives are often a compromise between conflicting interests, as for instance those of agriculture versus those of nature conservation, or allocation of groundwater for agriculture or for industry. In both examples, the key problem is the optimal groundwater table. If a compromise is not possible, a choice has to be made based on policy or even politics. The possibilities depend on the regional geohydrologic conditions as well as on the established land and water uses with the corresponding requirements, in the broadest sense.

The measures can be subdivided into two types, preventive and curative measures. Preventive measures should, in principle, be preferred, but this is not always possible or realistic. So, for instance, it is not possible to halt undesirable flow of saline or polluted groundwater in thick aquifers by sheetpile walls or grouting. Then, one should resort to curative measures, i.e. measures – such as selective pumping or water treatment after pumping – to cope with certain harmful effects, rather than to (be able to) eliminate the cause of such effects.

As groundwater flows slowly it may be necessary to take curative measures for short term effects, which may still last for many decades, and at the same time preventive measures aiming at restoration of the present situation and/or improved conditions in the long term. So, for instance, ongoing pollution by agricultural practices should be stopped by proper legislation, but even then it may take centuries before the present polluted groundwater has been replaced completely. Curative measures may still be needed.

Preventive measures for the various effects

In cases where the natural recharge of the groundwater is not enough or the drawdowns of the groundwater tables or piezometric levels are unacceptable, the solution might be artificial recharge of groundwater or induced recharge. Depending on the moment of realization, this measure can be considered as preventive or curative. The potential sources of water for artificial or induced recharge are rivers and canals with intakes from the rivers or closed estuaries. The quality of the water at the source should satisfy certain standards. Such standards depend mainly on the use of the water after infiltration and subsequent recovery but, in the case of artificial recharge, also on the requirements for transport by pipeline and subsequent infiltration. These two latter aspects may require some pretreatment. It is obvious that the water should be fresh. Therefore, the intakes should be beyond the reach of salt water intrusion in the rivers and estuaries (Savenije 1992).

Interesting examples are the public water supplies of the cities of The Hague and Amsterdam in the Netherlands. Both cities abstract water from the sanddunes along the Dutch coast. In the 1950s, the increased rates of abstraction could no longer be met from the natural recharge. The interface between fresh and saline groundwater moved upward rapidly. Therefore, artificial recharge was applied. Water from the river Rhine, and later on also from the river Meuse was transported by pipelines, of 60 and 80 km length, to the sanddunes where it was infiltrated in ponds and canals. So, the increased demand could be met and moreover the quality of the infiltrated water improved during its passage through the subsoil by breakdown of certain constituents and by smoothing out of the quality fluctuations of the infiltrated water during its passage through the subsoil.

The increased volumes of fresh groundwater, with thicknesses of the fresh water lenses of over 100 m, are good buffers in case of emergencies such as rupture of a pipeline or calamities making the river water for some time unfit for intake.

At present, there is a growing resistance, from the environmental point of view, against the infiltration in ponds and canals, and successful tests have been made to infiltrate the water at greater depths by means of injection wells. Therefore, the water should be pretreated and the rate of infiltration per well should not be too great.

A secondary result of the artificial recharge in the sanddunes was the fact that the interface between fresh and saline groundwater was pushed down with the effect that the inflow of saline groundwater underneath was reduced. More about this aspect further on in this subsection.

Artificial recharge by means of infiltration wells has also been proposed as a means to prevent irreparable damage to many historic buildings in a large area around the Markerwaard (54.000 ha.), the last polder to be reclaimed in the IJsselmeer (= lake IJssel), a large former tidal bay in the Netherlands, which is since its closure in 1932 a fresh water lake. The Markerwaard is located in an area with great characteristic length λ, determined by the presence of holocene clay and peat layers of high resistance against vertical groundwater flow on top of an aquifer of great transmissibility. When, after building a circular dike, the water level of that large part of the IJsselmeer is lowered by 5 to 6 m to create the Markerwaard, the piezometric levels would be drawn down in a large area around this future polder, with all its geotechnical consequences. The drawdown of the piezometric levels in the adjacent mainland can be compensated very precisely by a long line of infiltration wells of carefully adjusted capacities along the coastline. So, any damage to buildings and structures can be prevented.

Induced recharge by pumping a line of wells parallel to a river or estuary is possible only beyond the salt water intrusion length of the rivers and estuaries and there only at those places where the river or estuary intersects a good aquifer containing fresh and non–polluted groundwater. The flow in the rivers or estuaries prevents the deposition of fine sediments thus keeping the river bed open for infiltration without much resistance.

Lakes which came into being by dredging gravel and/or sand offer possibilities for recharge of aquifers, depending on their artificially controlled water level, and the rate of supply or discharge of water required to maintain such a level. The water level in the lake in relation to the surrounding groundwater tables, whether controlled or free, determines the boundaries of the groundwater flow system. In case the lake recharges the groundwater, it offers possibilities for abstraction of groundwater. In other cases, the recharge may bring about seepage in the surroundings of such a lake. This may be undesirable, especially in case saline or brackish groundwater is set in motion. However, in such cases, the seepage will in the long run become fresh. The lake level should be chosen with care.

So far the measures to recharge the groundwater, whether to enable sustainable abstraction of groundwater or to control the groundwater table or the piezometric level.

With respect to salt water intrusion, distinction has been made, in the previous section, between the local or small scale and the regional or large scale. Upconing of salt groundwater under abstraction wells can be prevented by proper location of the well field as a whole (van Dam 1992) and by pumping in many wells of small capacity and at great mutual distance rather than in a few wells of great capacity close together. This holds in particular there where the thickness of the pumped fresh water lens is only a few tens of meters.

In order to prevent or reduce salt water intrusion at regional or large scale, by inflow of saline groundwater under the coastline, several technical measures are possible. Their feasibility depends on the local or regional conditions. When dealing with artificial recharge in the sanddunes, mention was already made of the reduction of salt water inflow by a lowering of the interface between fresh and saline groundwater and the corresponding reduction of the height available for inflow of saline water between that interface and the impermeable base. The ideal situation is where the bottom of the fresh water lens can reach the impermeable base. As the thickness of the lens is roughly proportional to the root of a uniformly distributed rate of recharge, it will not always be technically or economically possible to achieve this.

Another solution to draw down the interface, and at the same time to reduce the inflow of saline groundwater at the coastline, is to pump the saline groundwater in a line of wells underneath the sanddunes. The pumped saline groundwater could either be discharged directly into the sea or be utilized after desalting and disposal of the brine into the sea. The cost of either of the two solutions will, in most cases, be prohibitive.

With respect to pollution of groundwater by diffuse or point sources there is only one remedy; it should be forbidden or restricted by law and the observance of the rules should be checked. New waste dumps should be installed in such a way that the underlying groundwater can not be polluted. This can be done by installing impermeable layers and/or drainage systems.

Curative measures for the various effects

As was mentioned at the beginning of the previous subsection some preventive measures which were taken too late can act in the beginning also as curative measures.

In cases where preventive solutions are not technically possible or economically feasible, one should resort to curative measures. So in case of seepage of saline or brackish water, the only proper solution might be to flush the receiving surface water system with a sufficient rate of fresh surface water taken from a river, when available. This flushing results both in dilution and removal of the salt carried by seepage. This is the realistic solution in the low-lying western and northern parts of the Netherlands, where the aquifers are thick. Pumping of saline groundwater at the coastline with the only objective to reduce saline seepage is not economically feasible, especially because a very long period of time must elapse before its favourable effects will be felt.

In case of salinization of abstraction wells by upconing, the only remedy is to give up the well for some time or even permanently. In the former case, abstraction may be resumed after some time, but at a lower pumping rate. The success depends on what is going on in the surroundings. Ongoing large scale salt water intrusion, due to overpumping, may spoil the well for ever; artificial recharge may improve the situation.

Groundwater pollution by waste dumps is in most cases still only a local effect. After tracing the distribution of the pollutant, it can be removed by selective pumping, i.e. pumping in the flow paths and at places and depths where it has arrived. The pumped water needs to be treated and/or discharged which can be either a technical or a financial problem. In many cases it is difficult to find a good place to discharge the effluent with permission. This pumping should be continued for a long period of time. The purified water can also be reinjected in the waste dump for continuous flushing of the dump. Another method is to isolate the waste dump by vertical walls, by grouting or with sheetpile walls down to some impermeable layer, if present at a reasonable depth. Even with this solution the surrounding groundwater which has already been polluted must be cleaned as described before.

REFERENCES

Badon Ghijben, W. & J. Drabbe 1889. *Nota in verband met de voorgenomen putboring nabij Amsterdam.* Tijdschrift van het Koninklijk Instituut van Ingenieurs, blz. 8–22.
Bridge, J.S. & M. R. Leader 1979. A simulation model of alluvial stratigraphy. *Sedimentology* 26: 617–644.
CHO-TNO (TNO Committee on hydrological research) 1980. *Research on possible changes in the distribution of saline seepage in the Netherlands.* Proceedings and information, No.26.
Custodio, E. 1987 (with the collaboration of G.A.Bruggeman). *Groundwater problems in coastal areas.* Studies and reports in hydrology, nr.45 UNESCO, Paris.
Dam, J.C. van 1976a. Partial depletion of saline groundwater by seepage. *Journal of Hydrology* 29 (3/4):315–339.
Dam, J.C. van 1976b. Possibilities and limitations of the resistivity method of geo–electrical prospecting in the solution of geo–hydrological problems. *Geoexploration* 14:179–193.
Dam, J.C. van 1983. The shape and position of the salt water wedge in coastal aquifers. *Proceedings of the Hamburg symposium of IAHS. IAHS publication* 146:59–75.

Dam, J.C. van 1986. Characterization of the interaction between groundwater and surface water: Salinity. *Keynote paper for the Budapest symposium of IAHS. IAHS publication* 156:165–179.

Dam, J.C. van 1992. Problems associated with saltwater intrusion into coastal aquifers and some solutions. *Proceedings of the ILT seminar on problems of lowland development. Institute of Lowland Technology (ILT), Saga university, Saga 840, Japan.*

Dam, J.C. van 1993a. Impact of sea–level rise on salt water intrusion in estuaries and aquifers. *Keynote lecture delivered at the international workshop "Seachange '93", Noordwijkerhout, 19–23 April 1993* (proceedings in press).

Dam, J.C. van 1993b. *Hydrologie, grondslag voor water– en milieubeheer.* Delft University Press, Delft.

D'Itri, F.M. & L. G. Wolfson 1987. *Rural groundwater contamination.* Lewis publishers, inc. Michigan.

Duijvenbooden, W. van & H. G. van Waegeningh (editors) 1987. Vulnerability of soil and groundwater to pollutants. *Proceedings of the international conference, Wageningen, the Netherlands. Proceedings and information No. 38. TNO committee on hydrological research, The Hague.*

Engelen, G.B. & G. P. Jones (editors) 1986. *Developments in the analysis of groundwater flow systems.* IAHS publication No. 163, Wallingford.

German IHP/OHP National Committee 1993. *International workshop*: Sea level changes and their consequences for hydrology and water management. *19–23 April 1993, Noordwijkerhout, Netherlands. A contribution to the UNESCO IHP–IV project H–2–2 Bundesanstalt für Gewässerkunde, IHP/OHP–Sekretariat, Koblenz.*

Herzberg, A. 1901. Die Wasserversorgung einiger Nordseebäder. *Journal für Gasbeleuchtung und Wasserversorgung, XLIV Jahrgang, nr. 44, pp. 815–819* en nr. 45, pp.842–844.

Hooghart, J.C. (editor) 1985. Water in urban areas. *Proceedings and information No. 33. TNO committee on hydrological research, The Hague.*

Hooghart, J.C. (editor) 1988. *Urban water '88,* Hydrological processes and water management in urban areas. *Netherlands national committee for the international hydrological programme, Zoetermeer, and TNO committee on hydrological research, Delft.*

IAHS (= International Association of Hydrological Sciences) 1988. *Consequences of spatial variability in aquifer properties and data limitations for groundwater modelling practice.* Report prepared by a working group of the international commission on groundwater. IAHS publication No. 175, Wallingford.

IAHS 1990. *Groundwater contamination risk assessment: A guide to understanding and managing uncertainties.* Report prepared by a working group of the international commission on groundwater. IAHS publication No. 196, Wallingford.

IPCC (= Intergovernmental Panel on Climate Change), Response strategies working group 1992. *Global climate change and the rising challenge of the sea.* Report of the coastal zone management subgroup. Ministry of transport, public works and water management, directorate general Rijkswaterstaat, The Hague.

Jelgersma, S. (et al.) 1993. Sea level changes and their consequences for hydrology and water management. *State of the art report at the international workshop "Seachange '93", Noordwijkerhout, 19–23 April 1993. Tidal waters division, Rijkswaterstaat, The Hague.*

Jinno, K., K. Momii & J.C. van Dam 1990. Activities of SWIM in the topic of groundwater salinization. *Journal of Groundwater Hydrology, Vol.32, nr.1, Febr. 1990. Japanese association of groundwater hydrology.*

Koltermann, C.E. & S. M. Gorelick 1992. Paleoclimatic signature in terrestrial floods deposits. *Science* 256:1775–1782, 25 June 1992. *American association for the advancement of science.*

Meisler, H., P. P. Leahy & L. L. Knobel 1984. Effect of eustatic sea–level changes on salt water–freshwater in the northern Atlantic coastal plain. *U.S. geological survey. Water–supply paper* 2255.

Savenije, H.H.G. 1992. *Rapid assessment technique for salinity intrusion in alluvial estuaries.* Ph.D. thesis Delft university of technology.

Stam, J.M.T., W. Zijl & A. K. Turner 1989. Determination of hydraulic parameters from the reconstruction of alluvial stratigraphy. *Proceedings of the 4th international conference on computational methods and experimental measurements, 23–26 May 1989, Capri, Italy.*

Stuyfzand, P.J. 1993. *Hydrochemistry and hydrology of the coastal dune area of the western Netherlands.* KIWA, Nieuwegein, the Netherlands.

Tetzlaff, D.M. & J. W. Harbaugh 1989. Simulating clastic sedimentation. *Computer methods in geosciences. Van Nostrand Reinhold Company, New York.*

Tjallingii, S.P., H. van Engen & D. Kampe (editors) 1993. Hydropolis, the role of water in urban planning. *Proceedings of the international workshop, March 29th – April 2nd, 1993, Wageningen, the Netherlands. Netherlands national committee for the international hydrological programme, Zoetermeer* (in press).

Ven, F.H.M. van de & J. C. Hooghart (editors) 1986. Urban storm water quality and effects upon receiving waters. *Proceedings and information No. 36. TNO committee on hydrological research, The Hague.*

12 GROUNDWATER POLLUTION

K. JINNO
Kyushu University, Fukuoka, Japan

INTRODUCTION

It is obvious from the various reports on the water quality problems that there are enormous complicated processes in the regime of the lowland water systems. Experiences and qualitative consideration for any problem are necessary in order not to mislead the direction for solving the problems. However at the same time, quantitative analysis is also indispensable when concrete counter-measures against pollution need to be planned. Besides, it should be noticed that quantitative approach does not only provide useful information on the processes which can only be known from the quantitative analysis, but also effectively instruct more universal and unbiased essences of the problems for the next generations who want to study the environmental issues. The following discussions demonstrate several problems which may be found in the lowland water quality systems and the results obtained herein should integrate our unlinked knowledge which have not yet been fully organized.

MODELLING OF GROUNDWATER FLOW AND MASS TRANSPORT

The processes related to the groundwater which an aquifer of lowland experiences are various and complicated. It is not always easy to quantify various problems of groundwater quality through numerical approaches, however, fundamental knowledge obtained by the numerical models should provide useful information when future predictions and engineering countermeasures are needed. From these points of view, the following sections demonstrate some possible approaches for quantifying the change in the groundwater quality.

Model equations

The equation of continuity for saturated and unsaturated flow is expressed as

$$\{C_W(h)+\beta S\}\frac{\partial h}{\partial t}=-\frac{\partial u_D}{\partial X}-\frac{\partial v_D}{\partial Y}\tag{12.1}$$

where the velocity components are given by the Darcy's law as

$$u_D=-k(h)\frac{\partial h}{\partial X}\ ,\quad v_D=-k(h)\left\{\frac{\partial h}{\partial Y}+\frac{\rho}{\rho_f}\right\}\tag{12.2}$$

The symbols are h as the hydraulic pressure, $C_W(h)$ as the specific moisture capacity, S as the specific storage coefficient, β as the dummy factor which takes either 1 in unsaturated or 0 in saturated zone, $k(h)$ as the permeability, ρ as the density dependent on the concentration of either salt or pollutant in the pore water, ρ_f as the density of pure water, X and Y as the coordinates in horizontal and vertical directions respectively. The last term in the vertical velocity of Equation(12.2) includes the gravitational effect of the density. For the expressions of the unsaturated properties between the hydraulic pressure and soil water content and permeability, the van Genuchten's model is applied herein as

$$S_\theta=\frac{\theta-\theta_r}{\theta_s-\theta_r}=\frac{1}{\{1+(\alpha|h|)^n\}^m}\tag{12.3}$$

$$k_r=\frac{k(h)}{k_0}=S_\theta^{1/2}\left\{1-\left(1-S_\theta^{1/m}\right)^m\right\}^2\tag{12.4}$$

$$C_W(h)=\frac{d\theta}{dh}=-\frac{\alpha m(\theta_s-\theta_r)S_\theta^{1/m}(1-S_\theta^{1/m})^m}{(1-m)}\tag{12.5}$$

The symbol θ is the soil water content. The non–reacting mass transport equation can be applied for the chloride concentration as follows

$$\begin{aligned}\frac{\partial(\theta C)}{\partial t}+\frac{\partial(u\theta C)}{\partial X}+\frac{\partial(v\theta C)}{\partial Y}&=\frac{\partial}{\partial X}\left(\theta D_{XX}\frac{\partial C}{\partial X}+\theta D_{XY}\frac{\partial C}{\partial Y}\right)\\&+\frac{\partial}{\partial Y}\left(\theta D_{YX}\frac{\partial C}{\partial X}+\theta D_{YY}\frac{\partial C}{\partial Y}\right)\end{aligned}\tag{12.6}$$

where the symbols used are u and v as the pore velocities, D with subscripts as the tensor of dispersivities which is dependent on the pore velocities u and v and the molecular diffusivity D_M as

$$\theta D_{XX} = \alpha_L \frac{u^2}{V} + \alpha_T \frac{v^2}{V} + \theta D_M$$

$$\theta D_{YY} = \alpha_T \frac{u^2}{V} + \alpha_L \frac{v^2}{V} + \theta D_M \qquad (12.7)$$

$$\theta D_{XY} = \theta D_{YX} = (\alpha_L - \alpha_T) \frac{u\,v}{V}$$

$$V = \sqrt{u^2 + v^2}$$

The symbols α_L and α_T are the longitudinal and transverse dispersivities, respectively. In the case of salt water intrusion, the chloride concentration $C(x,y,t)$ is related to the density of water as

$$C(x,y,t) = \frac{\rho - \rho_f}{\rho_s - \rho_f} \times 100 \quad (\%) \qquad (12.8)$$

Evaluation of transverse dispersivity

It is indispensable to evaluate both the longitudinal and transverse dispersivities in order to apply the numerical simulation model to practical problems. By carrying out the field observation for the change in the plume of the tracer material in bore holes, it is possible to evaluate these quantities. In this section, however, an alternative and conventional method for the evaluation of the transverse dispersivity is presented for the case of the steady state of salt water intrusion observed in the coastal aquifer (Hosokawa et al. 1990). When the assumption of the steady state salt water intrusion is valid, then the dominant direction of the dispersion is perpendicular to the shape of the salt–fresh water interface. Here, such an estimating scheme is demonstrated as illustrated in Figure 12.1. Taking the curvilinear coordinates parallel and perpendicular to the immiscible salt–fresh water interface, which is analytically obtained assuming the quasi–uniform flow (JSCE Handbook 1985), the mass transport Equation (12.6) expressed by the Cartesian coordinates can be approximated in the curvilinear coordinates as follows

$$U_x \frac{\partial C}{\partial x} = \alpha_T \frac{\partial}{\partial y} \left(U_x \frac{\partial C}{\partial y} \right) \qquad (12.9)$$

The analytical solution for the immiscible interface is given by

$$h(X) = \frac{|q|}{(\varepsilon k_0)} \left\{ \frac{2(\varepsilon k_0)}{|q|} X + 0.5 \right\}^{1/2}$$

(12.10)

The symbol $q(\text{cm}^2/\text{sec})$ is the fresh water discharge per unit width toward the sea. The similarity solutions for the velocity and chloride profiles are assumed as

$$U_x = U_\infty f(\eta)$$

(12.11)

$$C = 100 \{1 - f(\eta)\} ; \quad \rho = \rho_s - (\rho_s - \rho_f) f(\eta)$$

(12.12)

where the similarity function $f(\eta)$ with the variable η is obtained as

$$f(\eta) = \left[1 - \frac{1}{\sqrt{2\pi}} \int_\eta^\infty e^{-\xi^2/2} d\xi \right]^{1/2}, \quad \eta = \frac{y}{R(x)}, \quad R(x) = \sqrt{2\alpha_T x}$$

(12.13)

The detailed procedures of deriving the above solutions and accuracies for both concentration and velocity profiles are given by Momii et al. (1989) Hosokawa et al. (1990), Hosokawa (1992).

The similarity function is conveniently applied to the data set of the vertically measured concentration profile of salt water so that the unknown transverse dispersivity α_T is determined. By minimizing the square of fitting errors between the predicted and the measured values, the transverse dispersivity can be obtained. The criterion to be minimized is

$$J(\alpha_T) = \sum_{j=1}^N \left[C_M(X_0, Y_j) - C(x_i, y_j) \right]^2$$

(12.14)

where j is the number of a measurement point and N is the total measurement points along a bore hole. The notations used in Equation(12.14) are the location of a bore hole by X_0, the depth of measuring point by Y_j, the coordinates along and perpendicular to the curvilinear axes by x_i and y_j. Equation(12.14) can be plotted as a function of the transverse dispersivity α_T. Thus, the transverse dispersivity is obtained as to give a minimum value of Equation(12.14).

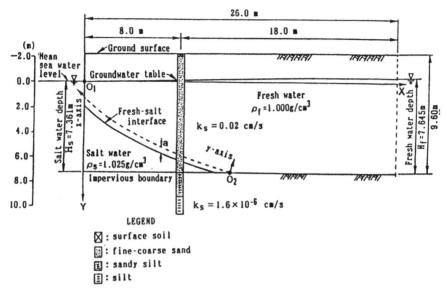

Figure 12.1. Cross–sectional view of the unconfined aquifer (field set–up).

NUMERICAL MODEL FOR SALT WATER INTRUSION

As an illustrative example, the steady state of salt water intrusion with the mixing zone developed along the salt–fresh interface is numerically simulated in order to check the result of the laboratory experiment. Both the velocity and concentration distributions are numerically obtained and compared with the vertical profiles of the similarity solution along a bore hole. Figure 12.2 shows the results of the velocity and concentration profiles by the numerical solutions. The similarity solutions agree well with the numerical solutions. Therefore, it could be confirmed that the similarity solution is valid when the steady state of salt water intrusion is observed in a homogeneous aquifer. Figure 12.3 demonstrates the result of the field application in which the bore hole was drilled 8 m apart from the coast as shown in Figure 12.1. We can observe that the mixing zone develops about 1 m thick. The results of the numerical and similarity solutions are in good agreement except the lower part of the mixing zone, which may be distorted due to the heterogeneity of the aquifer. Figure 12.4 is the sensitivity of the criterion J against α_T . The transverse dispersivity α_T can be estimated as 0.36 cm. The longitudinal dispersivity α_L is assumed to be equal to α_T since the longitudinal dispersion along the salt–fresh interface is not dominant compared with the transverse direction. Actually, the result of the numerical simulation for the steady state with the one order larger longitudinal dispersivity and with the obtained α_T shows almost same result compared with former case.

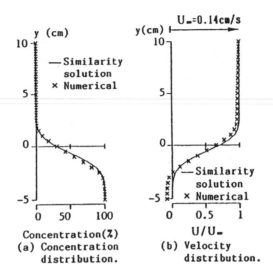

(a) Concentration distribution.

(b) Velocity distribution.

Figure 12.2. Validity of similarity assumption.

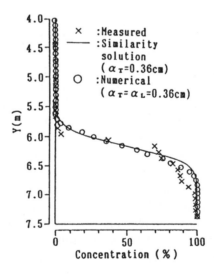

Figure 12.3. Concentration profile.

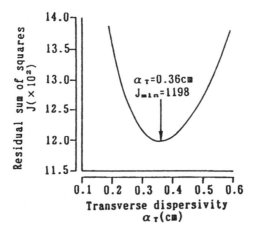

Figure 12.4. Sensitivity of the criterion J against α_T.

In Figures 12.5 and 12.6, the concentration distribution and the flow field are demonstrated. It can be observed that the salt–fresh immiscible interface given by Equation(12.10) tends to agree with the iso–concentration of 90% and the mixing zone develops along the interface. The salt water flow below the mixing zone takes place toward the fresh water region first and then the direction of the flow changes when the salt water approaches the interface. Besides, the unsaturated flow above the groundwater table is also calculated. From these results, a numerical simulation model can provide useful information in predicting the movement of salt water intrusion.

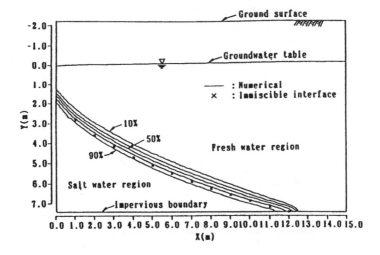

Figure 12.5. Concentration distribution by numerical solution.

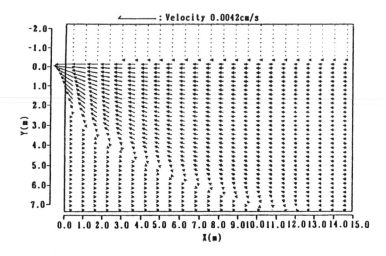

Figure 12.6. Calculated flow pattern.

NUMERICAL MODEL FOR GROUNDWATER POLLUTION BY CHLORINATED HYDROCARBONS

Groundwater pollution are also serious social concerns in lowland. Due to the agricultural and industrial activities, various types of groundwater pollution take place. An excess use of fertilizer and pesticides should cause eutrophication and contamination of toxic material in groundwater. On the other hand, the groundwater pollution caused by organic compounds are serious since early 1980s. The pollution by trichloroethylene (TCE) leaked from the storage tank in Silicon Valley, U.S.A. in 1981, implied that similar groundwater pollution by chlorinated hydrocarbons (CHC) could take place in many industrial countries also. According to the reports by the Japan Environment Protection Agency (JEPA) for the groundwater pollution surveyed at a total of 26,000 wells, 2.7% and 4.1% of the wells exceeded the guidelines standard for TCE and PCE. Simultaneously, 8.0% and 11.9% of additionally measured wells close to the polluted wells also exceeded the guidelines for TCE and PCE, respectively. JEPA set standards for TCE and PCE waste disposal at the concentration of 300 µg/l and 100 µg/l, respectively. In the law, for drinking water, 30 µg/l and 10 µg/l were also adapted for TCE and PCE. Because of these political actions taken against the polluted groundwater quality, it becomes indispensable not only to continue monitoring the pollution but also to establish a method of quantitative analyses so that appropriate countermeasures for the pollution can be made. In this section, examples of how the numerical simulation model for the CHC transport is applied to the practical problem, and the one-dimensional column experiment in which the concentrations in both gaseous and aqueous phases are measured, are shown. The numerical simulation model including the inherent transport processes of CHC is explained below.

Denoting the pollutant concentration of CHC in aqueous phase by $C_L(x,y,t)$, the mass transport equation is modelled as

$$\frac{\partial(\theta C_L)}{\partial t} + \frac{u}{R_d}\frac{\partial(\theta C_L)}{\partial x} + \frac{v}{R_d}\frac{\partial(\theta C_L)}{\partial y}$$

$$= \frac{1}{R_d}\frac{\partial}{\partial x}\left(\theta D_{xx}\frac{\partial C_L}{\partial x} + \theta D_{xy}\frac{\partial C_L}{\partial y}\right) + \frac{1}{R_d}\frac{\partial}{\partial y}\left(\theta D_{yx}\frac{\partial C_L}{\partial x} + \theta D_{yy}\frac{\partial C_L}{\partial y}\right) + \frac{Y}{R_d}$$

$$(12.15)$$

where R_d is the retardation factor when the adsorption of CHC to organic carbon in soil needs to be considered. The term Y in Equation(12.15) considers either the degradation by microorganisms and light or volatilization and dissolution with gaseous phase.

Adsorption model
The following expression is employed in the simulation of CHC adsorption to organic carbon as

$$R_d = 1 + \frac{(1-n_p)}{\theta}\cdot\sigma_s\cdot K_d \tag{12.16}$$

where n_p is the porosity, σ_s is the density of the soil, K_d is the partitioning coefficient between solid and aqueous phases. The coefficient K_d is related to organic carbons *(OC)* as

$$K_d = K_{OC}(OC) \tag{12.17}$$

where K_{OC} is the proportional constant correlated with the partitioning coefficient K_{OW} between octanol and water like

$$\log K_{OC} = \lambda\cdot\log K_{OW} - \beta \tag{12.18}$$

Symbols λ and β are the constants which are given by several researchers like Kauckhoff et al. Schwarzenbach et al. and Means et al.(all in National Research Council 1984).

Volatilization
Volatilization of CHC is another significant property which needs to be considered in modelling the transport. Applying the Henry's law for vapor of CHC, the following relation is used for the term Y as

$$Y = a \cdot \lambda_H (C_g - H \cdot C_l) \qquad (12.19)$$

where C_g is the vapor concentration, a is the volumetric fraction of air in pore which can be calculated from the soil water content, λ_H is the mass transfer coefficient, H is the Henry constant.

Similar to the transport equation for dissolved CHC in aqueous phase, the one–dimensional transport equation of volatiled vapor is expressed as follows:

$$\frac{\partial (a C_g)}{\partial t} = \frac{1}{R_d} \frac{\partial}{\partial y} \left(a D_g \frac{\partial C_g}{\partial y} \right) + \frac{a \cdot \lambda_H (H \cdot C_L - C_g)}{R_d} \qquad (12.20)$$

where D_g is the molecular coefficient in gaseous phase. While dispersed blobs of CHC descend through the unsaturated zone, some of the blobs remain in the pore space. Because the remaining blobs are easily volatiled, this process needs to be considered when it plays an important role for the spread of pollution. By adding the similar expression like Equation(12.19) to Equation(12.20), this process can be easily modelled. Details are given by Sleep & Sykes (1989). However, locating the blobs in the aquifer seems to be difficult.

Infiltration of dissolved CHC into aquifer through drainage channels

The area studied is located approximately 200 m downstream of the drainage channel from the camera factory where TCE is used for cleaning lenses. Infiltration was suspected from the unconcreted 108 m long reach of the channel. The cross sectional view of the aquifer below the channel is illustrated in Figure 12.7. Due to the long travel distance from the factory to the site, TCE at the effluent of the factory decreased to 14 % of the original concentration in the channel.

The decay of TCE in the channel is taken into account for the boundary condition at the bottom of the channel.

The dispersivities necessary for the numerical simulation are determined through the tracer test using boreholes B–2 and B–4 which are located 5.45 m apart. The longitudinal and transverse dispersivities are evaluated as 28 cm and 1.6 cm. The saturated permeabilities for the upper and lower layers below the channel bed are estimated as 0.00144 cm/sec and 0.167 cm/sec, respectively. The unsaturated properties for the upper and lower layers are referred to the soil catalogue (Mualem 1976) in which several soil size distributions are correlated to the unsaturated characteristics. The relations among hydraulic

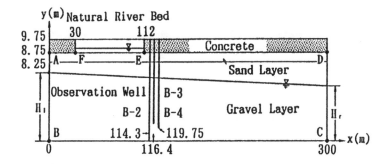

Figure 12.7. Cross sectional view below the drainage channel.

pressure, soil water content and unsaturated permeability given by van Genuchten are applied.

Several cases in which the groundwater level is changed are tested in order to simulate the effect of infiltration of water on the flux of TCE into the aquifer. Volatilization is not considered in this simulation.

Figure 12.8 illustrates the iso–concentration contours of TCE and the flow patterns where the groundwater is maintained at a relatively high level. We can observe that the infiltration of TCE takes place within the upper half of the aquifer. This can be explained by the fact that the vertical downward velocity is significant in the unsaturated zone below the channel bed. Figure 12.9 is the case where the groundwater is maintained at a lower level than the former case shown in Figure 12.8. Because less infiltration of channel water into the unsaturated zone, the infiltration of TCE also becomes less. When the groundwater rises from lower level to higher level, infiltration of TCE becomes limited within the shallow part of the aquifer due to the upward convection in the groundwater flow which is induced by the level changes. This fact can be seen in Figure 12.10.

On the other hand, significant infiltration takes place when the groundwater level changes from higher level to lower level as seen in Figure 12.11.

The vertical profiles of the pollutant concentration along the borehole B–4 obtained from the simulations are compared with several measurements in the same borehole. The calculated maximum values in the borehole range from 5% to 11 % of the original concentration for any cases and they are similar to the observed values. As a conclusion for the present simulation, the fluctuation in the groundwater level controls the TCE infiltration and the methodology used herein would be applicable to similar groundwater pollution.

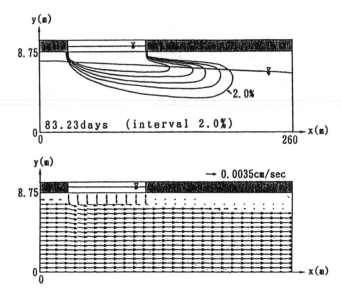

Figure 12.8. Iso-lines of dissolved TCE and flow pattern when the groundwater is maintained at a high level (Steady state).

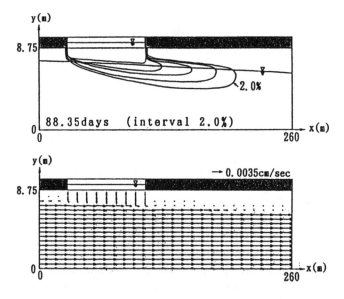

Figure 12.9. Iso-lines of dissolved TCE and flow pattern when the groundwater is maintained at a low level (Steady state).

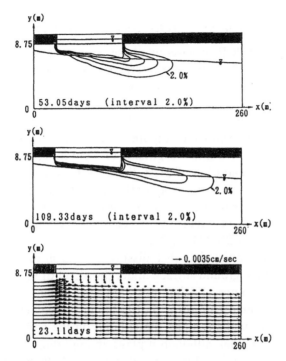

Figure 12.10. Iso-lines of dissolved TCE and flow pattern when the groundwater level ascends (Unsteady state).

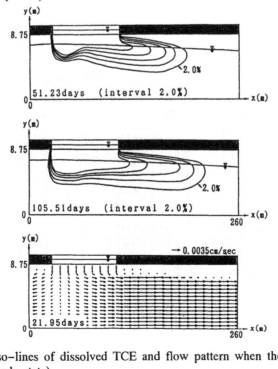

Figure 12.11. Iso-lines of dissolved TCE and flow pattern when the groundwater level descends (Unsteady state).

Transport of PCE vapor in unsaturated zone from polluted groundwater

Transport of TCE and PCE in gaseous phase is another important aspect in a regime of air pollution. Therefore, the interaction between aqueous and gaseous phases needs to be considered. A laboratory experiment in order to understand a fundamental transport has been carried out using columns of 40 cm in diameter of teflon resin as container which prevents adsorption and light degradation of PCE, as shown in Figure 12.12. Three columns of 5 cm in diameter, which are divided into 7 segments of 5 cm long, are used to measure water content and concentration of PCE dissolved in water. One column is set on the electric weight to measure evaporation from the soil. The column of 10 cm in diameter with 6 thermistors and 6 holes for sampling volatiled PCE is soaked in the glass box in which the concentration of dissolved PCE in the water was maintained at 28.044 mg/l.

The process to be simulated is that dissolved PCE is transported from the groundwater toward the ground surface under the affect of volatilization in pore space. The air temperature and humidity are also monitored. The observed temperature showed almost uniform profile along the column. Therefore, the effect of temperature gradient along the column on the soil water diffusion is not considered in the model, although it plays an important role at the real site.

Equation(12.20) is applied for the calculation of PCE concentration in gaseous phase. The equation for the dissolved PCE in aqueous phase is as follows

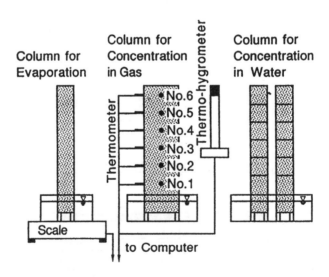

Figure 12.12. Experimental setup for vapor transport.

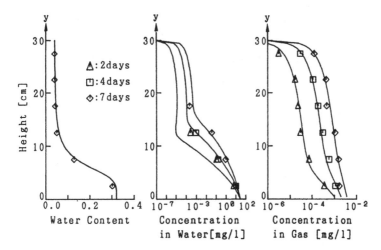

Figure 12.13. Vertical profiles of water content, PCE concentration in aqueous phase, and in gaseous phase.

$$\frac{\partial(\theta C_L)}{\partial t} + \frac{v}{R_d}\frac{\partial(\theta C_L)}{\partial y} = \frac{1}{R_d}\frac{\partial}{\partial y}\left(\theta D_L\frac{\partial C_L}{\partial y}\right) + \frac{a\cdot\lambda_H(C_g - H\cdot C_L)}{R_d} \qquad (12.21)$$

Figure 12.13 illustrates the profiles of soil water, dissolved concentration in water and volatiled vapor after 2, 4, and 7 days from the start of the experiment.

The symbols indicate the measured values. The profile of soil water did not change significantly for the whole period, even though slight evaporation occurred. On the other hand, the concentrations both in the aqueous and gaseous phases changed significantly. The solid lines in the figure represent the numerical solutions. In the simulation, the physical parameters used are 1.2 for the Henry constant H, 0.04 for the minimum water retention θ_r, 0.32 for the saturation water content θ_s, and 0.46 cm/sec for the saturated permeability. The unsaturated characteristics are represented by the van Genuchten equations.

The retardation factor is 1.0 because the soil does not contain organic carbon. In the simulation the sensitivity of the mass transfer coefficient λ_H is tested and determined as 3×10^{-7}/sec. Since this value seems to be very crucial for the prediction, the dependency on soil size, porosity and other unknown factors needs to be further analyzed. Figure 12.14 also shows the comparisons of volatiled PCE at each depth of the column. Reasonable agreement is obtained.

Figure 12.15 is the plots of the exchange rate of PCE between aqueous and gaseous phases along the column after 2, 4 and 7 days, through which we can observe that the volatilization occurs at the lower place close to the water table. The volatiled vapor of PCE is then transported upward through the molecular diffusion in gaseous phase. Since the upward transport of dissolved PCE through the aqueous phase is much slower than by gaseous molecular diffusion, redissolution of PCE from the gaseous phase to aqueous phase occurs at the place where the dissolved concentration in the aqueous is still below PCE saturation. As the vapor concentration increases and redissolution reaches the equilibrium, the location where redissolution takes place also moves upward. This fact can be seen in Figure 12.16.

The results obtained through the laboratory experiment and numerical simulation suggest the applicability of the present model for the practical problems.

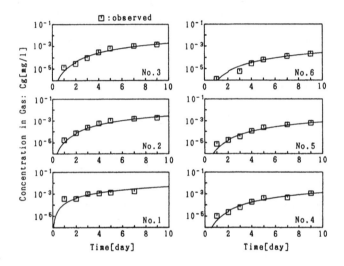

Figure 12.14. Time series plots of measured and calculated volatiled PCE at each depth.

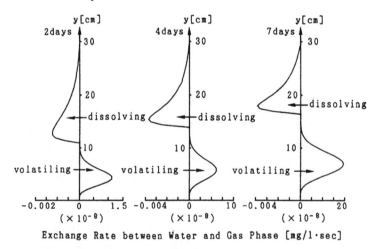

Figure 12.15. Vertical profiles of exchange rate between aqueous and gaseous phases (mg/l·sec).

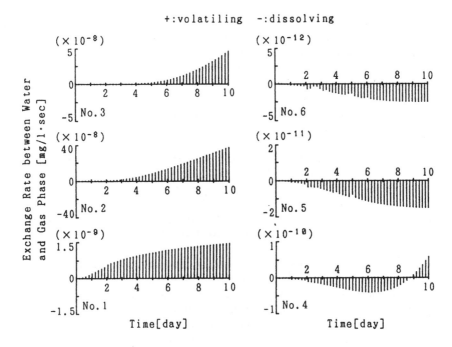

Figure 12.16. Time series plots of the exchange rate at each depth. Positive values indicate volatilization, and negative values dissolution into the water.

CONCLUSIONS

The concerns of groundwater quality as well as quantity need to be considered from the scientific point of view. Sufficient field surveys will enable the technical countermeasures more reliable. Then the numerical approach demonstrated herein as examples should provide fundamental information on various aspects of groundwater pollution and on the possible strategy in order to solve the problems. It is therefore emphasized that the long term monitoring of groundwater quantity and the groundwater hydrology is the key for preventing the groundwater resources from the hazards inherent to the human activities in the lowland.

REFERENCES

Egusa, N., K. Jinno, K. Momii & E. Sumi 1992a. Characteristics of infiltration of trichloroethylene from the natural river bed to groundwater. *Proceedings of hydraulic engineering, JSCE*, 36, 391–396 (in Japanese with English abstract).

Egusa, N., K. Jinno, H. Hironaka & M. Esaki 1992b. Mechanism of transport of volatile organics in the unsaturated zone. *Proceedings of the 2nd symposium on groundwater pollution and its management* (in Japanese).

Hosokawa, T., K. Jinno, & K. Momii, 1990. Estimation of transverse dispersivity in the mixing zone of fresh–salt groundwater. *International association of hydrological sciences publication*, No. 195:149–157.

Hosokawa, T. 1992. Fundamental study for the density flow and dispersion and the optimal fresh groundwater discharge. *Dissertation, Kyushu University, Fukuoka Japan* (in Japanese).

Japan Society of Civil Engineers. 1985. *Handbook of Hydraulics:382–384* (in Japanese).

Mendoza, C. A. & E. O. Frind 1990a. Advective–dispersive transport of dense organic vapors in the unsaturated zone 1. Model Development. *Water resources research*, 26(3), 379–387.

Mendoza, C. A. & E. O. Frind 1990b. Advective–dispersive transport of dense organic vapors in the unsaturated zone 2. Sensitivity Analysis. *Water resources research*, 26(3), 388–398.

Mendoza, C. A. & T. A. MaAlary 1990. Modeling of ground–water contamination caused by organic solvent vapors. *Ground water*, 28(2), 199–206.

Momii, K., T. Hosokawa, K. Jinno & T. Ito 1989. Estimation method of transverse dispersivity based on vertical salt concentration distribution in coastal aquifer. *Japan Society of Civil Engineers*. 411/12:45–53. (in Japanese with English abstract).

Mualem, Y. 1976. *A catalogue of hydraulic properties of unsaturated soils*.

National Research Council. 1984. Groundwater Contamination, Studies in Geophysics, *National academy press*.

Sleep, B. E. & J. F. Sykes 1989. Modelling the transport of volatile organics in variably saturated media. *Water resources research*, 25(1), 81–92.

13 NUMERICAL MODELLING OF MASS TRANSPORT PROBLEMS IN POROUS MEDIA

S. VALLIAPPAN, C. ZHAO and T. P. XU
University of New South Wales, Sydney, Australia

INTRODUCTION

The most common approach used for modelling the contaminant transport in the lowland range is the advection–dispersion equation. The advection–dispersion equation includes terms which describe the physical processes governing the movement, dispersion and transformation of a solute. Generally, the advection is due to bulk movement of water, either caused by differences in density (natural convection), by regional movement in the aquifer (advection), or by some artificial disturbance (forced convection). The dispersion is due to the irregular movement of water. On the pore scale, these irregularities are due to tortuosity of the flow paths and on a large scale they are due to the presence of zones of different permeabilities.

In the case of constant transport parameters with respect to time and position, the advection–dispersion equation is linear and the corresponding closed–form solutions can generally be derived. Many analytical solutions for the advection–dispersion equation are now available for a large number of initial and boundary conditions for one–dimensional transport problems (Van Genuchten & Alves 1982) and a smaller number of conditions for two and three dimensional transport problems (Cleary & Ungs 1987, Carnahan & Remer 1984, Javandel et al. 1984, Wexler 1989). Because of the large variability of low and mass transport properties in the field, the often transient nature of flow regime, and the non–ideal nature of applicable initial and boundary conditions, the usefulness of analytical solutions is often limited and numerical methods such as finite difference methods and finite element methods are needed to solve such problems.

In terms of numerical methods, a number of finite difference models (Bresler 1973, 1975, Davidson et al. 1975) and finite element models (Sykes 1975, Duguid & Reeves 1977, Segol 1977, Pickens et al. 1979, Bear & Verruijt 1987, Bear & Bachmat 1990, Valliappan & Zhao 1992, 1993, Zhao & Valliappan 1992) have been developed to deal with water and contaminant transport problems in porous media. Also, a number of finite difference models (Whisler & Klute 1965, Rubin 1967, Remson et al. 1967, Ibrahim & Brutsaert 1968, Hanks et al. 1969, Staple 1969, Giesel et al. 1973, Perrens & Watson 1977, Gillham et al. 1979) and a finite element model (Hoa et al. 1977) were presented to model unsaturated flow in porous media in which the hydraulic parameters are considered hysterestic. Considering the fact that numerical models have become the favored type of models for studying contaminant transport problems in engineering practice, the subject of this chapter is the use of numerical methods to simulate mass transport problems in porous media. Thus, a fundamental and practical introduction to finite difference and finite element methods are provided in great detail.

The arrangement of the main contents of this chapter is made as follows. In the second section, the governing equation of mass transport in porous media is derived from some physical principles. In the third section, the finite difference method for dealing with mass transport problems is outlined and the related formulas are given. In the fourth section, the finite element formulation for dealing with mass transport problems is derived and some problems associated with the method are also discussed. Finally, an example and discussions are given in the fifth section to illustrate how the finite element method is applied to solve the mass transport problems in porous media.

GOVERNING EQUATIONS OF MASS TRANSPORT PROBLEMS

Porous media and natural porous formation are very complex in reality because they display spatial variability of their geometric and hydraulic properties. Furthermore, this variability is of an irregular and complex nature. It generally defies a precise description, either because of insufficiency of information or because of the lack of interest in knowing the very minute details of the structure and flow field. Indeed, in most circumstances we are only interested in the behavior of a large portion of the formation, namely in averaging flow or transport variables over the space. This averaging process has a smoothing effect and filters out small scale variations associated with heterogeneity. Even in those cases in which this is not achieved, because of large scale variability, determining some gross features of the process may be quite satisfactory. Therefore, from the viewpoint of macroscopic mechanics, the porous media encountered in most of engineering practice can be regarded as homogeneous, isotropic media.

For a homogeneous, isotropic and flow-saturated porous medium, the governing advection dispersion equation for mass transport processes can be derived from the law of conservation of mass under the following assumptions: (1) the porous medium is homogeneous and isotropic; (2) the medium is saturated by flow; (3) the flow in the medium is a steady-state one; (4) Darcy's law applies to the flow in the medium. Under the Darcy assumption, the flow is described by the average linear velocity, which carries the dissolved solute by advection and forms one of the mechanism of mass transport processes in porous media. In reality, these is an additional mixing process, namely hydrodynamic dispersion, which is caused by variations in the microscopic velocity within each pore channel and from one channel to another. If the mass transport processes are described on a macroscopic scale using macroscopic parameters and taking into account the effect of microscopic mixing, it is necessary to introduce a second mechanism of mass transport, in addition to advection, to account for hydrodynamic dispersion.

As shown in Figure 13.1, the solute flux into and out of an infinitesimal elemental volume in the porous medium is considered to derive the governing equations of the problem. It should be noted that these governing equations, as a matter of fact, are certain mathematical statements of the conservations of mass. In a three-dimensional rectangular coordinate system, the specific discharge (superficial flow velocity) v has three components, v_x, v_y, and v_z, so that the average linear velocity $\bar{v} = \dfrac{v}{n}$ has also three components, $\bar{v}_x$, $\bar{v}_y$ and $\bar{v}_z$, where n is the porosity of the porous medium and is defined as the ratio of the volume of the voids V_v to that of the total volume of the representative element V_t. If the concentration of the solute C is defined as the mass of solute per unit volume of solution, the mass of solute per unit volume of porous media is nC. For a homogeneous medium, the porosity n is a constant. Thus, $\dfrac{\partial(nC)}{\partial x} = n\dfrac{\partial C}{\partial x}$. The mass of solute transported in the x direction by the two mechanisms of mass transport can be represented as

$$m_1 = \bar{v}_x nC\,dA$$

$$m_2 = nD_x\frac{\partial C}{\partial x}dA$$

(13.1)

where m_1 is the mass transported by advection; m_2 is the mass transported by dispersion and is derived from Fick's first law; D_x is the dispersion coefficient in the x direction and dA is the elemental cross-sectional area of the cubic element. The dispersion coefficient D_x is related to the dispersivity α_x and the diffusion coefficient D^* as follows:

$$D_x = \alpha_x \bar{v}_x + D^*$$

(13.2)

If F_x is defined to represent the total mass of solute per unit cross-sectional area transported in the x direction per unit time, then it can be written as

$$F_x = \bar{v}_x nC - nD_x\frac{\partial C}{\partial x}$$

(13.3)

It is noted that the negative sign before the dispersive term indicates that the contaminant moves toward the zone of lower concentration. Similarly, expressions in the y and z directions can be written as

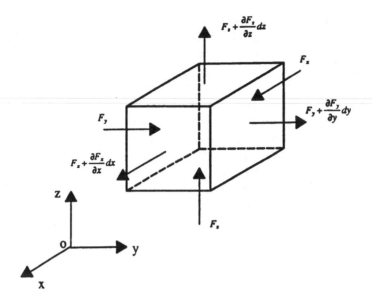

Figure 13.1. Mass balance in an infinitesimal element.

$$F_y = \bar{v}_y nC - nD_y \frac{\partial C}{\partial y}$$
$$F_z = \bar{v}_z nC - nD_z \frac{\partial C}{\partial z}$$

(13.4)

where

$$D_y = \alpha_y \bar{v}_y + D^*$$
$$D_z = \alpha_z \bar{v}_z + D^*$$

(13.5)

where α_y and α_z are dispersivity of the medium in the y and z directions respectively.

Therefore, the total amount of solute entering the cubic element in Figure 13.1 is

$$m_{in} = F_x dz\, dy + F_y dz\, dx + F_z dx\, dy$$

(13.6)

The total amount of solute leaving the cubic element is

$$m_{out} = (F_x + \frac{\partial F_x}{\partial x} dx) dz\, dy + (F_y + \frac{\partial F_y}{\partial y} dy) dz\, dx + (F_z + \frac{\partial F_z}{\partial z} dz) dx\, dy$$

(13.7)

where the partial terms represent the spatial change of the solute mass in the corresponding direction. The difference in the amount entering and leaving the element can be expressed as

$$m_{out} - m_{in} = (\frac{\partial F_x}{\partial x} + \frac{\partial F_y}{\partial y} + \frac{\partial F_z}{\partial z})dxdydz \tag{13.8}$$

If the dissolved substance is assumed to be nonreactive, the difference between the flux into the element and the flux out of the element equals the amount of dissolved substance accumulated in the element. The rate of mass change in the element is

$$m_{change} = -n\frac{\partial C}{\partial t}dxdydz \tag{13.9}$$

Using the law of conservation of mass, the governing equation of mass transport can be derived as

$$\frac{\partial F_x}{\partial x} + \frac{\partial F_y}{\partial y} + \frac{\partial F_z}{\partial z} = -n\frac{\partial C}{\partial t} \tag{13.10}$$

Substituting Equations 13.3 and 13.4 into Equation 13.10, the governing equation of mass transport in the three-dimensional porous medium can be written as

$$[\frac{\partial}{\partial x}(D_x\frac{\partial C}{\partial x}) + \frac{\partial}{\partial y}(D_y\frac{\partial C}{\partial y}) + \frac{\partial}{\partial z}(D_z\frac{\partial C}{\partial z})]$$
$$-[\frac{\partial}{\partial x}(\bar{v}_xC) + \frac{\partial}{\partial y}(\bar{v}_yC) + \frac{\partial}{\partial z}(\bar{v}_zC)] = \frac{\partial C}{\partial t} \tag{13.11}$$

For the homogeneous medium in which dispersion coefficients D_x, D_y and D_z are independent on the space variables and the velocity $\bar{v}$ is steady and uniform, Equation 13.11 can be further expressed as

$$[D_x\frac{\partial^2 C}{\partial x^2} + D_y\frac{\partial C}{\partial y^2} + D_z\frac{\partial^2 C}{\partial z^2}] - [\bar{v}_x\frac{\partial C}{\partial x} + \bar{v}_y\frac{\partial C}{\partial y} + \bar{v}_z\frac{\partial C}{\partial z}] = \frac{\partial C}{\partial t} \tag{13.12}$$

It should be noted that Equation 13.12 is derived from a three-dimensional porous medium. The solution to this equation provides the solute concentration C as a function of space and time.

Using the same procedure as above, the governing equation of mass transport in a two-dimensional porous medium can be derived as

$$[D_x\frac{\partial^2 C}{\partial x^2} + D_y\frac{\partial^2 C}{\partial y^2}] - [\bar{v}_x\frac{\partial C}{\partial x} + \bar{v}_y\frac{\partial C}{\partial y}] = \frac{\partial C}{\partial t} \tag{13.13}$$

Similarly, the governing equation of mass transport in a one-dimensional porous medium is

$$D_x\frac{\partial^2 C}{\partial x^2} - \bar{v}_x\frac{\partial C}{\partial x} = \frac{\partial C}{\partial x} \tag{13.14}$$

Moreover, it is possible to extend the governing equation of mass transport in a porous medium to include the effects of retardation of mass transport through adsorption, chemical reaction, biological transformations or radioactive decay. In this case, the mass balance carried out on the elemental volume must include a source-sink term. Taking retardation due to adsorption as an example, the governing equation of mass transport in a homogeneous medium in a one-dimensional system along the direction of flow can be written as

$$D_x \frac{\partial^2 C}{\partial x^2} - \bar{v}_x \frac{\partial C}{\partial x} + \frac{\rho_b}{n} \frac{\partial s}{\partial t} = \frac{\partial C}{\partial t} \tag{13.15}$$

where ρ_b is the bulk mass density of the porous medium; s is the mass of chemical constituent absorbed on a unit mass of the solid part of the porous medium. The first term of Equation 13.15 is the dispersion term, the second is the advection term and the third is the reaction term.

It should be pointed out that for certain simple boundary and initial conditions, there are some analytical solutions available for Equations 13.12, 13.13 and 13.14. However, in most field situations, the velocities are seldom uniform and the dispersivities are usually variable in space. For these situations, numerical methods such as the finite difference and the finite element methods must be used to obtain solutions.

FINITE DIFFERENCE METHODS FOR MASS TRANSPORT PROBLEMS

Finite difference and finite element methods are presently the most common numerical techniques for numerical modelling of mass transport problems in porous media. In these methods, not only the boundary conditions are approximated, but also the governing equation of the problem is discretized. One of the main characteristics for using both methods in their conventional forms is that the problem domain should be bounded. The advantages of both the finite difference and finite element methods lie in that: (1) hydraulic properties such as permeability and dispersivity can be easily varied throughout a system; (2) their formulations are suitable for modelling both steady-state and transient flow; and (3) they are comparatively simple and straightforward in implementation. Therefore, finite difference and finite element methods have found wide applications in engineering practice. In this section, the emphasis is on dealing with finite difference methods for mass transport problems in porous media in some detail. For the ease of brevity, a two-dimensional problem is considered to illustrate how the finite difference method is used to deal with such problems.

Finite difference methods are based on a discretization of the problem domain into a mesh that is usually rectangular, so that the contaminant concentration is computed at the grid points by solving the differential

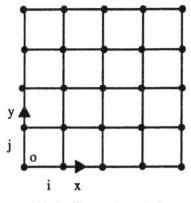

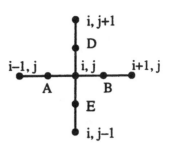

(A) An illustrative mesh   (B) A typical node

Figure 13.2. Finite difference mesh.

equation in the finite difference form throughout the mesh. There exist a variety of numerical techniques for solving the resulting linear equations of a system. Although the finite difference method may be applied with comparative ease to problems where the hydraulic properties vary from node to node, we will discuss the case where the hydraulic properties are constant, for the purpose of explaining the basic principles involved in the method for a square mesh. As mentioned in the last section, the governing equation of a two-dimensional mass transport problem in a porous medium is expressed as

$$D_x \frac{\partial^2 C}{\partial x^2} + D_y \frac{\partial^2 C}{\partial y^2} - \bar{v}_x \frac{\partial C}{\partial x} - \bar{v}_y \frac{\partial C}{\partial y} = \frac{\partial C}{\partial t} \qquad (13.16)$$

Using the finite difference method, the problem domain can be discretized by a square mesh as shown in Figure 13.2(A). The nodes are identified as (i, j), where i and j are integers, such that the coordinates of node (i, j) are defined by

$$x = (i - 1)a$$
$$y = (j - 1)a \qquad (13.17)$$

where a is the distance between grid points and can be expressed as

$$a = \Delta x = \Delta y \qquad (13.18)$$

The node (i, j) with its four surrounding grid points is shown in Figure 13.2(B), where A, B, D and E are middle points between the corresponding nodes. The derivative $\frac{\partial C}{\partial x}$ may be approximately expressed at points A and B as

$$[\frac{\partial C}{\partial x}]_A = \frac{C_{i,j} - C_{i-1,j}}{a}$$

$$[\frac{\partial C}{\partial x}]_B = \frac{C_{i+1,j} - C_{i,j}}{a}$$

(13.19)

Taking an average value of $[\frac{\partial C}{\partial x}]_A$ and $[\frac{\partial C}{\partial x}]_B$, the first derivative of C with respect to x at node (i,j) can be expressed as

$$[\frac{\partial C}{\partial x}]_{i,j} = \frac{C_{i+1,j} - C_{i-1,j}}{2a}$$

(13.20)

The second derivative of C with respect to x at node (i,j) may be written in the finite difference form as

$$[\frac{\partial^2 C}{\partial x^2}]_{i,j} = \frac{C_{i-1,j} + C_{i+1,j} - 2C_{i,j}}{a^2}$$

(13.21)

Similarly, the first and second derivatives of C with respect to y at node (i,j) can be derived as

$$[\frac{\partial C}{\partial y}]_{i,j} = \frac{C_{i,j+1} - C_{i,j-1}}{2a}$$

$$[\frac{\partial^2 C}{\partial y^2}]_{i,j} = \frac{C_{i,j-1} + C_{i,j+1} - 2C_{i,j}}{a^2}$$

(13.22)

The first derivative of C with respect to t can be also expressed in the finite difference form as

$$[\frac{\partial C}{\partial t}]_{i,j} = \frac{C_{i,j}^{t+\Delta t} - C_{i,j}^t}{\Delta t}$$

(13.23)

Substituting Equations 13.20 to 13.23 into Equation 13.16 and using the forward difference scheme for time, the finite difference equation of the system can be written as

$$C_{i,j}^{t+\Delta t} = C_{i,j}^t + \frac{\Delta t}{a^2}[D_x(C_{i-1,j}^t + C_{i+1,j}^t - 2C_{i,j}^t) + D_y(C_{i,j-1}^t + C_{i,j+1}^t$$

(13.24)

$$- 2C_{i,j}^t) - \frac{1}{2}\bar{v}_x a(C_{i+1,j}^t - C_{i-1,j}^t) - \frac{1}{2}\bar{v}_y a(C_{i,j+1}^t - C_{i,j-1}^t)]$$

where the superscripts refer to the time level.

It is noted that the forward difference scheme for time has the advantages of flexibility and simplicity, but gives meaningful answers only if Δt is less than some critical value, Δt_{crit},

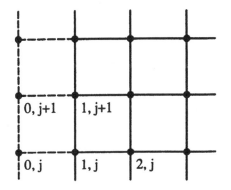

Figure 13.3. Treatment of the Neumann boundary condition.

$$\Delta t < \Delta t_{crit} \qquad\qquad (13.25)$$

If Δt exceeds Δt_{crit}, the solution will become unstable, giving results that are clearly in error. The value of the critical time step may be found by numerical tests or estimated by analyzing the simple problems.

Usually, there are three kinds of boundary conditions for mass transport problems in porous media. The first one is called the Dirichlet boundary condition in terms of C, where the contaminant concentration is known along the boundary, and therefore at each boundary node. The second is called the Neumann boundary condition in terms of $\frac{\partial C}{\partial n}$, where the first derivative of C with respect to the normal direction of the boundary n is specified along the related nodes. The third is called the mixed boundary condition, which is the combination of the Dirichlet and Neumann boundary conditions. In the finite difference method, the Dirichlet boundary condition can be directly applied to the boundary nodes, while the Neumann boundary condition needs to be considered using the following special technique. In order to approximate $\frac{\partial C}{\partial n}$ at boundary nodes, the mesh is extended one row of nodal points beyond the boundary as illustrated in Figure 13.3. The values for either i or j are zero for nodes (i,j) outside the boundary, which are called imaginary nodes. For the case illustrated in Figure 13.3, supposing $\frac{\partial C}{\partial n}$ is given along the boundary $x = 0$ $(i = 1)$, $\frac{\partial C}{\partial n}$ is approximated for node $(1, j)$ as

$$[\frac{\partial C}{\partial n}]_{1,j} = [\frac{\partial C}{\partial x}]_{1,j} = \frac{C_{2,j} - C_{0,j}}{2a} \qquad\qquad (13.26)$$

Therefore,

$$C_{0,j} = C_{2,j} - 2a[\frac{\partial C}{\partial x}]_{1,j} \tag{13.27}$$

which is used to calculate the value of $\dfrac{\partial^2 C}{\partial x^2}$ at the boundary nodes. Unlike the forward difference scheme which results in the explicit finite difference formulation, the central and backward difference schemes lead to the implicit finite difference formulation. The advantage of the implicit formulation is that the time step may be chosen much larger than for the explicit one. Disadvantages are due to a greater complexity, and that the Gauss-Seidel iteration must be applied at each time step. The explicit formulation is particularly attractive for three-dimensional problems. The formulation may be adapted to three-dimensional mass transport problems in porous media simply by generating a three-dimensional mesh and change the related expressions in Equation 13.24.

FINITE ELEMENT METHODS FOR MASS TRANSPORT PROBLEMS

The finite element method mainly differs from the finite difference method in the following two aspects: (1) the problem domain is discretized into a mesh of finite elements of any shape: (2) the governing differential equation of the problem is not solved directly but replaced by an approximate formulation. Although there exist various approximate procedures (Zienkiewicz 1977), the Galerkin weighted residual method may be the best one for dealing with mass transport problems in porous media.

As demonstrated in an earlier section (Equation 13.13), the governing equation of mass transport in a two-dimensional porous media can be written as

$$D_x \frac{\partial^2 C}{\partial x^2} + D_y \frac{\partial^2 C}{\partial y^2} - \bar{v}_x \frac{\partial C}{\partial x} - \bar{v}_y \frac{\partial C}{\partial y} = \frac{\partial C}{\partial t} \tag{13.28}$$

Using the Galerkin weighted residual method, Equation 13.28 can be rewritten for an element as

$$\iint_A [N]^T [D_x \frac{\partial^2 C}{\partial x^2} + D_y \frac{\partial^2 C}{\partial y^2} - \bar{v}_x \frac{\partial C}{\partial x} - \bar{v}_y \frac{\partial C}{\partial y}] dA$$

$$-\iint_A [N]^T \{\frac{\partial C}{\partial t}\} dA = 0 \tag{13.29}$$

where $[N]$ is the weighting function matrix of the element; C is the trial function of the element; A is the area of the element.

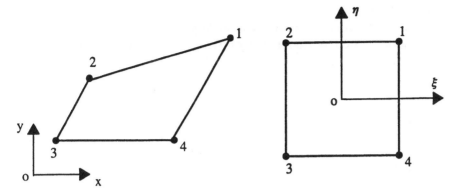

Figure 13.4. 4-node two-dimensional isoparametric finite element.

$$C = [N]\{C\}^e \tag{13.30}$$

where $\{C\}^e$ is the nodal concentration vector of the element at a given time.

Substituting Equation 13.30 into Equation 13.29 yields the following equation.

$$[\iint_A [N]^T (D_x \frac{\partial^2 [N]}{\partial x^2} + D_y \frac{\partial^2 [N]}{\partial y^2} - \bar{v}_x \frac{\partial [N]}{\partial x} - \bar{v}_y \frac{\partial [N]}{\partial y}) dA]\{C\}^e$$

$$- \iint_A [N]^T [N] \{\frac{\partial C}{\partial t}\}^e dA = 0 \tag{13.31}$$

Integrating by parts and applying Green's theorem to the second-order derivatives, Equation 13.31 can be written as

$$[G_e]\{C\}^e + [H_e]\{C\}^e + [R_e]\{\frac{\partial C}{\partial t}\}^e = \{f\}^e \tag{13.32}$$

where $[G_e]$, $[H_e]$ and $[R_e]$ are property matrices of the element; $\{f\}^e$ is a "load" vector of the element. They can be expressed as

$$[G_e] = \iint_A (D_x \frac{\partial [N]^T}{\partial x} \frac{\partial [N]}{\partial x} + D_y \frac{\partial [N]^T}{\partial y} \frac{\partial [N]}{\partial y}) dA$$

$$[H_e] = \iint_A (\bar{v}_x [N]^T \frac{\partial [N]}{\partial x} + \bar{v}_y [N]^T \frac{\partial [N]}{\partial y}) dA$$

$$[R_e] = \iint_A [N]^T [N] dA \tag{13.33}$$

$$\{f\}^e = \int_S [N]^T (D_x \frac{\partial C}{\partial x} n_x + D_y \frac{\partial C}{\partial y} n_y) dS$$

where S is the boundary of the element; n_x and n_y are the direction cosines of the outward unit normal vector on the boundary.

For a 4-node two-dimensional isoparametric finite element shown in Figure 13.4, the coordinate transform relationship between the global coordinate system xoy and the local coordinate system $\xi o\eta$ can be defined as

$$x = \sum_{i=1}^{4} N_i x_i$$

$$y = \sum_{i=1}^{4} N_i y_i \tag{13.34}$$

where

$$N_1 = \frac{1}{4}(1+\xi)(1+\eta)$$

$$N_2 = \frac{1}{4}(1-\xi)(1+\eta)$$

$$N_3 = \frac{1}{4}(1-\xi)(1-\eta) \tag{13.35}$$

$$N_4 = \frac{1}{4}(1+\xi)(1-\eta)$$

The contaminant concentration field within the element can be defined as

$$C = \sum_{i=1}^{4} N_i C_i \tag{13.36}$$

Comparing Equation 13.36 with Equation 13.30, the shape function matrix of the element can be written as

$$[N] = [N_1 \ N_2 \ N_3 \ N_4] \tag{13.37}$$

Following the differentiation rule in mathematics, the following relationships exist.

$$\frac{\partial [N]}{\partial \xi} = \frac{\partial [N]}{\partial x}\frac{\partial x}{\partial \xi} + \frac{\partial [N]}{\partial y}\frac{\partial y}{\partial \xi}$$

$$\frac{\partial [N]}{\partial \eta} = \frac{\partial [N]}{\partial x}\frac{\partial x}{\partial \eta} + \frac{\partial [N]}{\partial y}\frac{\partial y}{\partial \eta} \tag{13.38}$$

Equation 13.38 can be written in a matrix form as

$$
\left\{ \begin{array}{c} \dfrac{\partial [N]}{\partial \xi} \\[2mm] \dfrac{\partial [N]}{\partial \eta} \end{array} \right\} = \left[\begin{array}{cc} \dfrac{\partial x}{\partial \xi} & \dfrac{\partial y}{\partial \xi} \\[2mm] \dfrac{\partial x}{\partial \eta} & \dfrac{\partial y}{\partial \eta} \end{array} \right] \left\{ \begin{array}{c} \dfrac{\partial [N]}{\partial x} \\[2mm] \dfrac{\partial [N]}{\partial y} \end{array} \right\} \tag{13.39}
$$

Therefore,

$$
\left\{ \begin{array}{c} \dfrac{\partial [N]}{\partial x} \\[2mm] \dfrac{\partial [N]}{\partial y} \end{array} \right\} = \left[\begin{array}{cc} \dfrac{\partial x}{\partial \xi} & \dfrac{\partial y}{\partial \xi} \\[2mm] \dfrac{\partial x}{\partial \eta} & \dfrac{\partial y}{\partial \eta} \end{array} \right]^{-1} \left\{ \begin{array}{c} \dfrac{\partial [N]}{\partial \xi} \\[2mm] \dfrac{\partial [N]}{\partial \eta} \end{array} \right\} \tag{13.40}
$$

Finally, the property matrices of the element for mass transport in porous media can be expressed using the local coordinate as

$$
[G_e] = \int_{-1}^{1} \int_{-1}^{1} (D_x \frac{\partial [N]^T}{\partial x} \frac{\partial [N]}{\partial x} + D_y \frac{\partial [N]^T}{\partial y} \frac{\partial [N]}{\partial y}) |J| \, d\xi d\eta
$$

$$
[H_e] = \int_{-1}^{1} \int_{-1}^{1} (\bar{v}_x [N]^T \frac{\partial [N]}{\partial x} + \bar{v}_y [N]^T \frac{\partial [N]}{\partial y}) |J| \, d\xi d\eta \tag{13.41}
$$

$$
[R_e] = \int_{-1}^{1} \int_{-1}^{1} [N]^T [N] |J| \, d\xi d\eta
$$

where $|J|$ is the Jacobian determinant of the element.

It is noted that Equation 13.41 can be easily evaluated using numerical integration techniques. As a result, the global matrices of the system can be assembled using the standard technique in the finite element method. Thus, the global matrix equation for the mass transport in porous media can be expressed as

$$
[A]\{C\} + [R]\{\frac{\partial C}{\partial t}\} = \{f\} \tag{13.42}
$$

where

$$
[A] = [G] + [H] \tag{13.43}
$$

The solution of Equation 13.42 in time domain can be carried out by the finite difference approach for $\{\frac{\partial C}{\partial t}\}$.

$$
\{\frac{\partial C}{\partial t}\} = \frac{1}{\Delta t}(\{C\}^{t+\Delta t} - \{C\}^t) \tag{13.44}
$$

where the superscript represents the time level and Δt is the time step.

Having used the finite difference approach for the first derivative of C with respect to t, it is necessary to specify the time level in the time interval between t and $t + \Delta t$ at which another term $\{C\}$ is evaluated. In general, the nodal contaminant concentration vector, $\{C\}$, can be approximately anywhere between t and $t + \Delta t$.

$$\{C\} = (1 - \alpha)\{C\}^t + \alpha\{C\}^{t + \Delta t} \tag{13.45}$$

where $0 \le \alpha \le 1$. If $\alpha = 1$, the solution is fully implicit. If $\alpha = 0$, the solution is fully explicit. In the case of $\alpha = 0$, Equation 13.44 is called a forward difference approximation relative to time t. In the case of $\alpha = 1$, Equation 13.44 is called a backward difference approximation relative to time $t + \Delta t$. However, if $\alpha = 0.5$, Equation 13.44 is called a central difference approximation relative to time.

The formula for the forward difference scheme can be derived by letting $\alpha = 0$ in Equation 13.45 and substituting Equations 13.44 and 13.45 into Equation 13.42.

$$[A]\{C\}^t + [R]\frac{1}{\Delta t}(\{C\}^{t + \Delta t} - \{C\}^t) = \{f\}^t \tag{13.46}$$

Therefore,

$$[R]\{C\}^{t + \Delta t} = [R]\{C\}^t - \Delta t[A]\{C\}^t + \Delta t\{f\}^t \tag{13.47}$$

Letting $\alpha = 1$ in Equation 13.45 and substituting Equations 13.44 and 13.45 into Equation 13.42, the formula for the backward difference scheme can be expressed as

$$[A]\{C\}^{t + \Delta t} + [R]\frac{1}{\Delta t}(\{C\}^{t + \Delta t} - \{C\}) = \{f\}^{t + \Delta t} \tag{13.48}$$

Thus,

$$([A] + \frac{1}{\Delta t}[R])\{C\}^{t + \Delta t} = \frac{1}{\Delta t}[R]\{C\}^t + \{f\}^{t + \Delta t} \tag{13.49}$$

Similarly, letting $\alpha = 0.5$ in Equation 13.45 and substituting Equations 13.44 and 13.45 into Equation 13.42, the formula for the central difference scheme can be written as

$$[A](\frac{1}{2}\{C\}^t + \frac{1}{2}\{C\}^{t + \Delta t}) + [R]\frac{1}{\Delta t}(\{C\}^{t + \Delta t} - \{C\}^t) = \{f\}^{t + 0.5\Delta t} \tag{13.50}$$

Equation 13.50 can be further written as

$$(\frac{1}{2}[A] + \frac{1}{\Delta t}[R])\{C\}^{t + \Delta t} = (\frac{1}{\Delta t}[R] - \frac{1}{2}[A])\{C\}^t + \{f\}^{t + 0.5\Delta t} \tag{13.51}$$

Apparently, if the contaminant concentration distribution of a system is known for a time instant t, the contaminant concentration distribution of the system for another time instant $t + \Delta t$ can be found out using Equation 13.47, Equation 13.49 or Equation 13.51, depending on whichever difference scheme is chosen.

If the advection terms in Equation 13.28 play a predominant role in the mass transport process, the Galerkin finite element approach as mentioned above fails in solving the problem because its solution exhibits pronounced oscillatory behaviour and excessive numerical dispersion. These oscillations and numerical dispersion can only be avoided by using a drastic refinement of the finite element mesh, which makes the conventional Galerkin finite element approach inefficient in calculation. In order to overcome this problem, Huyakorn (1977), Huyakorn & Nilkuha (1979) presented an upwind finite element scheme. This scheme differs from the conventional Galerkin scheme in the following two aspects: (1) spatial discretization is performed through a general weighted residual technique which employs asymmetric weighting functions and yields upwind weighting effect for the advection term in the mass transport equation; (2) the time derivative term of the equation is weighted using the standard trial functions which are symmetric functions.

Using the upwind finite element scheme, Equation 13.28 can be written for an element as

$$\iint_A [W]^T [D_x \frac{\partial^2 C}{\partial x^2} + D_y \frac{\partial^2 C}{\partial y^2} - \bar{v}_x \frac{\partial C}{\partial x} - \bar{v}_y \frac{\partial C}{\partial y}] dA$$

$$-\iint_A [N]^T \{\frac{\partial C}{\partial t}\} dA = 0$$

(13.52)

where $[W]$ is the upwind weighting function matrix of the element; while $[N]$ is the conventional weighting function matrix as shown in Equation 13.29.

Using the same procedure as in the conventional Galerkin finite element approach, the property matrices of the upwind finite element can be derived as

$$[G_e] = \iint_A (D_x \frac{\partial [W]^T}{\partial x} \frac{\partial [N]}{\partial x} + D_y \frac{\partial [W]^T}{\partial y} \frac{\partial [N]}{\partial y}) dA$$

$$[H_e] = \iint_A (\bar{v}_x [W]^T \frac{\partial [N]}{\partial x} + \bar{v}_y [W]^T \frac{\partial [N]}{\partial y}) dA$$

$$[R_e] = \iint_A [N]^T [N] dA$$

(13.53)

$$\{f\}^e = \int_S [W]^T (D_x \frac{\partial C}{\partial x} n_x + D_y \frac{\partial C}{\partial y} n_y) dS$$

It is noted that from the computational point of view, all procedures employed in the Galerkin finite element approach can be directly used in the upwind finite element scheme so long as the weighting function matrix [N] is replaced by the upwind weighting function matrix [W] in the appropriate positions. Therefore, the following discussion is focused on the construction of the upwind weighting functions of the element.

For a one-dimensional transport problem, a linear element is shown in Figure 13.5. It is assumed that the flow velocity is in the positive direction from node 1 to node 2. Let the upwind weighting functions for nodes 1 and 2 be expressed in the form:

$$W_1(x) = N_1(x) - F(x)$$
$$W_2(x) = N_2(x) + F(x)$$

(13.54)

where

$$F(x) = a\frac{x^2}{h^2} + b\frac{x}{h} + c \qquad (0 \le x \le h)$$

(13.55)

where x is a coordinate and $F(x)$ is a piecewise quadratic function with a, b and c as undetermined coefficients; h is the length of the element. The function $F(x)$ must be chosen to satisfy the following requirements:

$$F(0) = F(h) = 0$$

(13.56)

Substituting Equation 13.56 into Equation 13.55 yields

$$F(x) = b(-\frac{x^2}{h^2} + \frac{x}{h})$$

(13.57)

where b is the remaining undetermined coefficient. For convenience in performing subsequent mathematical analysis, the following parameter is introduced.

$$\alpha = \frac{b}{3}$$

(13.58)

Substituting Equations 13.57 and 13.58 into Equation 13.54 yields

$$W_1(x) = N_1(x) + 3\alpha\frac{x^2}{h^2} - 3\alpha\frac{x}{h}$$

$$W_2(x) = N_2(x) - 3\alpha\frac{x^2}{h^2} + 3\alpha\frac{x}{h}$$

(13.59)

where $N_1(x)$ and $N_2(x)$ are the conventional shape functions and given by

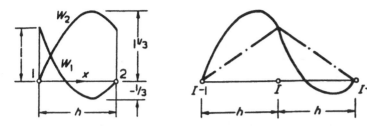

Figure 13.5. One-dimensional upwind finite element.

$$N_1(x) = 1 - \frac{x}{h}$$

$$N_2(x) = \frac{x}{h}$$

(13.60)

For a value of $\alpha = 1$, the corresponding upwind weighting functions are shown in Figure 13.5. It is evident that when α is non-zero, the upwind weighting function for any node I in the mesh is asymmetric with respect to the node. More weighting is given to the upstream portion of the line segment consisting of nodes $I-1$, I and $I+1$ when α is positive. For $\alpha = 1$, it can be proven that

$$\int_{x_{I-1}}^{x_I} W_I(x)dx = \int_0^h W_2(x)dx = h$$

$$\int_{x_I}^{x_{I+1}} W_I(x)dx = \int_0^h W_1(x)dx = 0$$

(13.61)

The numerical experience indicated that to dampen the numerical oscillation, the sign of α must be chosen in accordance with the sense of the direction of flow velocity, i.e. $\alpha > 0$ when $\bar{v}_x > 0$ and vice versa. Thus, for any node I in the mesh, greater weighting must always be given to the upstream element which contributes to that node, if oscillation is to be dampened. In view of this, the weighting functions so determined are termed upwind weighting functions. Besides, it has been demonstrated from a steady-state one-dimensional mass transport problem that the numerical solution is free of oscillation if $\alpha \geq 1$ or $(1 - \alpha)\dfrac{\bar{v}_x h}{D_x} < 2$ when $\alpha < 1$. Since α directly controls oscillation, it may be termed a damping factor.

To establish numerical stability with minimum loss of accuracy, the value of α must be carefully chosen. The expression for optimum α was obtained theoretically from a steady-state one-dimensional mass transport problem by

Christie et al. (1976). It can be written in the following form:

$$\alpha_{opt} = \coth(\frac{\bar{v}_x h}{2D_x}) - \frac{2D_x}{\bar{v}_x h} \tag{13.62}$$

where coth stands for hyperbolic cotangent.

For a quadrilateral element shown in Figure 13.6, let ξ, η represent a local isoparametric coordinate system. Since the element belongs to the Lagrangian family, its upwind weighting functions can be obtained by taking appropriate products of functions in each coordinate. In the process of forming such products, the fact that the damping factor α can vary from one element side to another should be taken into account. Thus, four different damping factors, namely $\bar{\alpha}_1$, β_1, $\bar{\alpha}_2$ and β_2, are assigned to sides 3-4, 4-1, 2-1 and 3-2 of the element in Figure 13.6, respectively. The sense of direction of flow velocity which corresponds to the positive values of the damping factors needs also to be considered for each side so that $\bar{\alpha}_i$ and β_i (i=1, 2) can be expressed as

$$\bar{\alpha}_i = \alpha_i^{opt} \text{sign}(\bar{v})$$
$$\bar{\beta}_i = \beta_i^{opt} \text{sign}(\bar{v}) \tag{13.63}$$

where α_i^{opt} and β_i^{opt} (i=1, 2) are the optimal values of the upwind parameters for the related element sides; $\bar{v}$ is the corresponding average scalar velocity of the element side and can be determined using the following formula.

$$\bar{v} = \frac{1}{2}(\{v_i\} + \{v_j\}) \cdot \{l_{ij}\} \tag{13.64}$$

where i and j are the node numbers of the element side; $\{v_i\}$ and $\{v_j\}$ are the velocities of nodes i and j respectively; $\{l_{ij}\}$ is a direction vector of this particular side in the local coordinate system and its positive direction is in coincidence with that of the local coordinate system of the element.

Taking node 1 as an example, the upwind weighting function for this node can be expressed as

$$W_1(\xi,\eta) = F_1(\xi,\bar{\alpha}_2)F_2(\eta,\bar{\beta}_1) \tag{13.65}$$

where the expressions for F_1 and F_2 can be derived directly from the one-dimensional upwind weighting function in Equation 13.59 in terms of ξ as follows.

$$F_1(\xi,\bar{\alpha}_2) = \frac{x}{h} - 3\bar{\alpha}_2\frac{x^2}{h^2} + 3\bar{\alpha}_2\frac{x}{h} \tag{13.66}$$

where

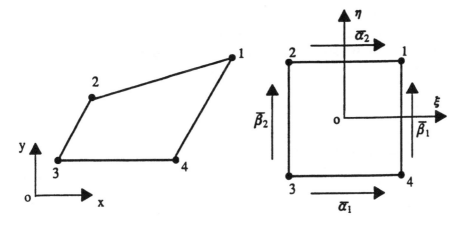

Figure 13.6. 4-node two-dimensional upwind finite element.

$$\frac{x}{h} = \frac{\xi + 1}{2} \tag{13.67}$$

Substituting Equation 13.67 into Equation 13.66 yields

$$F_1(\xi,\bar\alpha_2) = \frac{1}{4}[(1+\xi)(-3\bar\alpha_2\xi + 3\bar\alpha_2) + 2] \tag{13.68}$$

Similarly, $F_2(\eta,\bar\beta_1)$ can be derived as

$$F_2(\eta,\bar\beta_1) = \frac{1}{4}[(1+\eta)(-3\bar\beta_1\eta) + 3\bar\beta_1) + 2] \tag{13.69}$$

Thus, Equation 13.65 can be further written as

$$W_1(\xi,\eta) = \frac{1}{16}[(1+\xi)(-3\bar\alpha_2\xi+3\bar\alpha_2)+2][(1+\eta)(-3\bar\beta_1\eta+3\bar\beta_1)+2] \tag{13.70}$$

In a same manner, $W_2(\xi,\eta)$, $W_3(\xi,\eta)$ and $W_4(\xi,\eta)$ can be derived as follows.

$$W_2(\xi,\eta)=\frac{1}{16}[(1+\xi)(3\bar\alpha_2\xi-3\bar\alpha_2-2)+4][(1+\eta)(-3\bar\beta_2\eta+3\bar\beta_2+2)]$$

$$W_3(\xi,\eta)=\frac{1}{16}[(1+\xi)(3\bar\alpha_1\xi-3\bar\alpha_1-2)+4][(1+\eta)(3\bar\beta_2\eta-3\bar\beta_2-2)+4] \tag{13.71}$$

$$W_4(\xi,\eta)=\frac{1}{16}[(1+\xi)(-3\bar\alpha_1\xi+3\bar\alpha_1+2)][(1+\eta)(3\bar\beta_1\eta-3\bar\beta_1-2)+4]$$

It should be noted that in order to obtain satisfactory solutions, the derivatives, $\dfrac{\partial W_i}{\partial \xi}$ and $\dfrac{\partial W_i}{\partial \eta}$ (i=1, 2, 3, 4), of the element must be evaluated in such a way that when differentiation is taken with respect to one particular coordinate, the values of the upwind parameters along the remaining coordinate must be set to zero. Therefore, the derivatives of the upwind weighting functions of the element can be expressed as

$$\frac{\partial W_1}{\partial \xi} = -\frac{1}{4}(1 + \eta)(3\bar{\alpha}_2 \xi - 1)$$

$$\frac{\partial W_2}{\partial \xi} = \frac{1}{4}(1 + \eta)(3\bar{\alpha}_2 \xi - 1)$$

$$\frac{\partial W_3}{\partial \xi} = \frac{1}{4}(1 - \eta)(3\bar{\alpha}_1 \xi - 1)$$

$$\frac{\partial W_4}{\partial \xi} = -\frac{1}{4}(1 - \eta)(3\bar{\alpha}_1 \xi - 1)$$

(13.72)

and

$$\frac{\partial W_1}{\partial \eta} = -\frac{1}{4}(1 + \xi)(3\bar{\beta}_1 \eta - 1)$$

$$\frac{\partial W_2}{\partial \eta} = -\frac{1}{4}(1 - \xi)(3\bar{\beta}_2 \eta - 1)$$

$$\frac{\partial W_3}{\partial \eta} = \frac{1}{4}(1 - \xi)(3\bar{\beta}_2 \eta - 1)$$

$$\frac{\partial W_4}{\partial \eta} = \frac{1}{4}(1 + \xi)(3\bar{\beta}_1 \eta - 1)$$

(13.73)

Having obtained the upwind weighting functions of the element and their derivatives, the property matrices of upwind finite element can be straightforwardly evaluated using the same procedure as described in the conventional finite element method. Thus, the solution for the mass transport problems in a porous medium can be obtained.

NUMERICAL EXAMPLE AND DISCUSSIONS

In order to illustrate how the finite element method is applied to solve mass transport problems in porous media, a practical problem, which is sea water or contaminant water intrusion into an aquifer of fresh groundwater in a lowland area, is considered in this section.

Figure 13.7 shows the physical model for this illustrative example. This problem can be simplified as a one-dimensional solute transport problem. Supposing the solute is located on the surface at $x=0$ and that the unidirectional flow is along the horizontal direction, the problem is discretized into finite elements shown in Figure 13.8, where the solute will propagate from its source

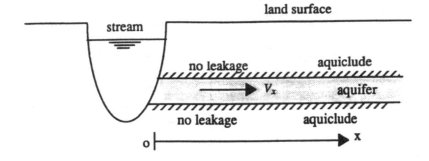

Figure 13.7. Physical model of a solute transport problem.

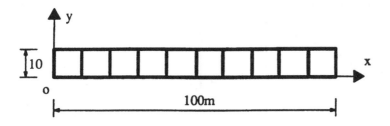

Figure 13.8. Discretized model of a solute transport problem.

(x=0) to the boundary (x=100 m) due to advection and dispersion. Since two-dimensional rectangular elements are used in the analysis, no-flow conditions along the two boundaries (y=0 and y=10 m) should be imposed for the calculation. To avoid the influence of the artificial boundary at x=100 m on the numerical results, the time range of interest and the location of the artificial boundary should be chosen appropriately. The problem is discretized by 10 two-dimensional upwind finite elements and the related boundary conditions are

$$C(x,t) = C_0 = 10 \ mg/cm^3 \quad at \ x = 0$$
$$C(x,t) = 0 \quad at \ x = 100 \ m$$

(13.74)

The analytical solution for this problem is available (Ogata & Banks 1961) and can be written as

$$C(x,t) = \frac{C_0}{2}[\exp(\frac{\bar{v}_x x}{D_x})\mathrm{erfc}(\frac{x + \bar{v}_x t}{2\sqrt{D_x t}}) + \mathrm{erfc}(\frac{x - \bar{v}_x t}{2\sqrt{D_x t}})]$$

(13.75)

where erfc is the complementary error function and can be expressed as

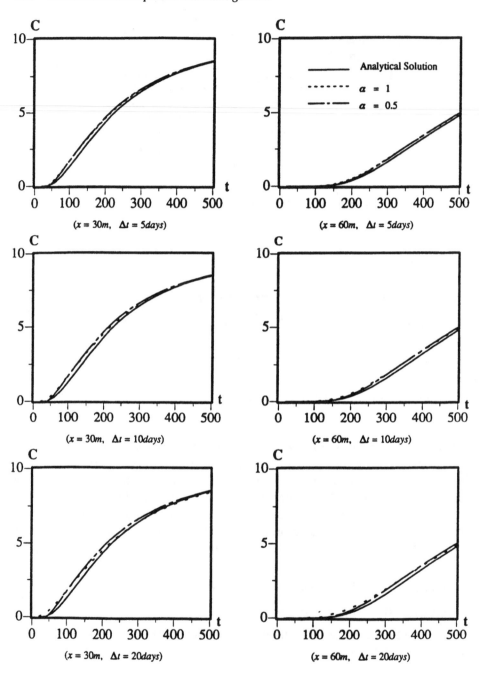

Figure 13.9. Concentration versus time for the solute transport problem.

$$\text{erfc}(z) = \frac{2}{\sqrt{\pi}} \int_z^\infty \exp(-u^2)du \qquad (13.76)$$

For the purpose of numerical calculation, the following parameters are used in the analysis: the average velocity of flow in the x direction $\bar{v}_x = 0.1\ m/day$; the dispersion coefficient $D_x = 1\ m^2/day$. Three different time intervals, namely $\Delta t = 5\ days$, 10 days and 20 days, are considered to examine the effect of time intervals on the numerical results. For the discretization of time, the forward difference scheme ($\alpha = 0$), the central difference scheme ($\alpha = 0.5$) and the backward difference scheme ($\alpha = 1$) are used in the calculation. Figure 13.9 shows the concentration through time for the solute transport problem. The detailed discussions about the numerical results and analytical solutions for this problem were reported in a previous paper (Valliappan & Zhao 1992). According to the related discussions, the following conclusions have been obtained: (1) the finite element method can provide valuable solutions for mass/contaminant transport problems in porous media; (2) either the central difference scheme or the backward difference scheme is suitable for the upwind finite element simulation of mass/contaminant transport problems since the CPU time can be reduced due to the use of large time increment; (3) compared to the central difference scheme or the backward difference scheme, the forward difference scheme is not accurate enough for the upwind finite element analysis except for using smaller time intervals.

REFERENCES

Bear, J. & A. Verruijt 1987. *Modelling groundwater flow and pollution.* Kluwer Academic Publisher.

Bear, J. & Y. Bachmat 1990. *Introduction to modelling of transport phenomena in porous fractured media.* Kluwer Academic Publisher.

Bresler, E. 1973. Simultaneous transport of solutes and water under transient unsaturated flow conditions. *Water Resour. Res.* 9: 975-986.

Bresler, E. 1975. Two-dimensional transport of solutes during nonsteady infiltration from a trickle source. *Soil Sci. Soc. Amer. Pro.* 39: 604-613.

Carnahan, C. L. & J. S. Remer 1984. Nonequilibrium and equilibrium sorption with a linear sorption isotherm during mass transport through an infinite porous medium: some analytical solutions. *J. Hydrol.* 73: 227-258.

Christie, I., D. F. Griffiths, A. R. Mitchell & O. C. Zienkiewicz 1976. Finite element methods for second order differential equation with significant first derivatives. *Int. J. Num. Meth. Eng.* 10: 1389-1396.

Cleary, R. W. & M. J. Ungs 1978. Analytical methods for ground water pollution and hydrology. *Water Resour. Prog. Rep.* 78-WR-15. Princeton Univ., Princeton, N. J.

Davidson, J. M., D. R. Baker & G. H. Brusewitz 1975. Simultaneous transport of water and absorbed solutes through soil under transient flow conditions. *Trans. Amer. Soc. Agr. Eng.* 18: 535-539.

Duguid, J. O. & M. Reeves 1977. A comparison of mass transport using average and transient rainfall boundary conditions. *Proc. of 1st int. conf. on finite elements in water resour:* 2.25-2.35.

Giesel, W., M. Renger & O. Strebel 1973. Numerical treatment of the unsaturated water flow equation: comparison of experimental and computed results. *Water Resour. Res.* 9: 174-177.

Gillham, R. W., A. Klute & D. F. Heermann 1979. Measurement and numerical simulation of hysterestic flow in a heterogeneous porous medium. *Soil Sci. Soc. Amer. J.* 43: 1061-1067.

Hanks, R. J., A. Klute & E. Bresler 1969. A numerical method for estimating infiltration, redistribution, drainage and evaporation of water from soil. *Water Resour. Res.* 5: 1064-1069.

Hoa, N. F., R. Gaudu & C. Thirriot 1977. Influence of the hysteresis effect on transient flows in saturated-unsaturated porous media. *Water Resour. Res.* 13: 992-996.

Huyakorn, P. S. 1977. Solution of steady state, convective transport equation using an upwind finite element scheme. *Appl. Math. Modelling* 1: 187-195.

Huyakorn, P. S. & K. Nilkuha 1979. Solution of transient transport equation using an upstream finite element scheme. *Appl. Math. Modelling* 3: 7-17.

Ibrahim, H. A. & W. Brutsaert 1968. Intermittent infiltration into soils with hysteresis. *ASCE J. Hydraul. Div.* 94: 113-137.

Javandel, I., C. Doughty & C. F. Tsang 1984. *Groundwater transport: Handbook of mathematical models, water resources monograph series.* AGU, Washington, D. C.

Ogata, A. & R. B. Banks 1961. A solution of the differential equation of longitudinal dispersion in porous media. *U. S. Geol. Survey Prof. Paper* 411-A.

Perrens, S. J. & K. K. Watson 1977. Numerical analysis of two-dimensional infiltration and redistribution. *Water Resour. Res.* 13: 781-790.

Pickens, J. F., R. W. Gillham & D. R. Cameron 1979. Finite element analysis of the transport of water and solutes in tile-drained soils. *J. Hydrol.* 40: 243-264.

Remson, I., A. A. Fungaroli & G. M. Hornberger 1967. Numerical analysis of soil-moisture systems. *ASCE J. Irrig. Drain. Div.* 93: 153-166.

Rubin, J. 1967. Numerical method for analyzing hysteresis-affected, post-infiltration redistribution of soil moisture. *Soil Sci. Soc. Amer. Proc.* 31: 13-20.

Segol, G. 1977. A three-dimensional Galerkin finite element model for the analysis of contaminant transport in saturated-unsaturated porous media. *Proc. of 1st conf. of finite elements in water resour:* 2.123-2.144.

Staple, W. J. 1969. Comparison of computed and measured moisture redistribution following infiltration. *Soil Sci. Soc. Amer. Proc.* 33: 840-847.

Sykes, J. F. 1975. Transport phenomena in variably saturated porous media. *Ph. D. thesis.* Univ. of Waterloo, Canada.

Valliappan, S. & C. Zhao 1992. Numerical simulation of contaminant transport in porous fractured media. *Proc. of the ILT seminar on problems of lowland development:* 107-119.

Valliappan, S. & C. Zhao 1993. Effect of medium porosity on contaminant transport in porous fractured media. *Proc. of conf. on geotech. management of waste and contaminant:* 473-478.

Van Genuchten, M. Th. & W. J. Alves 1982. Analytical solutions of the one-dimensional convection-dispersion solute transport equation. *Tech. Bull.* U. S. Dep. Agric.

Wexler, E. J. 1989. Analytical solutions for one two and three dimensional solute transport in groundwater systems with uniform flow. *U. S. Geol. Surv. Rep.* 89-56.

Whisler, F. D. & A. Klute 1965. The numerical analysis of infiltration, considering hysteresis, into a vertical soil column at equilibrium under gravity. *Soil Sci. Soc. Amer. Proc.* 29: 489-494.

Zhao, C. & S. Valliappan 1992. Numerical simulation of contaminant transport problems in infinite media. *Proc. of the ILT seminar on problems of lowland development:* 121-126.

Zienkiewicz, O. C. 1977. *The finite element method.* McGraw-Hill, London.

14 WATER QUALITY MANAGEMENT

K. KOGA[*], W. LIENGCHARERNSIT[**] and H. ARAKI[*]
* *Saga University, Saga, Japan*
** *Kasetsart University, Bangkok, Thailand & Saga University, Saga, Japan*

INTRODUCTION

In water resources planning and management, water quality should be considered together with water quantity, since the quality of water affects its functional uses. If a polluted water source is considered to be used as raw water source for domestic water supply, high investment and operational costs are needed for water quality improvement. The functional use of highly deteriorated water bodies in many regions is only limited to navigation or wastewater disposal. On the other hand, changes in water quantity evidently affect water quality, and so in the control and management of water quality, clear understanding of the hydrologic regime of the water body is necessary.

Water quality management is concerned with conservation and enhancement of water quality and ecosystems in water bodies by means of controlling pollutant discharge from human activities, taking into consideration various environmental factors such as physical, chemical and biological characteristics of the polluting materials, flow characteristics and assimilative capacity of the water body.

In the control and management of water quality problems, various sources of pollutants and their characteristics must be investigated. Dispersion patterns of the discharged pollutants in the water body as well as the manner in which they affect water quality and aquatic ecology must be studied. Then, a number of management plans or policies are formulated and the economic, environmental, political and social impacts of each alternative are evaluated. Often, some mathematical models are used as an analytical tool for identifying and evaluating the proposed management alternatives so that the best possible decisions will be made.

This chapter deals first with water quality characteristics and their significance. Then, goals and standards for water quality management are addressed. Some water pollution control measures which are applicable to the lowland areas are described including wastewater collection and treatment system, biomanipulation, reservoir mixing to limit algal growth, dredging and chemical treatment of sediments, river rehabilitation and floodplain restoration, etc. Then, mathematical models which are a useful tool in water quality management are explained. Finally, the strategies for water quality management in the lowlands are introduced.

WATER QUALITY CHARACTERISTICS AND THEIR SIGNIFICANCE

Water quality characteristics are generally classified into 3 main groups, i.e. physical, chemical and biological characteristics. However, in the examination of water quality, two types of measures can be categorized, i.e. gross and specific measures (Tchobanoglous & Schroeder 1987). The gross measures describe some combined characteristics of water quality with no distinct analysis on an individual constituent. These include turbidity, color, odor, solids, acidity, alkalinity, total hardness, biochemical oxygen demand (BOD), chemical oxygen demand (COD), total organic carbon (TOC), etc. The specific measures concern with a single characteristic or an individual constituent, such as pH, temperature, dissolved oxygen, some cations and anions, microorganism species, etc.

The physical characteristics of water include those apparent properties and some physical impurities such as turbidity, solids, color, odor, electrical conductivity, temperature, etc. The chemical characteristics show the amounts of chemical constituents present in water. These include pH, acidity, alkalinity, hardness, BOD, COD, TOC, some organic and inorganic compounds, dissolved gases such as oxygen, carbon dioxide, hydrogen sulfide, ammonia, etc., dissolved cations such as calcium, magnesium, iron, manganese, sodium, potassium, ammonium, mercury, lead, chromium, cadmium, zinc, arsenic, etc., dissolved anions such as hydroxide, sulfate, chloride, nitrite, nitrate, bicarbonate, carbonate, cyanide, bromide, fluoride, phosphate, etc. The biological characteristics indicate the presence of some microorganisms including bacteria, fungi, algae, protozoa, worms, viruses, etc.

In general, our first impression of water quality is based on its physical characteristics. Turbidity, color and odor indicate the presence of impurities in water. Though turbid water may not be dangerous for health, it is not attractive for direct consumption and people will try to find other water sources which may not be safe to drink. Therefore, in the drinking water quality standards, some limits are specified for these water quality parameters.

The solids in water can be classified as suspended solids and dissolved solids, or in more accurate terms, filterable and nonfilterable solids. Suspended solids include silt and clay from eroded ground and river banks or bottom. Dissolved solids usually occur from various sources, including contamination of saltwater from the sea, dissolution of carbonate rocks, as well as organic and inorganic matters in the soil, etc. The removal of solids is of great concern in water treatment, especially in processing industries which require water of high quality. The suspended solids in wastewater normally include organic matter which can decay and consume dissolved oxygen in water. When the concentration of suspended solids is high, it is more difficult to achieve effective bacterial disinfection since the solids provide the attached bacteria a protective barrier. Thus, the removal of suspended solids is also important in wastewater treatment.

The temperature of water is also an important parameter. Water temperature affects the natural ecosystems as well as chemical and biological reaction rates. The performance of biological wastewater treatment processes is related to the temperature of wastewater. In the cold regions, changes in water temperature in the fall and the spring may cause overturning of water and affect water quality in a lake.

The principal chemical constituents found in natural water include Ca^{+2}, Mg^{+2}, Na^+, K^+, Fe^{+2}, Mn^{+2}, HCO_3^-, CO_3^{-2}, Cl^-, SO_4^{-2}, and NO_3^-. Some heavy metals and toxic substances are found in water contaminated with industrial waste. Organic compounds from municipal wastewaters are a major cause of water pollution. Trace matters such as nitrogen and phosphorus in various forms, though present in small amounts, can cause eutrophication problems which subsequently affect water quality and ecosystems in the water bodies. Dissolved gases, especially oxygen, carbon dioxide, ammonia and hydrogen sulfide, are also an important factor in water quality management. These chemical constituents affect the functional uses of water.

The pH of water plays an important role in almost every phase of environmental engineering practice. In the field of water supply, it is a factor that must be considered in chemical coagulation, disinfection, water softening, and corrosion control. In wastewater treatment employing biological processes, pH must be controlled within a range favorable to the particular organisms involved. Usually, the chemical processes used to precipitate some dissolved ions or oxidize certain substances require that the pH be controlled within some favorable ranges. Most chemical equilibria are a function of pH. Changes in the pH value in natural waters caused by eutrophication or industrial wastewater disposal may have great impacts on the aquatic ecosystem.

Hardness of water is caused by divalent metallic cations, mainly Ca^{+2} and Mg^{+2}. These cations, together with some anions such as bicarbonates, may precipitate and form scales in boilers. Hard waters consume more soap before lather is formed. Therefore, hardness is an important factor to be considered in determining the suitability of water for domestic and industrial uses. Sanitary engineers must consider whether a softening process is needed or not in water treatment.

Though Fe^{+2} and Mn^{+2} are normally present in low concentrations in most natural waters, the oxidation to Fe^{+3} and Mn^{+4} forms colloidal precipitates which are unacceptable from the aesthetic viewpoint. These colloidal precipitates interfere with laundering operations and impart objectionable stains to plumbing fixtures. Removal of iron and manganese is normally practiced in water treatment, particularly when groundwater is used as a raw–water source.

Biochemical oxygen demand (BOD) is a water quality parameter used to determine the strength of wastewater. It is defined as the amount of oxygen required by bacteria while stabilizing decomposable organic matter under aerobic condition. The BOD value is the major criterion used in water pollution control where organic loading must be restricted to maintain desired dissolved oxygen levels (Sawyer & McCarty 1978). The BOD data are important in the design of a wastewater treatment plant. The BOD test is normally practiced to determine the performance of wastewater treatment systems.

Chemical oxygen demand (COD) is also an important parameter used in measuring the strength of domestic and industrial wastes. It is equivalent to the total amount of oxygen required for oxidizing all organic matter in wastewater. Though the COD test is based upon the action of strong oxidizing agents under acidic condition, it is very useful in determining the performance of wastewater treatment facilities due to the short time required for the analysis as compared with the BOD test.

Dissolved oxygen in water is vitally essential for all aquatic lives. It is one of the most important parameters used in water quality management. It is desirable to maintain the dissolved oxygen concentration at the level favorable for the growth and reproduction of fish and other aquatic organisms. In biological degradation, dissolved oxygen is the factor that determines whether the processes are brought about by aerobic or anaerobic microorganisms. The performance of aerobic treatment processes depends on the presence of dissolved oxygen. It also plays an important role in oxidation processes in water, e.g. oxidation of Fe^{+2} and Mn^{+2}, etc.

The compounds of nitrogen and phosphorus in domestic and industrial wastes as well as in natural waters are of great concern nowadays since they play an important role in the eutrophication processes. Nitrogen compounds also consume some dissolved oxygen in water during degradation. High nitrate content in drinking water will cause methemoglobinemia in infants (Sawyer & McCarty 1978). It is also found that ammonia is toxic to fish and some aquatic lives. Since phosphorus and nitrogen are essential elements for the growth of algae, control of their concentrations in the aquatic environment is important. So, data on nitrogen and phosphorus contents in wastewater are necessary for the design and operation of wastewater treatment facilities.

Chloride is commonly found in natural waters in widely varying concentrations. It is dissolved from the top soils and rocks and finally drained into the seas or oceans which serve as the main storages of chloride. Intrusion of sea water into estuaries and coastal groundwater aquifers and spray from the seas or oceans carry chloride back to inland areas. Chloride is present in human excreta resulting from the consumed food and water. Thus, domestic sewage effluent adds a remarkable amount of chloride to receiving waters. Usually, chloride at a reasonable concentration is not harmful to humans. However, high chloride concentration causes salty taste in water which is objectionable. Thus, the chloride content in water is one important parameter that determines suitability of the water body for being used as a source of water supply.

The principal forms of sulfur compounds encountered in the aquatic environment are sulfate and hydrogen sulfide. Sulfate is commonly found in most natural waters. It is utilized by plants and microorganisms to produce cell tissue. In turn, plants are consumed by humans and animals. Decaying of waste products as well as dead plants and animals under anaerobic condition will produce sulfide. In the presence of oxygen, sulfide will be oxidized by aerobic bacteria to form sulfite and sulfate. On the other hand, in the absence of dissolved oxygen and nitrate, sulfate serves as a source of oxygen for biochemical oxidations produced by anaerobic bacteria and is reduced to sulfide ion, which establishes an equilibrium with hydrogen ion to form hydrogen sulfide (Sawyer & McCarty 1978). At low pH the equilibrium will shift toward the formation of free hydrogen sulfide and can cause serious odor problems when it is present in a significant amount. Hydrogen sulfide also causes corrosion in concrete sewers when it escapes from the wastewater to collect at the walls and crown of sewers, and is oxidized to sulfuric acid which is a strong acid and corrosive to concrete.

Discharges of toxic metals and toxic organic compounds into natural waters have been significantly increased in the past few decades due to rapid industrialization and increasing use of chemicals in agriculture. Among these, heavy metals such as mercury, lead, chromium, zinc, cadmium, etc. and halogenated organic compounds such as dioxin, PCB (polychlorinated biphenyl), phenols, DDT and other pesticides have been recognized for their harmfulness.

In addition to the above mentioned principal chemical constituents, some minor minerals both ionic and non-ionic species are also found in natural waters. Some of them have a more or less significant role in the functional uses of water. These include aluminium, silica, boron, fluoride, arsenic, cyanide, and some radioactive substances.

The important biological parameters are measurements of coliform group of bacteria, including *Escherichia coli* which are found in the intestinal tract of humans and other warm-blooded animals. A few techniques have been used for the enumeration of microorganisms, e.g. standard plate count method, multiple-fermentation tube method, membrane-filter method, etc. In addition to the coliform group of bacteria, some biological parameters related to water-borne diseases such as pathogenic microorganisms and aquatic organisms that serve as hosts of some parasites, as well as ecologically related species such as floating algae are also determined for the purposes of disease prevention and water quality management.

SOURCES OF POLLUTANTS

The sources of pollutants that are discharged into natural waters can be classified into 2 main groups, namely point sources and non–point sources. The point sources include municipal and industrial wastewaters which are collected by sewerage systems and conveyed to discharge into the receiving waters, with or without treatment, at some locations. The non–point sources are normally referred to the polluted waters that flow into the natural waters at multiple locations, such as surface runoff from agricultural lands and urban areas. Usually, reduction of non–point source pollution by using treatment facilities is not economically feasible.

The discharged pollutants can be classified into some categories including oxygen–demanding material, nutrients, pathogenic organisms, suspended solids, salts, toxic metals, toxic organic compounds, and heat (Davis & Cornwell 1991). The decaying of biodegradable organic matter under the aerobic condition will consume dissolved oxygen in water. Thus the dissolved oxygen concentration of the water that contains organic wastes tends to reduce which will affect the survival of most aquatic organisms. The main sources of the oxygen–demanding materials are municipal wastewaters and most industrial wastes, particularly those from food–processing industries.

Among various nutrient substances, nitrogen and phosphorus are of primary concern since they are essential for the growth of algae. The major sources of these nutrients include municipal wastewaters and agricultural runoff. Most of the organic nitrogen present in the municipal wastewaters is in the form of proteins and their degradation products. The common forms of phosphorus compounds encountered in water environment are polyphosphates and orthophosphates. Polyphosphates are used for corrosion control in some public water supplies and industrial cooling water systems. They are also used as "builders" in some heavy–duty synthetic detergents. Orthophosphates occur from the hydrolysis of polyphosphates.

Many important communicable diseases are transmitted by water. These include cholera, typhoid fever, bacillary dysentery, helminthal infections, as well as some protozoal and viral infections. Discharge of excreta from sick persons or animals into the natural waters will cause spreading of the diseases. The main point sources of pathogenic organisms are hospitals and slaughterhouses.

Suspended solids entering the receiving waters come from many sources, either natural or man–made. These include soil erosion from watershed areas and river banks and bottom during heavy storms, discharges of wastewaters from strip mining, construction activity, logging operation, etc. The colloidal form of the suspended solids will result in turbid water while larger particles will deposit and may affect functional uses of the water bodies.

Natural water, either surface or groundwater, normally contains a certain amount of salts resulting from dissolution of soluble salts in soils or rocks which it flows through. Intrusion of saltwater from the seas or oceans into coastal groundwater aquifers and estuaries, as well as the discharge of wastewater from some human activities such as mining and industries are also the major sources of salts in inland waters. In some regions, the amount of dissolved salts has reached the levels which make the water unsuitable for certain usages.

Toxic substances including heavy metals and toxic organic compounds mainly occur from some industrial processes and runoff from agricultural lands which utilize pesticides in farming. Some heavy metals and toxic substances may enter groundwater aquifers from leaky hazardous waste landfill sites or from land disposal of wastewaters. Due to their high resistance to biodegradation, these toxic substances can persist and accumulate in the environment. In some areas, high contents of these substances have been found in river deposits and also in aquatic food chain including shellfish and fish, making them unsafe for consumption.

Thermal pollution is also recognized as one of the environmental problems. Discharge of cooling water from industries or thermal power plants can cause adverse impacts on the aquatic ecosystems. Many kinds of fish can live only in cold water. The increase in water temperatures in a portion of the river resulting from the discharge of hot water can completely block migration of these fish upstream and so affect their spawning. High water temperatures may affect reproduction of algae and other aquatic organisms. The rates of chemical and biological reactions in water are normally increased with temperature.

In water quality management, all the pollutant sources must be identified and pollutant loadings from these sources must be evaluated. Dispersion of the discharged pollutants and their reactions in water, as well as their effects on water quality and aquatic ecosystems should be clearly understood. To improve and maintain water quality in the receiving waters, control of pollutant loadings is necessary.

WATER QUALITY OBJECTIVES

Water quality requirements

In water quality management, requirements for water must be justified, basically depending upon the purposes of water uses, e.g. for drinking, domestic consumption, fishing, irrigation, industrial water, etc. Drinking water should not contain impurities in hazardous concentrations and it must be free from poisonous and undesirable substances including harmful microorganisms. It must also be significantly free from turbidity, color, taste and odor. The drinking water quality standards normally specify the acceptable limits of significant physical and chemical parameters as well as the required bacteriological quality of water delivered to the consumers. Evaluations of bacteriological quality are based upon quantitative tests for the presence of members of the coliform group of organisms, since the large number of coliform organisms in human excreta makes them excellent biological indicators of potential contamination by pathogenic enteric microorganisms.

The required industrial water quality depends upon purposes of water use. Water use for process purposes or for boiler feed must be of good quality. In some cases, industrial water must have a lower content of dissolved salts than that might be allowed in drinking water. On the other hand, water used for the cooling purposes may not be of good quality.

Irrigation water should not contain toxic substances including heavy metals which can accumulate in the plants and affect human health, and it should not contain too high levels of some organic and inorganic substances which may affect plant growth. Excessive salinity at root–zone level leads in succession to leaf burn, leaf drop, and plant destruction. Salinity effects are largely osmotic. High sodium exchange breaks down soil structures, seals pores, and interferes with drainage; in extreme cases of soil breakdown the pH of the soil may rise to the level of alkali soils. Although traces of boron are essential for plant growth, high concentrations are injurious. Chlorides are generally more injurious than sulfates, because chlorides are relatively more soluble and more toxic to some plants (Fair et al. 1968).

Nowadays, intensive attention on ecological aspects in water resources management is of great concern, especially in developed countries where high percentages of population are served by sewers and most domestic sewage, and industrial wastes are treated prior to discharging into receiving waters. In such countries, the primary concern of people shifts from the provision of food, shelter, hygiene and education to recreation, aesthetics, emotional well–being, etc. and there is room for protection of natural and ecological quality for its own sake, not just for economic reasons (Lijklema 1993). Presently, eutrophication is a key issue in the ecological aspect. This is caused by the discharge of large amounts of nutrients, particularly nitrogen and phosphorus, into surface waters. Though major point sources of these nutrients have been reduced by means of tertiary wastewater treatment systems and introduction of phosphate free detergents, discharge from non–point sources including agricultural lands is difficult to control. Furthermore, large stock of nitrate and phosphate adsorbed in the ground during the past few decades have been continuously released into surface waters. Thus, eutrophication problems in natural surface waters will continue for a remarkable period. Nevertheless, it should be noted that in water quality management, not only should organic matter and toxic substances in wastewaters be removed, but nutrient loading should also be reduced to a certain extent to minimize the eutrophication problems and enhance the biodiversity and ecological quality.

Water quality standards

To serve as regulations or guidelines for water quality management, a number of water quality standards have been established in various countries. Establishment of suitable water quality standards is a difficult task. Different types of water quality standards are set related to the intended use of water and objectives of the establishment. Not only the impacts on human health and environmental quality are considered, but also other factors such as economic and technical attainability, reliability of the analysis, legal enforceability, etc. are used as bases in setting the standards. In general, water quality standards can be classified into 3 main groups, i.e. drinking water quality standards, effluent standards, and stream standards or environmental quality standards.

The drinking water quality standards are set to serve as objectives for water treatment to protect the public health. These standards are concerned with constituents which have direct effects on human health and with those contaminants which affect the aesthetic qualities of drinking water. To serve as a guideline for water quality improvement for drinking purpose, the World Health Organization (WHO) has prepared the International Standards. The established standards were first published in 1958 and the current WHO standards were issued in 1984. In addition to establishing bacteriological, chemical, physical and radiological requirements for drinking water, the WHO standards also include standards of quality for water sources and approved methods for examination of water (ASCE & AWWA 1990).

The effluent standards are established to control the discharge of pollutants from domestic and/or industrial sources. In general, these standards are set in terms of permissible limits of harmful substances for protecting human health and items related to protection of living environment. All dischargers are required to reduce the strength of their wastewater to the specified limits before discharging, otherwise they will be subjected to some penalty.

The environmental quality standards are established to provide goals for water quality management based on the intended use of water. They provide the administrative targets to be reached as common objectives of a diversity of measures undertaken to recover or protect the quality of public waters. These standards must be such that ceaseless efforts can be made in pursuit of their attainment. The standards are required to be fulfilled in a certain time period and maintained thereafter. The period required for attainment of these standards is determined according to the degree of contamination at the time of designating each water area. Similar to the effluent standards, the environmental quality standards are set by taking into consideration those substances related to the protection of human health and living environment.

WATER QUALITY IMPROVEMENT AND POLLUTION CONTROL

Water quality improvement, either water treatment for domestic and industrial uses or wastewater treatment to reduce the strength of pollutants before final disposal, plays an important role in water quality management. There exist several methods for improving water quality which can be broadly classified as physical, chemical and biological processes. The design of treatment facilities depends on raw water/wastewater quality characteristics and the desired water quality to be managed. Regardless of many advanced methods developed for water and wastewater treatment nowadays, the water pollution problem is still one of the major environmental issues in many countries, since large budget is required for these treatment facilities. Over the years, the context of water pollution control has been based on the so-called 'water quality standards oriented'. In recent years, another context based on the 'goal oriented' has been introduced. The use of treatment facilities for water quality improvement can be considered as the 'hardware' implementation. On the other hand, the 'software' implementation is also applicable in water quality management for desired sustainable environment. The purpose of this section is to present an overview of various water quality improvement and pollution control systems that are applicable to the lowlands.

Wastewater collection and treatment

Contaminants in wastewater can be removed by a combination of physical, chemical and biological processes. Fundamental principles and performance of these processes are described extensively in available literature. The main purpose of this section is to outline specific context on wastewater collection and treatment systems that are appropriate for the lowlands. In general, the systems should be selected based on wastewater volume and characteristics, required performance, construction and operation costs, as well as the local conditions including topography, climate, etc. The steps in designing the wastewater collection and treatment systems include: 1) collection of data which include topographic maps, existing land use patterns and future plans, population, commercial establishments and industries, water use, existing systems, etc., 2) selection of the appropriate systems, 3) selection of design criteria, 4) evaluation of wastewater flow rates and characteristics, 5) selection of location for the treatment facilities, 6) design of the wastewater collection system, 7) design of the treatment processes, and 8) preparation of bidding documents including detailed drawings. In the lowlands, however, difficulties are often encountered in the implementation, especially in system selection and design of the wastewater collection system. This is because implementation in the lowlands have some specific characteristics including: 1) existing surface water flow is very slow, 2) limitation on the slope of sewers due to flat terrain, 3) less population density in agricultural lands, 4) lack of water resources due to small storage capacity, and 5) high groundwater table. With these characteristics, the so-called 'decentralized' systems which include an on-site treatment unit are required in some areas to reduce the high construction and operation costs of the sewerage system. High groundwater table will result in high seepage into sewers and diluted wastewater might affect operation of the treatment processes. Advanced treatment processes are sometimes needed when the treated effluent will be reused in agricultural lands to cope up with the high water demand and, in some cases, to control the eutrophication problem.

Improvement of polluted water

Pollutants discharged from various sources can spread widely in the receiving waters. By diluting with natural water, the concentrations will decrease. However, with high pollutant loading, the dilution effect alone cannot help solve the water pollution problems. Treatment of polluted water by means of the conventional treatment facilities is not practical due to large volume of water. In order to improve the quality of polluted water in a natural water body, some control measures have been applied. These include biomanipulation, enhancement of self–purification function, reservoir mixing to reduce algal growth, dredging and chemical treatment of sediments, etc. An overview of the above mentioned control measures are presented in the following sections.

Biomanipulation

Eutrophication problem in lakes, ponds and reservoirs is one of the most serious water pollution problems. In order to solve this problem, the amount of discharged nutrients particularly nitrogen and phosphorus must be reduced. However, the accumulated nutrients in water and sediment can continue the complete cycle for a remarkable period. Therefore, in addition to reduction in the discharge loading, it is necessary to remove the accumulated nutrients. One of the best ways is to perform dredging of the deposited materials. However, large budget is normally needed. Another technique which has recently been introduced for algal growth control and lake restoration is the so–called biomanipulation. In this method, intervention in the structure of ecosystems is deliberated by the introduction or removal of some plant or animal species. In many eutrophic lakes, the predominance of algae, notably cyanobacteria which is less susceptible to grazing than phytoplankton, reduces sunlight for macrophytes. The loss of macrophytes and the poor sunlight condition are detrimental for fish of prey. Thus, their victims including planktivorous fish can flourish which suppress the zooplankton. Figure 14.1 illustrates the simplified food web of this aquatic ecosystem. Removal of the planktivorous fish and introduction of fish of prey together with the provision of shelter opportunities has been shown to be very successful in a number of field experiments (Lijklema 1993). In recent years, several studies have been conducted in the Netherlands to assess the performance of the biomanipulation on lake restoration (e.g. Hosper & Jagtman 1990, Hosper & Meijer 1993, Jeppesen et al. 1990, Meijer et al. 1993a & b, Van Donk et al. 1990, etc.).

Enhancement of self–purification function

Generally, the self–purification capacity in natural environment is limited by the characteristic of the environment. The enhancement of self–purification function can be made in a number of ways, e.g. lengthening the time of passage through a given stretch of water, stepping up the rate of reaeration, introducing

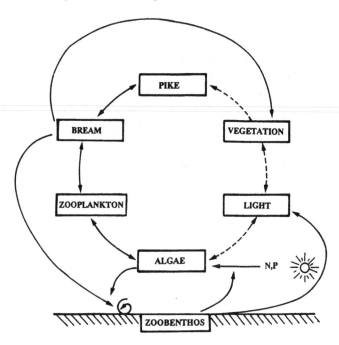

Figure 14.1. Simplified aquatic food web (after Lijklema 1993).

mechanical aeration, controlling the stratification of lakes, ponds, and impoundments, normalizing the accumulation of sludge deposits, etc. (Fair et al. 1968). Several experiments have been conducted to investigate the performance of some self–purification enhancement techniques. For example, Koga & Araki (1993) studied the performance of two methods for improving water quality in an open channel network, i.e. by inducing clean river water to dilute the polluted canal water and by installing biological contactors on the channel bed to increase the biological reaction rate. A significant improvement in water quality was reported from the dilution technique while a little improvement in water quality was observed from the introduction of the biological contactors. However, it is realized that the dilution method alone is not the best alternative for water pollution control, since it requires large volume of clean water and the polluted water is finally disposed into the river. A combination of pollutant reduction by some treatment facilities and dilution method should be considered. The use of mechanical aeration equipment has been implemented in many places to induce more oxygen from the atmosphere for maintaining aerobic condition, and to mix the upper layer of the water body for algal growth reduction.

Reservoir mixing

Undesirable eutrophication problem in lakes and reservoirs can be suppressed by vertical mixing which will reduce average light intensity and thus limit algal growth. Calculations and experiences have shown that, depending on temperature, irradiance and a few other factors, a mixing depth in the range of 10–12 meters is required (Lijklema 1993). Reservoir mixing is an economic algal control measure. The mixing can be readily be done by injection of air which has been found to be very effective in maintaining homogeneous conditions in the Biesbosch reservoirs in the Netherlands (Oskam 1993) and in some water supply reservoirs in Japan (Andou 1993).

Dredging and chemical treatment of sediments

For food web manipulations and reduced external loading to be effective, removal or treatment of lake sediments is necessary. In Sweden, the Netherlands and some other countries, a number of lakes have been dredged to remove soft and nutrient rich bottom material. Also, immobilization of phosphate by injection of calcium nitrate, aluminium or iron salts has been practiced, especially in water supply reservoirs (Lijklema 1993). Field experiments in the Petrusplaat reservoir in the Netherlands show that the use of copper sulfate and rooting up benthic cyanobacteria are successful methods for the control of geosmin production which has been indicated as the main cause of earthy–musty odor in reservoir water (Oskam 1993).

River restoration

Several regulation works have been implemented in lowland rivers for some purposes, such as straightening the channel to facilitate navigation or to bypass flood water, construction of dikes to prevent flooding or to reclaim more land, construction of weirs to divert water for agriculture and other uses. These activities have restricted the biodiversity and affected the river ecosystems. Nowadays, high attention is directed to the areas of environmental conservation and biodiversity. A number of river restoration projects have been implemented in some developed countries. These include the connection of previously cut–off river branches, diversion of peak flows from a meandering, ecologically attractive river through a somewhat higher lying straight channel to prevent undue scour and flooding of adjacent grounds, restoration of the more dynamic flooding, erosion and sedimentation patterns, etc. (Lijklema 1993). The main objective of river restoration is the conservation of nature. Successful reintroduction of plant and animal species typically found within a river ecosystem requires comprehensive studies and management of the channel, floodplain and tributaries (Kern 1992). Today, engineers and planners who are responsible for water resources development have a greater awareness of the impacts of the proposed projects on existing ecosystems. It is not only the aquatic community that benefits from a higher biodiversity and a more varied spatial distribution, but also terrestrial plants and animals living in the transition zone between water and land get better opportunities in the restoration projects.

Wetland treatment systems

A wetland is also a typical landscape in the lowlands where a variety of aquatic plants and animals are found. It is characterized by a comparatively shallow body of slow–moving water in which dense stands of water–tolerant plants such as water hyacinths or cattails are grown. The wetlands have been recognized regarding their significance on the aquatic ecosystems. In the context of water quality management, the wetlands are effective as wastewater treatment processes. Types of engineered wetlands used as aquatic treatment systems include artificial marshes, ponds, and trenches. Bacteria attached to the submerged roots and stems of aquatic plants growing in the wetlands are of particular importance in the removal of soluble and colloidal organic matters from a wastewater. The quiescent water conditions in the wetlands are suitable for the sedimentation of suspended solids. Nitrogen and phosphorus removal can be expected due to the absorption of aquatic plants. The ion exchange and adsorption capacity of the wetland sediments and the immersed parts of aquatic plants that reduce the perturbating effects of climatic variables contribute to the effectiveness of wetland treatment systems. In the design of the wetland treatment systems, three important aspects should be considered. These include pretreatment of wastewater, wastewater distribution, and mosquito control. It should be recognized that the wetlands have high self–purification capacity and are of great value in biodiversity and ecological development (Tchobanoglous & Schroeder 1987).

Groundwater quality management

Groundwater and surface water are closely related both in terms of water quantity and water quality. In the rainy season, surface water will percolate into underlying groundwater aquifers either from overland runoff or from water bodies. On the other hand, stored water in the aquifers will be gradually released when the piezometric head of groundwater is higher than surface water level which helps supply base flows in rivers during the dry periods. As regards water quality, polluted surface water may easily reach and pollute groundwater. Sea water intrusion depends on the rate of fresh water flow into the sea. It is thus obvious that the management should always include both resources simultaneously, incorporating each of them in the overall system according to their individual features.

In general, surface water is much more susceptible to man–made pollution than groundwater. However, when groundwater has been polluted, restoration of water quality either by removal of contaminated pollutants or by leaching with clean groundwater is a lengthy process and often practically impossible. This is because of the slow motion of groundwater and adsorption phenomena on the surface of the solid matrix, especially when fine grained materials like clay are present in an aquifer.

Saltwater intrusion into the coastal aquifers is a main groundwater pollution issue. Control of this phenomenon can be made by 1) reducing the rate of groundwater abstraction or relocating abstraction wells, 2) artificial recharge to increase the safe yield, to raise the piezometric level, and to increase the volume of fresh groundwater available for emergencies, etc., and 3) abstraction of saline and brackish groundwater to decrease the piezometric head of the saltwater and, thus, the interface between saltwater and fresh water is forced back to the sea side (Van Dam 1992).

In those areas with high possibility of groundwater contamination such as the areas around the solid waste and hazardous waste landfill sites, a properly planned groundwater monitoring program is necessary. The purpose of groundwater monitoring is to ensure that groundwater remains uncontaminated from the human activities. If contamination has been detected, early warning to the water users can be given and some measures to cleanup the contaminated aquifer or to prevent spreading of the contaminants can be undertaken at the beginning stage.

In groundwater management, the entire groundwater aquifer should be regarded as a single basin from the view point of its water balance. Withdrawal of groundwater from one well will affect the piezometric level in the adjacent wells. Pollutants reaching the aquifer may be transported and reach areas at large distances from the original source. Thus, the management of groundwater should be centralized and requires an appropriate legal and institutional framework. One cannot leave individual land owners to pump according to their needs or to dump pollutants on their land (Bear 1979).

MATHEMATICAL MODELS FOR WATER QUALITY MANAGEMENT

In an attempt to study the distribution patterns of the discharged pollutants in the receiving water body and to identify the optimal water quality management plans which best meet the specified objectives, several mathematical models have been formulated. The water quality models consist of a set of mathematical expressions which are developed to represent physical, chemical and biological processes concerning dispersion and interaction of substances in water. The principle of conservation of mass is the basis for these models. The purpose of this section is to present an overview of water quality models which have been utilized as an analytical tool in water quality management. More details about water quality modeling can be found in Bird et al. (1960), Biswas (1976), Connor & Brebbia (1977), Pinder & Gray (1977), Bear (1979), Loucks et al. (1981), Orlob (1983), Thomann & Mueller (1987), and others.

Types of water quality models

Two major groups of water quality models can be classified, namely simulation models and optimization models (Loucks et al. 1981). The simulation models are concerned with determination of spatial and temporal dispersion of the selected water quality constituents while the optimization models deal with evaluation of the most suitable management alternatives. The optimization models normally include a set of objective functions together with a number of constraints.

Besides the above mentioned classification, the water quality models can also be classified based on 1) the water quality parameters that are involved, 2) time–scale condition, i.e. steady–state or unsteady–state condition, 3) spatial dimensionality, i.e. one–, two–, or three–dimensional models, 4) the particular aspects of the hydrologic system that are simulated, and 5) the numerical techniques used in model formulation or the computational methods used to obtain the model solution.

The complexity of the water quality models depends on the constituents included in the models which can vary from a single constituent such as chloride, some heavy metals, etc. to a set of interrelated constituents such as BOD–DO, nitrogen compounds, phosphorus compounds, or the aquatic ecosystem which includes some significant nutrient elements and planktons. The multi–constituent models normally require more data and computer time, but they can provide a more detailed description of water quality characteristics resulting from various management policies.

The models can be classified as steady–state and unsteady–state models. The steady–state models are used in the case that hydrologic conditions, and water quality do not change with time. These models are usually simpler and require less computer time, and are normally applied for long–term planning. On the other hand, the unsteady–state models are used to evaluate dynamic or time–varying conditions. These models permit an evaluation of transient phenomena but normally require more input data and computer time.

The spatial dimensionality of the models depends on the assumptions pertaining to the mixing patterns in water bodies. Although all real systems are three–dimensional, sufficiently accurate results may be obtained from one– or two–dimensional models. In the one–dimensional model, complete mixing is assumed in the vertical and lateral directions. The two–dimensional models may assume complete mixing in the vertical direction as in the case of shallow water bodies, or they may assume complete mixing in the lateral direction as in the case of stratified narrow channels.

The models can be developed as deterministic or stochastic models. In the deterministic models, the mean values of input parameters are fed and the models yield estimates of mean values of various water quality constituents. In the stochastic models, the randomness or uncertainty of model parameters is taken into account.

Several computational methods have been used in model development, ranging from hand computation or using a simple calculator to advanced numerical techniques and optimization methods which normally require a high–speed computer. Though the simple models which can be solved manually are preferable, they are usually limited to very simplified problems which may not be applicable to many real conditions. Regarding the numerical techniques, two important methods, namely the finite difference method and the finite element method, have been extensively used in the development of simulation models in recent years. In the development of optimization models for identification and evaluation of management alternatives which satisfy economic and water quality goals, some different optimization algorithms have been used, including linear programming, dynamic programming, integer programming, quadratic programming, etc. Selection of the optimization algorithm depends on the characteristics of the problems and the approximations used in model formulation.

General concepts in model development and application

In order to use the mathematical models in water quality management successfully, one should be aware of some important concepts in model developments and applications. These include:

1) A simple model should be considered if it can be used to solve the problem and provides sufficiently accurate results. It is not necessary to include all the parameters and variables in the analysis, since this would make the model unnecessarily complicated with many assumptions and huge data are required. However, interrelationships among various water quality parameters must be clearly understood, otherwise the presence of some constituents in water that affect the production or reaction rates of the studied water quality parameters may not be taken into consideration. This may result in an incomplete and unreliable model.

2) It is preferable to use a specific model in problem solving rather than a generalized multipurpose model, since the generalized model is expensive, difficult to understand, and requires a lot of data. In addition, most generalized models cannot provide adequate details at the micro level.

3) The modeler should not only emphasize on the mathematical method used in model formulation, but the validity of each parameter included in the model, e.g. dispersion coefficients, reaction rates, etc. should also be concentrated. Too many assumptions on the model parameters will result in unreliable and useless outputs.

4) Modeling and data collection process should proceed in parallel. Modeling often gives a better insight to the type of data that should be collected. The mere existence of data is not enough; its accessibility, accuracy, and usability are important criteria. Raw data often require a great deal of massaging before it can be converted to a form where it can be used as input to a model (Biswas 1976).

5) The model should be developed in such a way that the results obtained can be linked to the current decisions. It should be able to provide answers to the decision makers' questions such as what decisions should be made at present to meet the planned water quality goals, or what would be the effects on water quality resulting from the current decisions, etc.

6) The developed model should be updated from time to time as more information is available or understanding of the processes including interrelationships among various parameters and variables gets improved. This will make the model up–to–date and useful for current applications.

7) The model user should understand clearly the principal concepts and assumptions used in model development. He should realize the applications and limitations of the model being used, otherwise the results obtained might lead to a wrong decision.

8) A good document manual is very helpful for users which will facilitate the use of the model. In addition to the steps in using the model and details of input data to be fed, the document manual should clearly identify the principal basis and assumptions used in model development, as well as limitations of the model.

Surface water flow characteristics and models

Normally, there exist many types of surface water bodies in the lowland areas, e.g. canal, tidal river, estuary, open channel network, storage pond, lake, reservoir, wetlands, bay and coastal waters, etc. Physical characteristics of these water bodies, which include flow pattern, depth, etc. are more or less different. These physical properties are closely related to water quality. Dispersion of a pollutant is mainly influenced by water flow pattern. On the other hand, stratified flows in lakes and estuaries are normally caused by difference in water density which might be due to the difference in water temperatures or due to the high content of suspended solids or more saline water in the bottom layer. Therefore, in water quality management, the characteristics of water flow and relationship between hydrodynamic properties and water quality should be clearly understood.

In general, surface water flow in the lowlands is characterized by slow flow pattern due to gentle slope of the areas. The flow pattern might be influenced by tidal fluctuations in the sea. Also, flow in a lake or reservoir with large surface area is usually affected by wind motion. The flow in wetlands is similar to water movement at front line of a shallow lake although definition of wetlands is very wide. As regards an open channel network, which has been developed for water supply, irrigation, navigation, drainage, etc. and becomes a typical landscape in lowlands, the flow is more or less complicated due to its various connecting branches, particularly when this open channel network is connected to a tidal river.

In recent years, a large number of mathematical models have been developed to simulate the hydrodynamic characteristics of the surface waters. These hydrodynamic models are also very useful in the planning and management of surface water quality, since they provide more insight into the physical properties of the water body which, as previously mentioned, are closely related to the dispersion pattern of water quality constituents. These hydrodynamic models are mainly based on the principles of conservation of mass and momentum. The continuity and momentum equations can be written in the vectorial form as (Mauersberger 1983):

$$\frac{\partial \rho}{\partial t} + \nabla \cdot (\rho v) = 0 \tag{14.1}$$

and

$$\frac{Dv}{Dt} + 2\omega \times v = \sum_i F_i - \nabla\phi - \frac{1}{\rho}\nabla p + \frac{1}{\rho}\nabla \cdot P \tag{14.2}$$

where ω is the angular velocity of the Earth's rotation,

F_i represents the nonconservative external forces (per unit mass),

ϕ is the time–independent potential of external forces (e.g. gravity),

v is the velocity vector,

p is the pressure,

P is the dissipative part of the stress tensor, i.e. the friction tensor including turbulent friction.

Equation 14.1 is the continuity equation which describes the rate of change of density at a fixed point resulting from the change in mass velocity vector. Equation 14.2 is the momentum equation, in which the first term on the left–hand side represents the rate of change of flow velocities, while the second term accounts for the coriolis force. The first two terms on the right–hand side account for the external forces, while the third and the fourth terms represent the pressure force and the friction force, respectively.

From Equations 14.1 and 14.2, the physical parameters, viz. viscosity and density, are the fluid properties. The pressure force and external gravitational force are functions of fluid density, which is in turn influenced by temperature and content of impurities in the fluid. The friction force is dependent on the fluid viscosity which is also a function of temperature. In addition, density variation in the water body will affect turbulent diffusion. The diffusion transport is reduced by strong stratification, thus affecting dispersion patterns of pollutants in water. This indicates the effect of water quality on flow phenomenon and vice versa. Often, the hydrodynamic models are developed together with the water quality models to get more accurate results, for example, in the study of stratified flows in lakes and estuaries. Figure 14.2

illustrates the relationship between water density and flow phenomenon. However, in many cases, the density effect is small compared with other factors and can be ignored, for example in the shallow water with strong mixing by wind or tidal current.

In the computation of hydrodynamic circulation in a shallow water body with only little vertical variation of the flow, the vertically averaged model can be formulated by averaging the above mentioned three–dimensional equations over the depth. This vertically averaged model is less complicated and requires less computer time. It has been applied to determine the circulation patterns in several vertically well mixed water bodies and provides satisfactory results. However, since this two–dimensional model ignores the effects of velocity and density variations in the vertical direction, it is not applicable to simulate the problems of stratified flow in a lake due to temperature gradients or at the estuary mouth due to saline water intrusion.

Besides the vertically averaged two–dimensional model, another type of two–dimensional model, the laterally averaged model, has been developed to simulate the circulation patterns in a deep and narrow channel. This model ignores variation of flow in the lateral direction. It has been applied to determine the density and viscosity effects on flow patterns in an estuary and a narrow lake or reservoir.

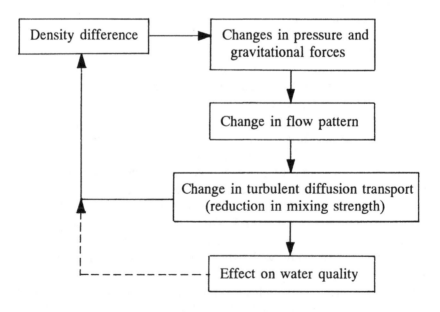

Figure 14.2. Diagram of relationship between water density and flow pattern.

A one-dimensional flow model can be developed by averaging over the whole cross-sectional area of the flow. This model has been used to find approximate flow properties in a stream. It requires less effort in formulation and requires much less computer time and input data.

Conservation of mass equation

The principle of conservation of mass is the basis of most water quality models. The conservation of mass equation can be written in the vectorial form as (Mauersberger 1983):

$$\frac{\partial C}{\partial t} + \nabla \cdot (vC - D \cdot \nabla C) = G \qquad (14.3)$$

where C is the substance concentration,

v is the velocity vector,

D is the diffusion coefficient tensor,

G is the source–sink term.

The first term denotes the time rate of change of concentration (mass per unit volume) of substance. The second and the third terms represent the advective and diffusive transports, respectively. The last term is the source–sink term which accounts for the net production rate per unit volume of that substance. The turbulent diffusion coefficient tensor D consists of six independent components. If three axes of the spatial coordinates coincide with principal axes of diffusion flux, these six components can be reduced to three components. In real flow phenomena, however, such coincidence rarely occurs, especially in the case of non–isotropic turbulent diffusion.

To obtain a unique solution, appropriate boundary conditions must be specified. For the pollutant dispersion problems, there are two types of boundary conditions, i.e. 1) substance concentration at the open boundary, and 2) inflow/outflow mass transport across the boundary.

Similar to the hydrodynamic models, the vertically averaged two–dimensional substance balance equation can be derived by averaging over the depth, i.e. variation in substance concentration and mass transport, either by advection or diffusion, in the vertical direction are ignored. The two–dimensional equation in the (x,y) coordinates can be written as:

$$\frac{\partial c_i}{\partial t} + u\frac{\partial c_i}{\partial x} + v\frac{\partial c_i}{\partial y} - \frac{1}{h}\{\frac{\partial}{\partial x}(hK_x\frac{\partial c_i}{\partial x}) + \frac{\partial}{\partial y}(hK_y\frac{\partial c_i}{\partial y})\} - R_i = 0 \qquad (14.4)$$

where c_i is the vertically averaged concentration of the water quality constituent i,

u and v are vertically averaged flow velocities in the x- and y- directions, respectively,

h is water depth,

K_x and K_y are diffusion coefficients in the x- and y- directions, respectively,

R_i is the source–sink term of the constituent i.

For the dispersion problems in a small stream or river with complete mixing over the whole cross-sectional area, the one–dimensional model can be adopted. The one–dimensional form of the substance balance equation can be written as:

$$\frac{\partial AC}{\partial t} + \frac{\partial}{\partial x}(QC - AE\frac{\partial C}{\partial x}) = S \qquad (14.5)$$

where C is the averaged concentration over the cross–section,

        A is the cross–sectional area,

        Q is the river discharge rate,

        E is the longitudinal dispersion coefficient,

        S is the source–sink term.

The one–dimensional water quality model requires much less computer time and input data. Though the vertical and lateral variations in pollutant concentration are ignored, it is still very useful in many cases. It can provide approximate results on the pollutant distribution patterns which can be used in the first screening of the proposed management alternatives.

Diffusion

Diffusion is the process by which the dissolved or suspended substance in the water moves when a concentration gradient exists. The diffusion process may be the result of molecular activity, the so–called 'molecular diffusion', or the action of turbulence, the so–called 'turbulence diffusion'. For a channel with non–uniform velocity distribution over the cross–section, there is a spreading of substance in the longitudinal direction. While the fluid elements in the same cross–section travel at different speeds, they will separate and tend to spread the substance greatly in the longitudinal direction. This causes a transverse gradient of the substance concentration, and the molecular and turbulent diffusion will act to make the concentration more uniform over the cross–section. This process is called 'longitudinal dispersion' (Streeter & Wylie 1983).

Diffusion plays an important role in energy, momentum and mass transport phenomena. In the investigation of substance dispersion pattern in a water body, adequate knowledge on the diffusion process under various flow conditions taking place in that water body is necessary. In the mass balance equation which is the basis for all water quality models, the transport by diffusion is represented by the diffusion term which is usually expressed in terms of the second-order derivative of substance concentration with respect to distance. This term includes a parameter called 'diffusion coefficient', or 'longitudinal dispersion coefficient' for the one-dimensional equation.

Several empirical formulae have been developed to express the value of diffusion coefficient in terms of flow properties such as water depth, flow velocity, etc. Taylor (1954) expressed the diffusion coefficient in terms of velocity variations as:

$$K_x = \overline{u_L^2} \int_0^\infty R_{Lu}(\tau)d\tau \qquad (14.6)$$

and

$$K_y = \overline{v_L^2} \int_0^\infty R_{Lv}(\tau)d\tau \qquad (14.7)$$

where K_x and K_y are the diffusion coefficients in the $x-$ and $y-$ directions, respectively,

$\overline{u_L^2}$ and $\overline{v_L^2}$ are time average of the Lagrangian squared velocity variations,

R_{Lu} and R_{Lv} are Lagrangian autocorrelations in the $x-$ and $y-$ directions, respectively.

In the Eulerian system, the diffusion coefficient can be expressed as:

$$K_x = \overline{u_E^2} \cdot \beta \cdot \theta_{Eu} \qquad (14.8)$$

and

$$K_y = \overline{v_E^2} \cdot \beta \cdot \theta_{Ev} \qquad (14.9)$$

where $\overline{u_E^2}$ and $\overline{v_E^2}$ are the values of the Eulerian time average of the squared velocity variations,

β is a parameter the value of which depends on the scale of turbulence,

θ_{Eu} and θ_{Ev} are the integral time scale in the Eulerian system.

From the hourly recorded data of flow velocities, the velocity variations, u_E' and v_E' can be obtained by removing the periodic components of the flow velocities. These velocity variations, which are random in nature, are then used to estimate the values of the diffusion coefficients (Taylor 1954).

Elder (1958) extended the Reynolds analogy and analysis used by Taylor (1954) to describe the diffusion of marked fluid in the turbulent flow in an open channel. The longitudinal and lateral diffusion coefficients were expressed in terms of channel depth and friction velocity as:

$$K_l = 5.93 \ h \ v_* \tag{14.10}$$

and

$$K_n = 0.23 \ h \ v_* \tag{14.11}$$

where K_l is longitudinal diffusion coefficient,

K_n is lateral diffusion coefficient,

h is total water depth,

v_* is friction velocity which is related to the mean velocity by:

$$v_* = \sqrt{\frac{\tau_b}{\rho}} = \frac{V\sqrt{g}}{C_h} \tag{14.12}$$

in which

τ_b is friction shear stress,

V is mean velocity,

C_h is De Chezy's coefficient.

When the diffusion coefficients in the longitudinal and lateral directions are known, the values in the $x-$ and $y-$ directions can be computed using the following expressions:

$$K_x = K_l\cos^2\alpha + K_n\sin^2\alpha \tag{14.13}$$

and

$$K_y = K_l\sin^2\alpha + K_n\cos^2\alpha \tag{14.14}$$

where α is the angle between flow direction and the $x-$ axis.

In the study on lateral spreading of substance in natural channels, many investigators assumed that the lateral mixing coefficient was proportional to the product of shear velocity and average depth (Yotsukura & Sayre 1976):

$$E_z = \alpha h U_*$$ (14.15)

where E_z is the transverse mixing coefficient,

h is the average depth,

U_* is the shear velocity,

α is an empirical coefficient that increases slightly with the width-to-depth ratio.

Okoye (1970) reported the α values ranging from about 0.1 to 0.2. In curved channels, the α values are considerably larger owing to secondary flows (Yotsukura & Sayre 1976). In flumes, the values ranging from 0.5 to 2.5 for a bend were reported by Fisher (1969) and Chang (1971), while Yotsukura et al. (1970) and Sayre & Yeh (1973) reported the values ranging from about 0.6 for a gradually curving reach up to about 10 for a very sharp bend in the Missouri river.

Transverse mixing in natural channels

In studying the dispersion pattern of a pollutant in a river, knowledge of the mixing characteristics is important. The assumption that the discharged pollutant is completely mixed over the water depth or the whole cross-section of the channel at the discharging point often leads to a wrong estimation.

From a hydraulic viewpoint, the river downstream from the continuous source may be divided into three zones according to the vertical and transverse mixing conditions. The first zone extends from the discharging point downstream to a section where the concentration of the discharged pollutant is considerably uniform over the channel depth. For a neutrally buoyant soluble effluent discharged without excess momentum at the surface, the length of this zone is of the order of 50 to 100 times the depth of the channel. For an effluent that is either lighter or heavier than the ambient water this length may be much higher. The second zone extends farther downstream to a section where the concentration distribution becomes considerably uniform over the channel width. For a point source located at one river bank, the length of the second zone may vary from a few kilometers to a hundred kilometers depending on the width of the channel. The third zone begins where the second zone ends and extends as far as the concentration is detectable (Yotsukura & Sayre 1976).

Normally, the geometry and flow properties in a natural river are not uniform. These properties affect the concentration distribution of the discharged pollutant and should be taken into consideration when water quality data or the results of plume studies are evaluated. In studying the characteristics of plumes from a steady point source near the center line of a meandering river, it is often noticed that the plumes always shift from the center line and that the distribution curves tend to be skewed to one side or the other. It is recognized that fluctuation in river discharge has a significant effect on this oscillating pattern. To normalize the plume data by taking the river discharge into account, Yotsukura & Cobb (1972) introduced a different format of result presentation.

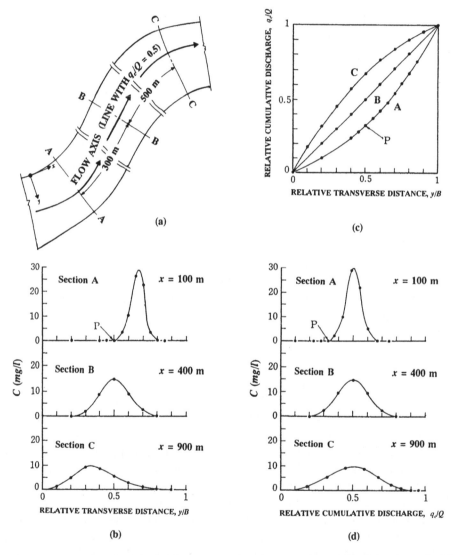

Figure 14.3. Transverse distributions of substance concentration. (a) channel with nonuniform section; (b) concentration plotted against relative transverse distance; (c) relative cumulative discharge vs. relative transverse distance; and (d) concentration plotted against relative cumulative discharge.

In their format, the abscissa is changed from the relative transverse distance to the relative cumulative discharge (q_c/Q), in which q_c is the cumulative discharge from one bank to the corresponding transverse distance and Q is total discharge over the whole cross section. An example is shown in Figure 14.3. Figure 14.3a illustrates a meandering channel with nonuniform cross section, where the flow axis, the vertical plane which divides the total discharge into 2 halves, i.e. q_c/Q = 0.5, always shifts from the center line of the channel. Figure 14.3b illustrates the transverse distribution of substance concentration discharged from a steady point source located upstream near the center line of the channel. It can be seen that the substance concentration distribution over each section does not follow the Gaussian distribution but its peak has shifted to the left or to the right. Figure 14.3c shows the plots between relative cumulative discharge and relative transverse distance at three different sections, A, B and C. When the abscissa of the concentration curves is changed to the relative cumulative discharge (Figure 14.3d), the plumes exhibit much less transverse shifting and tend to be more symmetrical, resembling the Gaussian distribution curves. The lateral shift in the concentration versus distance curve is due to transverse convective transport (Chang 1971, Holley 1971) and the asymmetry is mostly due to nonuniform distribution of flow across the channel (Yotsukura & Sayre 1976).

In developing a mathematical model for transverse mixing study in a meandering river with variable width, some investigators, e.g. Chang (1971), Fukuoka & Sayre (1973), Yotsukura & Sayre (1976), etc., have adopted an orthogonal curvilinear coordinate system. This system consists of three mutually orthogonal sets of coordinate surfaces, namely longitudinal, transverse, and horizontal coordinate surfaces. A simplified diagram showing the plan view of this orthogonal curvilinear coordinate system is illuatrated in Figure 14.4. To cope up with the differences in horizontal distances measured along different longitudinal coordinate surfaces from one transverse coordinate surface to another, the so-called metric coefficients m_x and m_z are introduced. With this system and metric coefficients, the following two-dimensional continuity and mass transport equations are derived (Yotsukura & Sayre 1976):

$$m_x m_z \frac{\partial h}{\partial t} + \frac{\partial}{\partial x}(m_z h v_x) + \frac{\partial}{\partial z}(m_x h v_z) = 0 \qquad (14.16)$$

and

$$m_x m_z \frac{\partial}{\partial t}(hc) + \frac{\partial}{\partial x}(m_z h v_x c) + \frac{\partial}{\partial z}(m_x h v_x c)$$

$$= \frac{\partial}{\partial x}(\frac{m_z}{m_x} h E_x \frac{\partial c}{\partial x}) + \frac{\partial}{\partial z}(\frac{m_x}{m_z} h E_z \frac{\partial c}{\partial z}) \qquad (14.17)$$

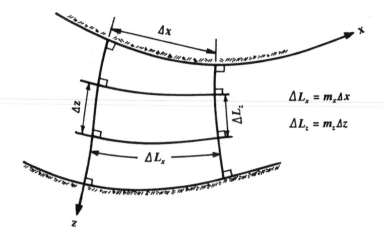

Figure 14.4. A simplified (plan view) orthogonal curvilinear coordinate system for a
natural channel.

where v_x　and v_z are the depth–averaged flow velocities in the longitudinal
and transverse directions, respectively,
h　　is water depth,
E_x　is the longitudinal dispersion coefficient,
E_z　is the transverse dispersion coefficient.

It should be noted that the coefficients E_x and E_z include the combined effects
of depth–averaged turbulent diffusion and convective dispersion. Assuming that
the longitudinal dispersive transport is small compared with the convective
transport, the steady–state continuity and mass transport equations become:

$$\frac{\partial}{\partial x}(m_z h v_x) + \frac{\partial}{\partial z}(m_x h v_z) = 0 \tag{14.18}$$

and

$$m_z h v_x \frac{\partial c}{\partial x} + m_x h v_z \frac{\partial c}{\partial z} = \frac{\partial}{\partial z}\left(\frac{m_x}{m_z} h E_z \frac{\partial c}{\partial z}\right) \tag{14.19}$$

To obtain a more compact and convenient form, Yotsukura & Sayre (1976)
replaced the independent variable z by the cumulative discharge q_c, which is
defined in the orthogonal curvilinear coordinate system as:

$$q_c = \int_{Z_L}^{z} m_z h v_x dz \qquad (14.20)$$

By integrating Equation 14.18 with respect to z from the left bank Z_L to z, and substituting the integral by q_c as defined in Equation 14.20, and by making use of Leibnitz's rule and the conditions that $v_x = v_z = 0$ at Z_L and that z is independent of x, the following equation is obtained:

$$m_x h v_z = -\frac{\partial}{\partial x} \int_{Z_L}^{z} m_z h v_x dz = -\frac{\partial q_c}{\partial x} \qquad (14.21)$$

Substituting Equation 14.21 into 14.19, the following equation is obtained:

$$m_z h v_x \frac{\partial c}{\partial x} - \frac{\partial q_c}{\partial x}\frac{\partial c}{\partial z} = \frac{\partial}{\partial z}(\frac{m_x}{m_z} h E_z \frac{\partial c}{\partial z}) \qquad (14.22)$$

After replacing $c(x,z)$ by $c(x,q_c)$ and making use of the chain rule of partial derivatives, Equation 14.22 is reduced to:

$$\frac{\partial c}{\partial x} = \frac{\partial}{\partial q_c}(m_x h^2 v_x E_z \frac{\partial c}{\partial q_c}) \qquad (14.23)$$

For an approximate solution, the quantity $m_x h^2 v_x E_z$ in Equation 14.23 may be treated as a constant factor of diffusion D. Then Equation 14.23 reduces to the form similar to the classical Fickian diffusion equation:

$$\frac{\partial c}{\partial x} = D\frac{\partial^2 c}{\partial q_c^2} \qquad (14.24)$$

Equations 14.23 and 14.24 are very useful in studying the two-dimensional dispersion phenomena in a meandering river. It can be solved by an analytical method or some numerical techniques. Yotsukura & Sayre (1976) also described several solution techniques and methods for estimating parameters with examples from some field investigations.

Analytical solutions of the substance balance equation

For simple dispersion problems, the substance balance equation can be solved analytically. Some examples are shown as follows:

1) Simplified one-dimensional substance balance equation: For a channel with uniform cross-sectional area, water flow and dispersion coefficient, the substance balance equation can be written as:

$$\frac{\partial C}{\partial t} + U \frac{\partial C}{\partial x} - E \frac{\partial^2 C}{\partial x^2} + KC = 0 \qquad (14.25)$$

where C is the averaged concentration over the cross-section,

U is the averaged flow velocity,

E is the longitudinal dispersion coefficient,

K is the first-order decaying rate of the substance.

2) Long channel with specified substance concentration at the upstream end: For a uniform channel with specified substance concentration at the upstream end $(x = 0)$, the boundary and initial conditions are given by:

$$
\begin{aligned}
C(0,t) &= C_0 &&\text{for } t \geq 0 \\
C(\infty,t) &= 0 &&\text{for } t \geq 0 \qquad (14.26)\\
C(x,0) &= 0 &&\text{for } x \geq 0
\end{aligned}
$$

The following solutions are obtained (Ippen 1966):

$$\frac{C}{C_0} = \frac{1}{2} \exp(\frac{xU}{2E}) \ [\ \exp(\frac{x}{2E}\sqrt{U^2 + 4EK}) \ \text{erfc}(\frac{x + \sqrt{U^2 + 4EK \cdot t}}{\sqrt{4Et}})$$

$$+ \exp(-\frac{x}{2E}\sqrt{U^2 + 4EK}) \ \text{erfc}(\frac{x - \sqrt{U^2 + 4EK \cdot t}}{\sqrt{4Et}}) \ ] \qquad (14.27)$$

For the steady-state case, we have:

$$\frac{C}{C_0} = \exp [\ \frac{x}{2E} (U - \sqrt{U^2 + 4EK}) \ ] \qquad (14.28)$$

3) Long channel with specified substance concentration at the downstream end: For the case that the substance concentration at the downstream end $(x = 0)$ is specified while the value at the upstream end $(x = -\infty)$ is equal to zero, which is similar to the case of saltwater intrusion from the sea to an estuary, the

boundary and initial conditions are:

$$
\begin{aligned}
C(0,t) &= C_0 && \text{for } t \geq 0 \\
C(-\infty,t) &= 0 && \text{for } t \geq 0 \\
C(x,0) &= 0 && \text{for } x \leq 0
\end{aligned}
\tag{14.29}
$$

The following solutions are obtained (Ippen 1966):

$$
\frac{C}{C_0} = \frac{1}{2}\exp(\frac{xU}{2E})\ [\ \exp(-\frac{x}{2E}\sqrt{U^2+4EK})\ \mathrm{erfc}(\frac{-x+\sqrt{U^2+4EK}\cdot t}{\sqrt{4Et}})
$$

$$
+\ \exp(\frac{x}{2E}\sqrt{U^2+4EK})\ \mathrm{erfc}(\frac{-x-\sqrt{U^2+4EK}\cdot t}{\sqrt{4Et}})\ ]
\tag{14.30}
$$

For steady–state case, we have:

$$
\frac{C}{C_o} = \exp\{\frac{x}{2E}(U+\sqrt{U^2+4E\cdot K})\}
\tag{14.31}
$$

4) Long channel with specified substance loading at a fixed section: For a long channel with instantaneous substance loading M (total mass) at a fixed section ($x = 0$), the boundary and initial conditions are:

$$
\begin{aligned}
C(\infty,t) &= 0 && \text{for } t \geq 0 \\
C(x,0) &= \frac{M}{A}\delta(x)
\end{aligned}
\tag{14.32}
$$

where A is the cross–sectional area of the channel,

$\delta(x)$ is the Dirac delta function with the property of $\int_{-\infty}^{\infty}\delta(x)\mathrm{d}x = 1$.

The following solution is obtained (Ippen 1966):

$$
C = \frac{M}{A\sqrt{4\pi Et}}\ \exp[-\frac{(x-Ut)^2}{4Et}-Kt]
\tag{14.33}
$$

For a continuous loading with the rate m (mass per unit time), the solution can be obtained by considering concentration increment dC introduced at time t, and integrating over the period from $t = 0$ to t. The following solutions are obtained:

$$C = \frac{m \cdot \exp(\frac{xU}{2E})}{2A\sqrt{U^2+4EK}} \; [\; \exp(\frac{x\sqrt{U^2+4EK}}{2E}) \; \{ \; \mathrm{erf}(\frac{x+\sqrt{U^2+4EK}\cdot t}{\sqrt{4Et}})-1 \; \}$$

$$+\exp(\frac{-x\sqrt{U^2+4EK}}{2E}) \; \{ \; \mathrm{erf}(\frac{-x+\sqrt{U^2+4EK}\cdot t}{\sqrt{4Et}})+1 \; \} \;] \qquad \text{for } x \geq 0$$

$$(14.34)$$

and

$$C = \frac{m \cdot \exp(\frac{xU}{2E})}{2A\sqrt{U^2+4EK}} \; [\; \exp(\frac{x\sqrt{U^2+4EK}}{2E}) \; \{ \; \mathrm{erf}(\frac{x+\sqrt{U^2+4EK}\cdot t}{\sqrt{4Et}})+1 \; \}$$

$$+\exp(\frac{-x\sqrt{U^2+4EK}}{2E}) \; \{ \; \mathrm{erf}(\frac{-x+\sqrt{U^2+4EK}\cdot t}{\sqrt{4Et}})-1 \; \} \;] \quad \text{for } x \leq 0$$

$$(14.35)$$

For the steady-state case, we have:

$$C = \frac{m}{A\sqrt{U^2+4EK}} \; \exp \{ \; \frac{x}{2E}(U-\sqrt{U^2+4EK}) \; \} \quad \text{for } x \geq 0 \quad (14.36)$$

and

$$C = \frac{m}{A\sqrt{U^2+4EK}} \; \exp \{ \; \frac{x}{2E}(U+\sqrt{U^2+4EK}) \; \} \quad \text{for } x \leq 0 \quad (14.37)$$

5) Simplified two-dimensional substance balance equation: For a basin with constant depth and flow velocities, the substance balance equation can be written as:

$$\frac{\partial C}{\partial t}+U\frac{\partial C}{\partial x}+V\frac{\partial C}{\partial y}-K_x\frac{\partial^2 C}{\partial x^2}-K_y\frac{\partial^2 C}{\partial y^2}+KC = 0 \qquad (14.38)$$

where C is the vertically averaged substance concentration,

U and V are vertically averaged flow velocities in the x- and y-directions, respectively.

K_x and K_y are diffusion coefficients in the $x-$ and $y-$ directions, respectively,

K is the first-order decaying rate of the substance.

6) Basin with specified substance loading at a fixed point. Consider a constant-depth basin of B in width, with flow velocities U in the longitudinal direction (the $x-$ direction) and $V = 0$ in the lateral direction (the $y-$ direction), and continuous substance loading Q/H (mass per unit depth per unit time) discharged at a fixed point ($x = 0$, $y = 0$) on one side of the basin. If the substance is conservative ($K = 0$) and the diffusion in the longitudinal direction is neglected as compared with the advection term, and with the upstream boundary condition $C = f(y)$ ($0 \le y \le B$), then the steady-state distribution pattern of the substance can be computed from (Nanbu 1958):

$$C = \frac{1}{B}\int_0^B f(y)\,dy + \frac{2}{B}\sum_{n=1}^{\infty} \exp\{-\frac{K_y}{U}(\frac{n\pi}{B})^2 x\}\cos\frac{n\pi y}{B}\cdot\int_0^B f(y)\cos\frac{n\pi y}{B}\,dy \quad (14.39)$$

If the specified upstream boundary concentration is $C = 0$ ($0 \le y \le B$), the following equation is obtained (Awaya 1968):

$$C = \frac{Q}{BHU}\{1 + 2\sum_{n=0}^{\infty}\exp[-\frac{K_y}{U}(\frac{n\pi}{B})^2 x]\cos\frac{n\pi y}{B}\} \quad (14.40)$$

Figure 14.5 illustrates the substance concentration distribution pattern as a function of $x/(B^2 U/K_y)$ and y/B (Awaya 1968). For $x \ll B^2 U/K_y$, Equation 14.40 is reduced to:

$$C = \frac{Q}{H}\frac{1}{\sqrt{\pi U K_y x}}\exp(-\frac{Uy^2}{4K_y x}) \quad (14.41)$$

Although the above mentioned analytical solutions are obtained for the simple cases of dispersion problems which rarely occur in nature, they can provide some information, at least the order of magnitude, on pollutant dispersion. They are also useful in verifying the developed numerical models and in determining the sensitivity of some model parameters.

Numerical models

For complicated dispersion problems such as dispersion in a water body with irregular shape and with multiple pollutant sources or the problem of transient dispersion with variations in flow and pollutant loading, the analytical solutions are not available, so the computer-based numerical methods are used in solving

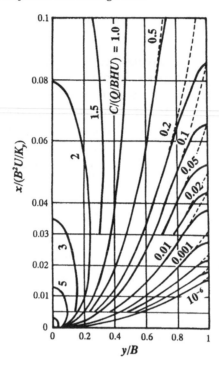

Figure 14.5. Distribution pattern of conservative substance in a constant–depth basin
(Awaya 1968).

the substance balance equation. In the past few decades, parallel to the advance
in computer technology, much effort has been devoted to the development of
more effective numerical techniques which can be used in solving the partial
differential equations. Among the various numerical techniques, the finite
difference and the finite element methods are most well–known and have been
widely used in the development of water quality models in recent years.

In the finite difference method, the partial differential equation is transformed
to a set of algebraic difference equations in which the continuous variable is
replaced by a discrete variable. The spatial derivative appearing in the partial
differential equation is approximated by the difference in the values of
dependent variable at two adjacent grid points divided by the distance between
the grid points, while the time derivative is approximated by the difference in
the variable values at two subsequent time steps divided by the time interval.
This set of finite difference equations is then solved numerically to yield values
of dependent variable at a predetermined number of grid points in the study
domain.

Normally, the finite difference method makes use of orthogonal grids (Figure 14.6) which may have some difficulties when applied to the natural water bodies with irregular boundaries. Some other types of grids which are more suitable for the irregular water bodies have been developed. For example, Bauer (1979) developed a three–dimensional finite difference model with an irregular grid to simulate flow patterns in a homogeneous lake. Tatom & Waldrop (1978) developed a finite difference model for long, narrow reservoirs using orthogonal curvilinear coordinates taking into consideration reservoir inflows and outflows, cooling water circulation, and surface wind stress. In addition, several techniques have been developed to get more accurate and stable finite difference models, e.g. the use of staggered grid (Heaps 1969, Simons 1971, etc.) and the use of alternate–direction implicit method (Peaceman & Rachford 1955, Pinder & Bredehoeft 1968, Leendertse 1967 & 1970, etc.).

The finite element method is a very powerful numerical technique which has been applied to solve various engineering problems nowadays. It was first used in solving the structural engineering problems and later applied to other fields. It is an extremely effective and flexible method which can easily handle any shape of boundary including moving boundary, and any combination of boundary conditions. Several types of element configurations are available (Figure 14.7), and combination of different element configurations in a model is possible without difficulties. The shape and size of the elements can be varied to suit the study domain.

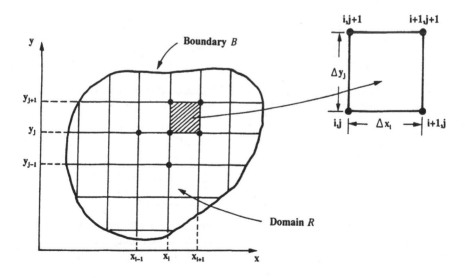

Figure 14.6. An orthogonal grid normally used in the finite difference models.

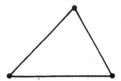

Triangular element with
nodes at the corners

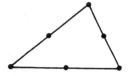

Triangular element with nodes
at the corners and mid–sides

Triangular element with nodes at
the center and corners

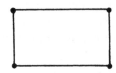

Rectangular element with
nodes at the corners

Rectangular element with nodes
at the corners and mid–sides

General quadrilateral element
(linear isoparametric)

Curved side element
(quadratic element)

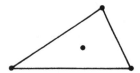

Tetrahedron element with
nodes at the corners

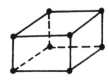

Cube element with nodes
at the corners

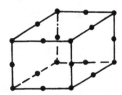

Cube element with nodes at
the corners and mid–sides

Figure 14.7. Example of element configurations used in the finite element models.

Several techniques have been used in the formulation of the finite element model. For the fluid flow and dispersion problems, however, the Galerkin's weighted residual method is normally used. In the weighted methods, dependent variables in the governing equations, which are in the form of partial differential equations, are approximated by some trial functions which are expressed in terms of their values at nodal points identified in the study domain. With this approximation, some errors or residuals will occur. These residuals are forced to be zero, in an average sense, by setting the weighted integrals of the residuals to zero. In the Galerkin's technique, the weighting functions are the same as the trial functions. With this technique, the original partial differential equations will be transformed to a set of algebraic equations with dependent variables at the nodal points as unknown variables. With appropriate boundary and initial conditions, this set of algebraic equations can be solved by some numerical methods, e.g. the Gauss elimination method. More details about the finite element method applied to the fluid flow problems can be found in available literature (e.g. Zienkiewicz 1971, Strang & Fix 1973, Connor & Brebbia 1977, Pinder & Gray 1977, Gallagher et al. 1975a,b & 1978, Chung 1978, Zienkiewicz & Morgan 1983, Taylor & Hughes 1981, Wendt 1992, and others)

In addition to the above mentioned finite difference and finite element methods, some modified numerical schemes have been developed in the formulation of water quality models. The main purposes of the modification and development are to make the numerical models more realistic which best suit the real physical phenomena and to minimize the errors resulting from the numerical approximations. For example, Futawatari & Kusuda (1993) developed a two–layered sediment transport model using a Lagrangian reference frame. This frame travels up and down the estuary with water movement, so the advection term can be eliminated and numerical dispersion can be reduced. A number of multi–layered finite element models have been developed for stratified flow problems (e.g. Wang & Connor 1975, Simons 1973 & 1980, Cheng et al. 1976, McNider & O'Brien 1973, etc.). In these models, the vertically averaged continuity and momentum equations are obtained by integrating the three–dimensional equations over each constant–density layer. Mass and momentum transfers across the interfaces are also taken into account. Use of these models is limited to strongly stratified water bodies.

In spite of the ignorance of vertical and lateral variations in the flow and substance concentration, a number of one–dimensional numerical models have been successfully applied in many cases. The one–dimensional models can be developed by using the finite segment, finite difference or finite element method. Among these, the so–called 'node-branch model', which has been developed by using the finite element method, is very useful in analyzing the unsteady flow and substance dispersion in a hydraulic network system (Booij 1980, Koga et al. 1988, Spaans et al. 1989, etc.). The schematization of the node–branch network is shown in Figure 14.8. Each branch connects two nodes.

The nodes which are the junction points of adjacent branches are considered as no–volume points. The model formulation is based on the one–dimensional continuity and momentum equations for flow computation and mass balance equation for computation of substance concentration. Flow properties and substance concentration in each branch are expressed in terms of the values at both ends. At the node which is the downstream end of two or more branches, there is a possibility that the substance concentration is discontinuous if the dispersion is small. Therefore, a distinction is made between the concentrations at the nodes (C_i and C_j) and at the ends of the connecting branch (c_1 and c_2) as follow (see Figure 14.9):

If flow is directed inward to the branch (from a node), the concentration is considered continuous, so

$$c_1^+ = C_i^+ \quad \text{if} \quad Q_1^+ \geq 0 \tag{14.42}$$

and

$$c_2^+ = C_j^+ \quad \text{if} \quad Q_2^+ < 0 \tag{14.43}$$

Note that the superscript '+' denotes the value to be computed in the next time step. If flow is directed outward from the branch to the node, the concentration may be discontinuous. The transport flux S is computed from the products of flow and nodal concentration, i.e.

$$S_1^+ = Q_1^+ c_1^+ \quad \text{if} \quad Q_1^+ < 0 \tag{14.44}$$

and

$$S_2^+ = Q_2^+ c_2^+ \quad \text{if} \quad Q_2^+ \geq 0 \tag{14.45}$$

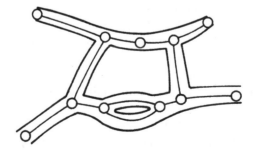

Figure 14.8 Schematization of network.

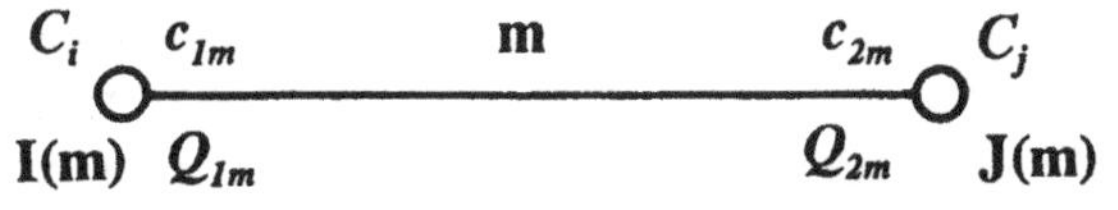

Figure 14.9 One branch and its adjacent nodes (after Booij 1980).

Equations 14.42 to 14.45 are then coupled with the unsteady–state mass balance equation in the discretized form which is derived by applying the Galerkin's weighted residual method to the one–dimensional mass balance equation. This results in the following branch equations:

$$S_1^+ = N_{m1}C_i^+ + N_{m2}C_j^+ + N_{m3} \qquad (14.46)$$

and

$$S_2^+ = N_{m4}C_i^+ + N_{m5}C_j^+ + N_{m6} \qquad (14.47)$$

where $N_{mi}(i=1,2,..6)$ are functions of cross–sectional area and length of the branch, discharge, dispersion coefficient, reaction rate of substance, time interval, as well as the nodal concentrations and transport fluxes computed from the previous time step.

The equations of all branches in the network are then assembled to form the system equations which can be solved to obtain the values of substance concentrations at the ends of each branch.

Ecological models

In the past few decades, more attention has been paid to the impacts of development projects on the existing ecosystems. A number of ecological models have been developed to simulate the aquatic ecosystems which include some aquatic organisms, their nutrient cycles, as well as some other water quality parameters. In these models, the transport processes due to convective and diffusive flows are coupled with the kinetic processes which describe chemical and biological transformations occurring in water. The parameters normally included in the models of this type are phytoplankton, zooplankton, nitrogen and phosphorus in various forms, dissolved oxygen, BOD, etc. The ecological and physical processes such as photosynthesis, respiration, zooplankton grazing, fish predation, nutrient uptaking and recycling, decomposition of organic matter, sedimentation, atmospheric reaeration, etc. are represented by mathematical functions. These functions are assembled and operated together to simulate the behaviors of the ecosystems. In recent years, attempts have been made to include more biological realism, taking into consideration feedback mechanisms and couplings to other constituents and to environmental factors. More details on the ecological models can be found in the available literature including Chen (1970), Thomann et al. (1970, 1974, 1975 & 1985), Larsen et al. (1974), O'Connor et al. (1969, 1973 & 1976), Chen & Orlob (1975), Di Toro et al. (1971 & 1975), Di Toro & Connolly (1980), Najarian & Harleman (1975), Walsh (1976), Williams & Hinwood (1976), Baca & Arnett (1976), Jorgensen (1976, 1979 & 1983), Jorgensen & Harleman (1978), Harleman (1978), Orlob (1983), Thomann & Mueller (1987), and others.

Integrated models for water management

From the fact that water quality is closely related to water quantity, the management of water quality needs the control and management of water quantity in addition to reduction in the discharged pollutants. Dispersion of pollutants depends mainly on the configuration and flow pattern of the water bodies. Therefore, in using mathematical models for water quality prediction and planning, data on water quantity are necessary. These data can be obtained either from the past records or from water quantity simulation models. That is the water quality models can be integrated with the related water quantity models which can provide complete analysis of the water system.

Nowadays, the eutrophication and ecological problems are of great concern in the field of water quality management. Many experts from various fields have been involved in the investigations and researches in these problems. Usually, these experts need a flexible and effective tool for good communication with others. The so–called 'integrated water quality management model' is considered as a bridging/communicating tool for these experts of diversified fields. At present, many pilot studies are conducted to develop this integrated computer model.

Models for water policy analysis

The study on water policy analysis is very important especially in the lowlands. In order to solve the complicated water resources problems from the viewpoint of integrated water management, at first, the water functions should be analyzed taking into consideration various concerned aspects including economic, environmental, social and political aspects. The structured information is then composed to analyze and evaluate the water management policies, and some strategic plans for water management are established.

Comprehensive studies on water policy analysis have been conducted in the Netherlands under the so–called PAWN (Policy Analysis of Water Management for the Netherlands) project with an aim at developing a methodology for assessing the multiple consequences of national water management policies and applying this methodology in generating and analyzing alternative water management policies (Koudstaal et al. 1986). This kind of policy analysis study is found to be very useful in the planning for integrated water resources management and should also be considered in other lowlands taking into account their own traditional water use categories, climate, geographical conditions, etc.

STRATEGIES FOR LOWLAND WATER QUALITY MANAGEMENT

The starting point for water quality management is the analysis of water system. This analysis does not only include the physical, chemical and biological qualities of a drainage basin up to its watershed, but also includes the functional uses of water, its ecological values and potentials, as well as the managerial, legislative and financial structures of the water management field.

In water quality management it is necessary to use some parameters to measure the strength and harmfulness of the discharged wastes. All the factors which affect physical, chemical, and biological processes of the selected water quality parameters must be taken into account. For organic pollution in surface water, the biochemical oxygen demand (BOD) and dissolved oxygen are mainly used as pollution indicators. Dispersions of some toxic substances such as heavy metals, chlorinated hydrocarbons, etc. are of interest when contaminations from industries and hazardous waste landfill sites are evaluated. Salinity or chloride concentration is used as an indicator in investigating the problems of saltwater intrusion in estuaries and coastal groundwater aquifers. Some ecological parameters, such as planktonic species, benthic organisms, fish, etc. are considered in the investigation of the effects of waste discharge on the aquatic ecosystems. Selection of appropriate parameters that best suit for the tackled problem is very important in water quality management.

A national policy on water quality management in lowlands should be established. This policy document is a strategic plan for solving water resources problems. The policy document should set goals for the functional uses of the water system and also specify the period of time for these goals to be achieved, set priorities and standards and initiate changes, if necessary, in the organizational, legislative and financial structures to facilitate the management.

In order to be able to formulate these goals in a realistic way, their feasibility must be assessed. Numerous tactics can lead, or at least contribute, to a certain goal. It is the role of policy analysis to select the better tactics and formulate feasible policies. This can be done by a systematic screening procedure, in which the promising tactics are evaluated using a number of models and databases with some scenario assumptions for external input variables such as growth rates of population and industrial production. The impacts on the functions of the water system are evaluated in terms of macro- and micro-economic and ecological effects.

A number of water quality standards, criteria, regulations and guidelines must be set up and applied in water quality management so that the established goals can be achieved within the specified period and maintained thereafter. There are many types of standards depending upon the functional uses of water. These include drinking–water standards, stream standards, effluent standards, and water quality standards for some other types of water uses. These standards are considered as the objectives for water quality management.

Surface water and groundwater are closely related especially in the lowlands. This is particularly obvious when the quality of surface water decreases with the rapid economic growth, causing at the same time an increase in groundwater use for domestic and industrial purposes. Therefore, in water quantity management, conjunctive use of both types of water resources should be considered. This concept is also applicable to water quality management, since the polluted surface water and some methods of wastewater disposal can cause deterioration in groundwater quality. Moreover, excessive use of groundwater in the coastal areas will result in intrusion of saltwater to freshwater aquifers which is a major groundwater pollution issue in many regions. From the fact that the water system is one integrated system and division into separate parts may create cross–boundary problems, the integrated water management concept should be emphasized.

REFERENCES

American Society of Civil Engineers & American Water Works Association (ASCE & AWWA) 1990. *Water treatment plant design*. 2nd ed., McGraw–Hill, New York.

Andou, N. 1993. Environmental and ecological considerations in water management in Japan. *Proc. integrated water management seminar:* 269–294, Tokyo.

Awaya, Y. 1968. Water pollution phenomena in rivers. *JSCE special lecture on hydraulics.* B–13 (in Japanese).

Baca, R.G. & R.C. Arnett 1976. A finite element water quality model for eutrophic lakes In W.G. Gray, G. Pinder & C. Brebbia (eds.), *Finite element in water resources*. 4:125 147, Pentech, Plymouth, Devon.

Bauer, S.W. 1979. Three–dimensional simulation of time–dependent elevations and currents in a homogeneous lake of arbitrary shape using an irregular–grid finite–difference model. In W.H. Graf & C.H. Mortimer (eds.), *Hydrodynamics of lakes:* 267–276, Elsevier, Amsterdam.

Bear, J. 1979. *Hydraulics of groundwater*. McGraw–Hill, Israel.

Bird, R.B., W.E. Stewart & E.N. Lightfoot 1960. *Transport phenomena*. John Wiley & Sons, New York.

Biswas. A.K. (ed.) 1976. *Systems approach to water management*. McGraw–Hill, Tokyo.

Booij, N. 1980. Report on the ICES subsystem FLOWS. *Report No.78–3,* Delft University of Technology, Delft, the Netherlands.

Chang, Y.C. 1971. Lateral mixing in meandering channels. *Ph.D. dissertation*. University of Iowa, Iowa City.

Chen, C.W. 1970. Concepts and utilities of ecological models. *J. ASCE, San. Eng. Div.*

96(5):1085–1097.

Chen, C.W. & G.T. Orlob 1975. Ecologic simulation of aquatic environments. In B.C. Patten (ed.), *Systems analysis and simulation in ecology*. vol.3, Academic Press, New York.

Cheng, R.T., T.M. Powell & T.M. Dillon 1976. Numerical models of wind–driven circulation in lakes. *Applied Mathematical Modelling*. 1(3):141–159.

Chung, T.J. 1978. *Finite element analysis in fluid dynamics*. McGraw–Hill, New York.

Connor, J.J. & C.A. Brebbia 1977. *Finite element techniques for fluid flow*. Butterworth & Co., London.

Davis, M.L. & D.A. Cornwell 1991. *Introduction to environmental engineering*. 2nd ed., McGraw–Hill, Singapore.

Di Toro, D.M., D.J. O'Connor & R.V. Thomann 1971. Nonequilibrium systems in natural water chemistry. *American Chemical Society, Advances in chemistry series*. 106:131–180.

Di Toro, D.M., D.J. O'Connor, R.V. Thomann & J.L. Mancini 1975. Phytoplankton–zooplankton–nutrient interaction model for western Lake Erie. In B.C. Patten (ed.), *Systems analysis and simulation in ecology*. 3:423–474, Academic Press, New York.

Di Toro, D.M. & J.P. Connolly 1980. *Mathematical models of water quality in large lakes, Part 2: Lake Erie*. ERL, ORD, USEPA, Duluth, 231 pp.

Elder, J.W. 1958. The dispersion of marked fluid in turbulent shear flow, *J. Fluid Mechanics*. 5:544–560.

Fair, G.M., J.C. Geyer & D.A. Okun 1968. *Water and wastewater engineering, vol.2: Water purification and wastewater treatment and disposal*. John Wiley & Sons, New York.

Fisher, H.B. 1969. The effects of bends on dispersion in streams. *Water Resources Research*. 5(2):496–506.

Fukuoka, S. & W.W. Sayre 1973. Longitudinal dispersion in sinuous channels. *J.ASCE., Hydraulics Div*. 99(1):195–217.

Futawatari, T. & T. Kusuda 1993. Modeling of suspended sediment transport in a tidal river. In A.J. Mehta (ed.), *Nearshore and estuarine cohesive sediment transport*. Coastal and estuarine studies vol. 42, American Geophysical Union.

Gallagher, R.H., J.T. Oden, C. Taylor & O.C. Zienkiewicz 1975a. *Finite elements in fluids. vol.1*. John Wiley & Sons, London.

Gallagher, R.H., J.T. Oden, C. Taylor & O.C. Zienkiewicz 1975b. *Finite elements in fluids. vol.2*. John Wiley & Sons, London.

Gallagher, R.H., O.C. Zienkiewicz, J.T. Oden, M.M. Cecchi & C. Taylor 1978. *Finite elements in fluids. vol.3*. John Wiley & Sons, London.

Harleman, D.R.F. 1978. A comparison of water quality models for the aerobic nitrogen cycle. *Research memorandum RM–78–34*. International Institute for Applied Systems Analysis, Laxenburg, Austria.

Heaps, N.S. 1969. A two–dimensional numerical sea model. *Philosophical transaction*. Series A 265(1160):93–137.

Holley, E.R. 1971. Transverse mixing in rivers. *Report S132*. Delft Hydraulic Lab., Delft, the Netherlands.

Hosper, S.H. & E. Jagtman 1990. Biomanipulation additional to nutrient control for restoration of shallow lakes in the Netherlands. *Hydrobiologia*. 200/201: 523–534.

Hosper, S.H. & M.L. Meijer 1993. A simple test for the assessment of chances for clear water, following drastic fish–stock reduction in shallow, eutrophic lakes. *Ecological Engineering*: 63–72, Elsevier, Amsterdam.

Ippen, A.T. 1966. *Estuarine and coastal hydrodynamics*. McGraw–Hill, New York.

Jeppesen, E., J.P. Jensen, P. Kristensen, M. Sondergaard, E. Mortensen, O. Sortkjaer & K. Olrik 1990. Fish manipulation as a lake restoration tool in shallow, eutrophic temperate lakes 2: threshold levels, long–term stability and conclusions. *Hydrobiologia*. 200/201:219–227.

Jorgensen, S.E. 1976. A eutrophication model for a lake. *Ecological Modelling.* 2:147–165.
Jorgensen, S.E.(ed.) 1979. *Handbook of environmental data and ecological parameters.* International Society for Ecological Modelling, Pergamon, Oxford.
Jorgensen, S.E. 1983. Ecological modeling of lakes. In G.T. Orlob (ed.), *Mathematical modeling of water quality in lakes and reservoirs.* John Wiley & Sons, New York.
Jorgensen, S.E. & D.R.F. Harleman (eds.) 1978. Hydrophysical and ecological modelling of deep lakes and reservoirs. *Summary report of a IIASA workshop: Collaborative proc. CP–78–7.* International Institute for Applied Systems Analysis, Laxenburg, Austria.
Kern, K. 1992. Restoration of lowland rivers: the German experience. In P.A. Carling & G.E. Petts (eds.) *Lowland floodplain rivers:* 279–297, John Wiley & Sons, London.
Koga, K. & H. Araki 1993. Enhancement and control of self–purification mechanism on water quality in open channel network systems. In N. Soga (ed.), *Circulation and control of man–made substances in environment (1990–1992):* 234–237, Man–environment system research report, Ministry of Education, Culture and Science, Japan.
Koga, K., N. Booij, P. Ankum, W. Segeren, W. Schuurmans, K. Inomae & H. Araki 1988. Numerical mode of water quality in hydraulic network systems. *Report of the faculty of science and engineering.* 16:91–100, Saga University, Japan.
Koudstaal, R., H.A. Pennekamp & J. Wesseling (Eds.) 1986. *Planning for water resources management in the Netherlands.* Institute for Land and Water Management Research (ICW) and Delft Hydraulics (WL), the Netherlands.
Larsen, D.P., H.T. Mercier & K.W. Malveg 1974. Modeling algal growth dynamics in Shagawa Lake, Minnesota. In E.J. Middlebrooks, D.H. Falkenberg & T.E. Maloney (eds.), *Modeling the eutrophication process:* 15–33, Ann Arbor Science, Ann Arbor.
Leendertse, J.J. 1967. Aspects of a computational model for long–period water wave propagation. *CA memorandum RM–5294–PR.* Rand Corporation Santa Monica.
Leendertse, J.J. 1970. A water quality simulation model for well mixed estuaries and coastal seas: vol.I, Principles of computation. *CA memorandum RM–6230–PR.* Rand Corporation Santa Monica.
Lijklema, L. 1993. Ecosystem considerations in water management. *Proc. integrated water management seminar:* 295–310, Tokyo.
Loucks, J.R., J.R. Stedinger & D.A. Haith 1981. *Water resource systems planning and analysis.* Prentice Hall, New Jersey.
Mauersberger, P. 1983. General principles in deterministic water quality modeling. In G.T. Orlob (ed.), *Mathematical modeling of water quality: streams, lakes, and reservoirs.* International Institute for Applied Systems Analysis, John–Wiley & Sons, London.
McNider, R.T. & J.J. O'Brien 1973. A multilayer transient model of coastal upwelling. *J. Physical Oceanography.* 3(3):258–273.
Meijer, M.L., E. Jeppesen, E. Van Donk, B. Moss, M. Scheffer, E.H.R.R. Lammens, E.H. Van Nes, B.A. Faafeng & J.P. Jensen 1993a. Long–term response to fish–stock reduction in small, shallow lakes: interpretation of 5 year results of 4 biomanipulation case studies in the Netherlands and Denmark. *Proc. 'Shallow Lakes 1992' conference.* Denmark.
Meijer, M.L., E.H. Van Nes, E.H.R.R. Lammens, R.D. Gulati, M.P. Grimm, J.J.G.M. Backx, P. Hollebeek, E.M. Blaauw & A.W. Breukelaar 1993b. The consequences of a drastic fish stock reduction in the large and shallow Lake Wolderwijd, the Netherlands. Can we understand what happened? *Proc. 'Shallow Lakes 1992' conference.* Denmark.
Najarian, T.O. & D.R.F. Harleman 1975. A real time model of nitrogen cycle dynamics in an estuarine system. *Report No. 204.* Ralph M. Parsons Lab. for Water Resources and Hydrodynamics, M.I.T., Cambridge, Massachusetts.
Nanbu, S. 1958. Basic study on dispersion and dilution phenomena of discharged

wastewater. *J.JSCE.* 59:26–31 (in Japanese).

O'Connor, D.J., J.P. St. John & D.M. Di Toro 1969. Water quality analysis of the Delaware river estuary. *J. ASCE., San. Eng. Div.* 95(6):1225–1252.

O'Connor, D.J., R.V. Thomann & D.M. Di Toro 1973. *Dynamic water quality forecasting and management.* Office of research and development, USEPA.

O'Connor, D.J., R.V. Thomann & D.M. Di Toro 1976. Ecologic models. In A.K. Biswas (ed.), *Systems approach to water management:* 294–333, McGraw–Hill, Tokyo.

Okoye, J.K. 1970. Characteristics of transverse mixing in open–channel flows. *Report KH R–23.* Keck Laboratory, California Institute of Technology.

Orlob, G.T. (ed.) 1983. *Mathematical modeling of water quality: streams, lakes, and reservoirs.* International Institute for Applied Systems Analysis, John Wiley & Sons, London.

Oskam, G. 1993. Water quality management in the Biesbosch reservoirs with special reference to anti–eutrophication measures. *Proc. integrated water management seminar:* 243–268, Tokyo.

Peaceman, D.W. & H.H. Rachford Jr. 1955. The numerical solution of parabolic and elliptic differential equations. *J. Society of Industrial Applied Mathematics.* 3(4):28–41.

Pinder, G.F. & J.D. Bredehoeft 1968. Application of the digital computer for aquifer evaluation. *Water Resources Research.* 4(5):1069–1093.

Pinder, G.F. & W.G. Gray 1977. *Finite element simulation in surface and subsurface hydrology.* Academic Press, New York, 295 pp.

Sayre, W.W. & T.P. Yeh 1973. Transverse mixing characteristics of the Missouri river downstream from the Cooper Nuclear Station. *IIHR Report 145.* Iowa Institute of Hydraulic Research, University of Iowa, Iowa City.

Sawyer, C.N. & P.L. McCarty 1978. *Chemistry for environmental engineering.* 3rd ed., McGraw–Hill, Tokyo.

Simons, T.J. 1971. Development of numerical models of Lake Ontario. *Proc. 14th conference on Great Lakes research:* 655–672, International Association of Great Lakes Research, Ann Arbor.

Simons, T.J. 1973. Development of three–dimensional numerical models of the Great Lakes. *Ontario science series No.12.* Canada Center for Inland Waters, Burlington.

Simons, T.J. 1980. Circulation models of lakes and inland seas. *Canadian bulletin of fish and aquatic sciences 203.* 146 pp.

Spaans, W., N. Booij, N. Praagman, R. Noorman & J. Lander 1989. *Duflow – A micro computer package for the simulation of one–dimensional unsteady flow in channel systems.* International Institute for Hydraulic and Environmental Engineering, Rijkswaterstaat Dienst Getijdewateren & Delft University of Technology, Delft, the Netherlands.

Strang, G. & G.J. Fix 1973. *An analysis of the finite element method.* Prentice–Hall, New Jersey.

Streeter, V.L. & E.B. Wylie 1983. *Fluid mechanics.* McGraw–Hill, Singapore.

Tatom, F.B. & W.R. Waldrop 1978. Curvilinear unsteady two–dimensional depth–averaged hydrodynamic model. *Proc. conference on environmental effects of hydraulic engineering works:* 329–339, International Association of Hydraulics Research, Knoxville.

Taylor, C. & T.G. Hughes 1981. *Finite element programming of the Navier–Stokes equations.* Pineridge Press, Swansea, 244 pp.

Taylor, G.T. 1954. Dispersion of matter in turbulent flow through a pipe. *Proc. The Royal Society, series A.* London.

Tchobanoglous, G. & E.D. Schroeder 1987. *Water quality.* Addison–Wesley, Menlo Park, California.

Thomann, R.V., D.J. O'Connor & D.M. Di Toro 1970. Modeling of the nitrogen cycle in estuaries. *Advances in water pollution research.* 2(III):9/1–9/14, Pergamon Press.

Thomann, R.V., D.M. Di Toro & D.J. O'Connor 1974. Preliminary model of the Potomac estuary phytoplankton. *J. ASCE., Env. Eng. Div.* 100:699–715.

Thomann, R.V., D.M. Di Toro, R.P. Winfield & D.J. O'Connor 1975. Mathematical modeling of phytoplankton in Lake Ontario. *OR report EPA 660/3–75–005.* USEPA, Carvallis.

Thomann, R.V., N.J. Jaworski, S.W. Nixon, H.W. Paerl & J. Taft 1985. *The 1983 algal bloom in the Potomac estuary.* Prepared for Potomac Strategy State/EPA Management Committee, Washington, D.C.

Thomann, R.V. & J.A. Mueller 1987. *Principles of surface water quality modeling and control.* Harper & Row, New York.

Van Dam, J.C. 1992. Problems associated with saltwater intrusion into coastal aquifers and some solutions. *Proc. ILT seminar on problems of lowland development.* Saga, Japan.

Van Donk, E., M.P. Grimm, R.D. Gulati, P. G.M. Heuts, W.A. de Kloet & L. Van Liere 1990. First attempt to apply whole–lake food–web manipulation on a large scale in the Netherlands. *Hydrobiologia.* 200/201:291–302.

Walsh, J.J. 1976. Models of the seas. In D.H. Cushing & J.J. Walsh (eds.) *The ecology of the seas:* 388–407, Blackwell Scientific Publications, London.

Wang, J.D. & J.J. Connor 1975. Mathematical modeling of near–coastal circulation. *Report No. W152.* Ralph M. Parsons lab. for water resources and hydrodynamics, M.I.T., Cambridge, Massachusetts.

Wendt, J.F. 1992 (ed.). *An introduction to computational fluid dynamics.* Springer–Verlag New York, 291 pp.

Williams, B.J. & J.B. Hinwood 1976. Two–dimensional mathematical water quality model. *J. ASCE., Env. Eng. Div.* 102(1):149–163.

World Health Organization (WHO) 1984. *Guidelines for drinking water quality.* vol.1., WHO, Geneva.

Yotsukura, N. & E.D. Cobb 1972. Transverse diffusion of solutes in natural streams. *U.S.G.S. Prof. paper: 582–C.*

Yotsukura, N., H.B. Fisher & W.W. Sayre 1970. Measurement of mixing characteristics of the Missouri river between Sioux city, Iowa, and Plattsmouth, Nebraska. *U.S.G.S. Water supply paper: 1899–G.*

Yotsukura, N. & W. Sayre 1976. Transverse mixing in natural channels. *Water Resources Research:* 12(4):695–704.

Zienkiewicz, O.C. 1971. *The finite element method in engineering science.* McGraw–Hill, London, 521 pp.

Zienkiewicz, O.C. & K. Morgan 1983. *Finite elements and approximation.* John Wiley & Sons, New York, 327 pp.

15 WATER MANAGEMENT FOR AGRICULTURE

V. V. N. MURTY
Asian Institute of Technology, Bangkok, Thailand

INTRODUCTION

Lands affected by fluctuating surface water levels are generally referred to as lowlands. Lowlands could exist due to natural topographical conditions or come into existence due to human activities like excessive groundwater withdrawal, extraction of oil and gas or land reclamation in coastal areas in the form of polders. In the development and management of lowlands a wide range of issues relating to engineering, ecology and agricultural production need to be considered. In respect of agricultural production, lowlands around the world support large cropped areas and appropriate management of these areas is an important aspect in lowland development. In most of the Asian countries, rice, an important food crop, is grown under lowland conditions. The lowland rice cultivated areas are relatively flat in topography and have standing water for a considerable period of time during a year. In managing the lowlands for agricultural production, there are several issues and problems which are to be handled. Crop production requires large quantities of fresh water and efficient water management will reduce the demands on fresh water resources. As there is a continuous demand on land resources for industrial and urban development, an efficient utilization of the available land resources for agriculture is required. Lowland crop production systems, especially rice fields, are known to be causative factors in certain environmental hazards (Mather 1984). Lowland management should be able to recognize the environmental hazards and initiate steps for mitigating the same. In this section different issues to be considered in water management for agriculture in lowlands with special reference to management of areas under rice cultivation are outlined.

Demand for water for agriculture in lowlands

Despite the fact that large parts of lowlands could be subjected to periodic inundations, agriculture in lowlands require fresh water resources.

The future population growth, the resulting demand for food production, availability of water resources and sustainable development are interrelated. The projections of population in the world as made by the UN (1991) indicates that of the population increase up to the year 2000, about 90 percent will happen in the developing countries. Such an increase in the population will naturally require higher agricultural production in the respective countries of the region. Agriculture will continue to be the major consumer of the world's water resources even though in future it can be expected that there will be increasing demand from the industry and municipal sectors (Tables 15.1 and 15.2). The growing scarcity of water is likely to result in intense competition between the agricultural and non–agricultural sectors and if market forces are allowed to prevail, water availability for agriculture will decrease.

The Food and Agriculture Organization emphasise the importance of water in crop production by stating: "not withstanding the fact that land is indispensable for agricultural production, it is water rather than land which is the binding constraint. It is only when this water constraint is released that other technical constraints such as nutrients and pests become important" (Rydzewski & Abdullah 1992).

Wherever large irrigation systems supply water for agriculture in lowlands, the management of the irrigation systems and the development of the lowlands are to be considered together. Efficient management of irrigation systems involve appropriate water deliveries to match crop water requirements, control of seepage from the conveyance system and land grading for attaining higher application efficiencies. As the volume of fresh water handled by the irrigation systems is considerable, improvements in the management of irrigation systems will help in conserving the fresh water resources and make the irrigated agriculture sustainable.

Compared to most of the agricultural crops, lowland rice requires large amount of water for successful crop production. Closely related with water are the environmental concerns in rice cultivation. In addition to some health hazards, with increasing use of fertilizers, insecticides and pesticides, contamination of water resources can be expected. In almost all Asian countries, at present it is practically impossible to allocate more land resources for rice cultivation. The same can also be said about water resources. It therefore appears that the existing rice fields need to be managed in an efficient manner with a view to minimize the water requirements as well as environmental hazards. Structural rehabilitation of rice fields and modernization of rice field management are important in the development of lowlands for sustainable agriculture. Table 15.3 shows the extent of rice cultivated areas in some Asian countries.

Table 15.1. Estimated changes in water use (WRI 1990).

	1980s use (m^3 x 10^9 y)			Estimated use in 2000 (m^3 x 10^9 y)		
	Consumptive Use	Waste Water	Total Use	Consumptive Use	Waste Water	Total Use
Agriculture	1623.0	583.0	2206.0	1920.0	665.0	2585.0
Domestic	110.2	152.9	263.1	174.5	282.0	456.5
Industry	98.1	661.8	759.9	225.5	993.0	1218.5
Total	1831.3	1397.7	3229.0	2320.0	1940.0	4260.0

Table 15.2. World water demand (km^3 y^1) according to use (Shiklomanov 1991).

Water users	1900	1940	1950	1960	1970	1980	1990	2000
Agriculture								
A	525	893	1130	1550	1850	2290	2680 (68.9)	3250 (62.6)
B	409	679	859	1180	1400	1730	2050 (88.7)	2500 (86.2)
Industry								
A	37.2	124	178	330	540	710	973 (21.4)	1260 (24.7)
B	3.5	9.7	14.5	24.9	38.0	61.9	88.5 (3.1)	117 (4.0)
Municipal Supply								
A	16.1	36.3	52.0	82.0	130	200	300 (6.1)	441 (8.5)
B	4.0	9.0	14	20.3	29.2	41.1	52.4 (2.1)	64.5 (2.2)
Reservoirs								
A	0.3	3.7	6.5	23.0	66.0	120	170 (3.6)	220 (4.2)
B	0.3	3.7	6.5	23.0	66.0	120	170 (6.1)	220 (7.6)
Total								
A	579	1060	1360	1990	2590	3320	4130 (100)	5190 (100)
B	417	701	894	1250	1540	1950	2360 (100)	2900 (100)

A: Total water consumption, B: Irretrievable water losses.
Percentage figures in parenthesis

Table 15.3. Rice areas in selected regions of Asia.

Region	Rice Ecosystem Areas (thousands of hectares)				
	Irrigated	Rainfed lowland	Upland	Deep Water	Total
NW & S India, Pakistan, Sri Lanka	71.0% (10,640)	16.2% (2429)	9.1% (1367)	3.7% (560)	100.0% (14,996)
East India, Nepal, Myanmar, Bangladesh	18.9% (8566)	48.5% (21,927)	14.4% (6510)	18.2% (8252)	100.0% (42,255)
Thailand, Laos, Kampuchea, Vietnam	22.1% (3845)	52.9% (9204)	8.1% (1401)	16.9% (2935)	100.0% (17,385)
Philippines, Malaysia, Indonesia	58.2% (8130)	23.6% (3290)	13.4% (1873)	4.8% (669)	100.0% (13,962)
China, Dem. Rep. of Korea, Rep. of Korea	92.3% (32,196)	5.5% (1910)	2.2% (779)	0.0% (0)	100.0% (34,885)
–Total	50.1% (63,377)	30.5% (38,760)	9.4% (11,930)	9.8% (12,416)	100.0% (126,483)

–(Percentages are relative to total area of each region; modified from IRRI 1991)

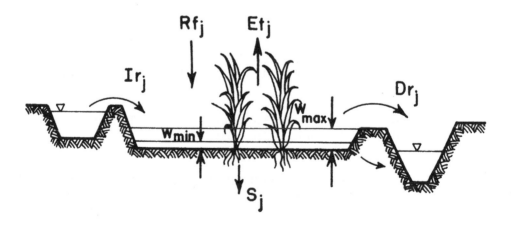

Figure 15.1. Water balance components in a lowland rice field.

WATER BALANCE COMPONENTS IN RICE FIELDS

In some of the lowlands, where the natural precipitation is adequate and can meet the crop water requirements, rice is grown under rainfed conditions.

At locations where rainfall is inadequate or ill distributed, crop water requirements are to be supplemented by irrigation. In either case, the water balance components need to be understood for an efficient utilization of the water.

Referring to Figure 15.1 the water balance components in a paddy field can be expressed as:

$$W_j = W_{j-1} + Rf_j - Et_j - S_j + Ir_j - Dr_j \tag{15.1}$$

where W_j = water depth in the field at the end of the given period, W_{j-1} = water depth in the field at the beginning of the given period, Rf_j = effective rainfall during the period, Et_j = crop evapotranspiration, S_j = mean seepage and percolation for the period, Ir_j = depth of the irrigation, and Dr_j = surface drainage.

The water balance equation is applicable to any selected period of time like a day, a week or a month. If *Wmax* is designated as the maximum depth of water possible in the rice field, *Wopt*, the optimum depth and *Wmin*, the minimum depth at which irrigation is to be given, the water balance equation can be used for determining the irrigation schedules and the depth of water to be applied for each irrigation. The values of Et_j and S_j are estimated for the given period. Rainfall occurring during the period under consideration will add to the water balance to the extent that the field is capable of retaining the rainfall, based on the initial depth of water. Excess rainfall will go as drainage.

Evapotranspiration

Evaporation from soil and water surfaces, and transpiration from the plant leaves are combined, and is known as evapotranspiration (ET). It is also referred to as consumptive water use. Evapotranspiration depends upon climatic parameters like solar energy, wind velocity, relative humidity and temperature, plant characteristics and soil–water regime. Maintaining ET at the potential rate i.e. the rate which is not hindered by water shortage is essential for high yields of rice. Crop yields are likely to decline with reduced rate of ET.

In order to plan for the irrigation deliveries as well as other water management practices, an estimation of the evapotranspiration values is needed. The approaches that are available for estimating evapotranspiration values could be broadly divided into two groups.

Hydrological or water balance approaches: These methods include hydrological approaches for determining the water balance components, soil moisture sampling and use of lysimeters. Detailed description of the methods are given in ILRI (1973) and Jensen et al. (1990). Among these methods, lysimeters are the most accurate and are usually used to calibrate other methods. Lysimeters are isolated soil masses installed in the field either with appropriate weighing mechanisms or with water table control. For upland crops, weighing type of lysimeters are used for evapotranspiration studies (Allen et al. 1991). For lowland rice, however, a different principle is used. Circular or square tanks are embedded in the cropped area and the same crop is planted in the lysimeter as well as in the field where it is installed (Figure 15.2). Evapotranspiration values are measured by measuring changes in the water levels in the lysimeter. Care should be taken that the soil and the crop in the lysimeter are similar to the field in which the lysimeter is installed.

Methods based on Meteorological approaches: From a large number of methods based upon different meteorological approaches, Doorenbos & Pruitt (1977) selected four methods and outline procedures for using these methods for estimating the evapotranspiration values. These are:
  i)   Radiation method,
  ii) Blaney–Criddle method,
  iii) Penman method, and
  iv) Pan evaporation method.
Among these methods, the Penman method requires the most climatological data, but supposed to give the best accuracy. The pan evaporation method is the most convenient as it requires only the pan evaporation data in the calculations.

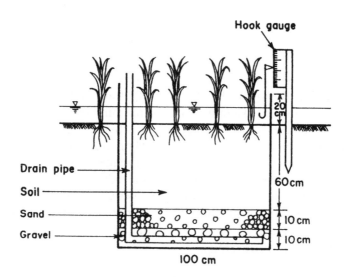

Figure 15.2. Lysimeter for measuring evapotranspiration.

In the four methods mentioned above, the reference crop evapotranspiration is calculated first and using the values of crop coefficients, the evapotranspiration values for the particular crop are calculated. In respect of lowland paddy, it is reported that in tropics the evapotranspiration values during wet season are 4–5 mm/day, while in dry season values of 6–7 mm/day can be expected (De Datta 1981). Tomar & O' Toole (1979) estimated the regional evapotranspiration values of rice as given in Figure 15.3. Table 15.4 shows the range of values of ET for different crops estimated under different agro–climatological conditions.

In a given location, knowing the evapotranspiration of the crops and the other components of the water balance equation, the crop water requirements can be estimated. Application and conveyance efficiencies are to be considered when demands on the irrigation systems are to be determined.

Seepage and percolation

Seepage losses refer to the loss of water through the borders of the rice fields and are particularly significant when the fields are located adjoining natural or artificial drainage channels. Seepage losses are influenced by the rice field layout and condition of the perimeter embankment. Improving the condition of the embankments could reduce the seepage losses, but certain amount of seepage losses mainly occurring due to drying and cracking of the soil are unavoidable.

N–Type Apparatus (Figure 15.4) is used for the combined measurement of vertical percolation and actual evapotranspiration. Rubber bags are used in order to maintain the same hydraulic head inside and outside the instrument for uniform downward percolation of water. Bags are to be replaced (generally every alternate day) to avoid any leakage of water from inside or vice versa. Daily water subsidence is measured by micrometer hook gauge which gives differences in water depth. Subtracting the amount of water lost from the lysimeter, the N–type apparatus readings give the vertical percolation from the respective fields.

Inclined sloping gauges (Figure 15.5) can be used for the measurement of total water requirements. This instrument is designed in such a way that for one unit of water level loss, the instrument shows five units of loss. Hence, relatively accurate readings can be obtained. The change in water level in the field using a sloping gauge will give the sum of ET, percolation and seepage provided that there is no rainfall or surface drainage during the period of measurement.

Percolation losses refer to the vertical movement of water through the soil profile. In lowland rice fields, percolation losses could vary from 1 mm/day to values of 30 mm/day or higher depending upon the soil type. Puddling of the top soil is done in order to reduce the percolation losses. There is no clear indication of the desirable rate of percolation required for optimum crop yield. Research in Japan has indicated that a percolation rate of 10–15 mm/day was

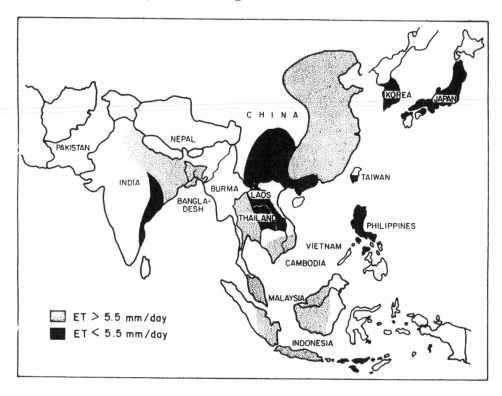

Figure 15.3. Regional evapotranspiration values of rice (after Tomar & O'Toole 1979; with permission from IRRI).

Table 15.4. Approximate range of seasonal ET for selected crops.

Seasonal ET Crop	(mm)	Seasonal ET Crop	(mm)
Bananas	700–1700	Potatoes	350–625
Beans	250–500	Rice	500–900
Cotton	550–950	Sorghum	300–650
Deciduous trees	700–1050	Soybeans	450–825
Grains (small)	300–450	Sugarbeets	450–850
Maize	400–750	Sugarcane	1000–1500
Oil seeds	300–600	Sweet potatoes	400–675
Onions	350–600	Vegetables	250–500
Oranges	600–950		

(Modified from Doorenbos & Pruitt 1977)

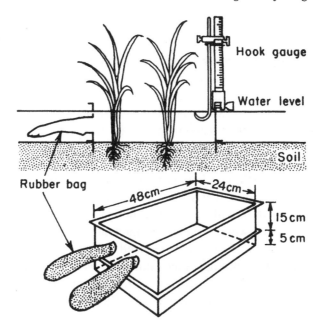

Figure 15.4. N–type instruments for rice fields (IRRI 1989).

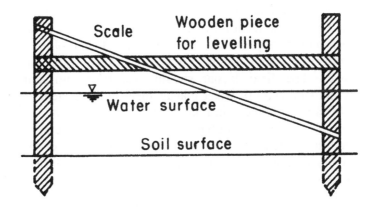

Figure 15.5. Sloping gauge (scale 1:5 ratio).

favorable for supply of dissolved oxygen, the removal of harmful substances, and the maintenance of root activity. However, there was little benefit on yield under good soil conditions. In fact, with some situations, the loss of plant nutrients may be serious if the percolation rate is high (De Datta 1981).

PRACTICES FOR EFFICIENT WATER USE

An understanding of the water balance components in rice fields at a given location not only helps in proper planning of the irrigation schedules but also helps in initiating appropriate measures for controlling losses if they are far excess of permissible values. It should also be realized that high yields of rice do not require continuously standing water in the field. High yields could be obtained with saturated soil regime maintained continuously in the field to allow evapotranspiration to take place at the potential rate. In case of the problem of weed growth, continuous flooding up to panicle initiation and then continuous saturation after that seems to be the most appropriate technique of water–efficient irrigation without yield reduction (Tabbal et al. 1992). Different factors influencing the water balance components are shown in Table 15.5.

Excessive percolation losses are reduced either by puddling or by soil compaction. Seepage through embankments can be reduced by proper maintenance and compaction. Irrigation schedules are to match the crop water requirements. Several of these practices for efficient water use in rice fields are outlined by Koga (1992).

FIELD LAYOUT AND CONSOLIDATION

The objective of field layout design in relation to water management is to improve the fields to an ideal form and function for water utilization conditions such as distribution, application and drainage. Other objectives consist of achieving effective use of rainfall, reuse of surface water and reduction in system losses. The field layout should also take into consideration the economics and simplicity of water management in the system and the working efficiency of farm machinery in relation to the given layout.

The layout of rice fields in a given area initially depend upon the topography but gets modified over a period of time. Several factors like social conditions, land ownership, methods of cultivation, water supply, etc. influence the layout. An examination of the field layouts in many of the Asian countries indicate the following problems or difficulties.

i) Because of the irregular layouts followed, considerable land area is lost in bunds (low embankments).

ii) Movement of water either for irrigation or drainage becomes difficult.

iii) Utilization of rainfall is not up to the maximum possible extent.

The rice field layouts could broadly be classified as (i) plot–to–plot layout and (ii) individual plot layout. There are varied forms and mixed layouts consisting of both of these systems (Figure 15.6).

In individual plot layout irrigation and drainage are easily facilitated, whereas in plot–to–plot layout water has to move from one field to another. Each of

Table 15.5. Factors influencing evapotranspiration and percolation in rice fields (Modified from Watanabe 1992).

Evapotranspiration	Meteorological conditions; a) temperature b) humidity c) solar radiation d) wind direction e) others and velocity Rice growing; a) variety b) soil c) meteorology d) cultivation management e) others
Vertical (deep) percolation	Hydraulic conditions; a) lay of land b) water level of drainage c) groundwater table canal d) rice transpiration e) others Permeability of soil; a) soil profiles b) cracks and porosity c) oxidation–reduction potential d) generation of gas e) tillage f) water management g) others
Horizontal (levee) percolation	Hydraulic conditions; a) altitude of ground surface from next lots b) ponding water depth c) water level in drainage canal d) water management Permeability of levee; a) soil of levee b) water content in levee c) levee management d) small animals and e) field layout insects in levee

these systems need to be examined within a given hydrological and topographical condition. The objective of the system layout is to obtain efficient movement of water both for irrigation and drainage. Movement of machinery from field to field also need to be considered. The objective of rice field layout design in relation to water management is to improve the fields to an appropriate form and function for water utilization conditions such as distribution, application and drainage. At the same time the goal is to achieve highest possible effective use of rainfall, reuse of surface runoff water and reduction of system losses like conveyance and seepage losses.

Rice field consolidation in Japan

In the area of land consolidation, particularly with reference to lowland rice fields, Japan has made considerable progress.   In the land consolidation projects, the fields have been laid with irrigation and drainage channels, roads and changed ownership as necessary.   This has improved the land and labor

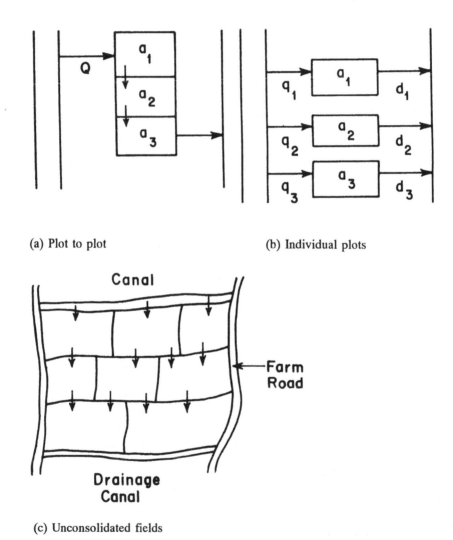

(a) Plot to plot   (b) Individual plots

(c) Unconsolidated fields

Figure 15.6. Different forms of field layouts.

productivity and also made the introduction of modern farm machinery. Table 15.6 shows the progressive changes and relations in respect of rice farms and farming systems in Japan.

In order to evaluate paddy fields from the view point of land consolidation, Yamaji (1992) suggests the following standards.

i) Accessibility to roads and canals: In one cultivation unit, road, irrigation canal and drainage canal are all to be accessed. One cultivation unit is usually one field lot in Japan.

ii) Surface soil: The surface soil has appropriate properties, such as water holding capacity, drainage, bearing capacity and nourishment, and is not polluted.

iii) Land levelling: Land levelling of surface soil should be sufficient. The standard deviation of the measured data from average height of surface ups and downs should be less than 28 mm in the stage of consolidation, and less than 19 mm in the stage of farming.

iv) Shape and size: Present agricultural methods do not use human power, but the power of livestock and machinery. So it is desirable to work straight along longer side of the field. Therefore, the shape is desirably to be rectangular and long and slender because a small number of turning makes machinery operation efficient.

Rice fields in other countries

In USA, six states namely Arkansas, California, Louisiana, Mississippi, Missouri and Texas produce rice. It has been reported that in Texas about 20% of rice fields are consolidated into flat level type, whereas in California the fields have been consolidated and given a mild grade to help in surface drainage (Yamaji 1992).

In Thailand, the land consolidation projects have been termed as the intensive and extensive types. In the intensive type, attempts are made to change the structure of roads and canals radically whereas in the extensive type some minimum improvements in the layout of roads and canals are made.

In respect of the existing rice fields, the following points need to be considered.

i) In a given area, the layout of the rice field need to be examined both from irrigation and drainage view point. Irrigation and drainage channels need to be provided to optimize the time required for these operations.

ii) The paddy field embankments need to be examined and straightened wherever possible providing suitable height. Peripheral embankments are to be suitably compacted to reduce seepage.

iii) Farm roads as appropriate need to be provided for movement of farm produce and machinery.

Table 15.6. Changes and relation between rice farming and farm system in Japan (Yamaji 1992).

	19th Century	Recent	Present	Pioneer
Reclamation	Manpower & Cattle	Manpower & Cattle	Present	Laser–controlled bulldozer
Fieldwork				
Seeding	Transplant	Transplant	Transplant	Direct seeding
Nursery	Field	Field	Nursery Box	(none)
Plowing	Hoe & plow	Hoe & plow	Tractor	Tractor (large)
Irrigation	Manpower	+ pump	+ pump	+ pump
Puddling	Manpower	+ Cattle	Tractor	Tractor (L) or none
Transplanting	Manpower	Manpower	Machine	(none)
Fertilizer Application	Manpower	Manpower	Machine	+ Water Solution
Pest Control	(agronomical)	Pesticide	Sprayer (pesticide) Helicopter	+ Small Helicopter
Weeding	Manpower	Manpower	Sprayer (herbicide)	+ Small Helicopter
Water Management	Gate operation	Gate operation	+ Auto Irrigator	+ Auto Irrigator
Drainage	Natural	+ Tile drain	+ Mixed Tile Drain	Both–way U.G. Pipe
Harvest	Sickle	(+ binder)	Combine Harvester	Combine Harvester (L)
Land Owning	Landlord	Owner Farmer	Owner Farmer	Gather Culti. Right
Typical Field Lot				
Length of Longer Side	depend on topographic condition	54 m	100 m	150 – 200 m
Length of Shorter Side		18 m	30 m	100 – 500 m

POLDERS

The term polders refer to certain special types of drained lands established in low–lying coastal areas, flood plains, shallow lakes and in upland depressions. Luijendijk et al. (1988) provide the most widely used definitions of polders as:

i) A tract of lowland reclaimed from the sea, or other body of water, by dikes, etc. In the polder, the runoff is controlled by sluicing or pumping, and the water table is independent of the water table in the adjacent areas.

ii) A level area which was originally subjected to a high water level, either permanently or seasonally due to either ground water or surface water. It is referred to as polder when it is separated from the surrounding hydrological regime in such a way that its water level can be controlled independently of its surrounding regimes.

In terms of water management for agriculture, demand, utilization, drainage and groundwater control, polders and lowland rice fields have several similarities. Water balance equation similar to that of Equation 15.1 can be formulated taking into consideration leaching requirements and groundwater contribution wherever applicable. Evapotranspiration requirements of crops will be similar. Polders situated in coastal areas will be subject to the influence of sea water and suitable provision for leaching of salts has to be provided. Soil reclamation could be part of development of the polders in many locations. Subsequent water management practices are developed taking into consideration the requirements of soil management as well. Luijendijk et al. (1988) outline the various stages of development of polder areas relating to water management infrastructure and the kind of land use as follows:

1. First stage: open uncontrolled drainage system; rain–fed agriculture; one food crop annually; subsistence farming; main transportation by boats;

2. Intermediate stages: semi–closed drainage system; semi–controlled water management based on flood and saline water protection; flushing in the case of acid sulphate soils and water conservation by surface and subsurface water control; mainly rain–fed agriculture; more than one crop annually; intermediate economic land farming; agro–processing industry and internal road system; trans–area transport;

3. Final stage: closed system/fully controlled water management/pumping; irrigation system; multiple cropping and integrated farming; commercial farming; sustained economic growth; industries and complete road infrastructure system.

DRAINAGE OF AGRICULTURAL AREAS IN LOWLANDS

Drainage of agricultural areas situated in lowlands is a difficult job mainly because of finding an outlet for the drainage water. In spite of this difficulty wherever possible drainage of agricultural lands should be attempted as it will help in improving the productivity of these areas.

The drainage of upland crops and lowland rice differ in many ways. Lowland rice requires a shallow water depth during most of its crop period. Standing water is initially needed for land preparation and transplanting. Rice is able to absorb nutrients when the soil is saturated and water is standing on the soil surface. Despite these differences, provision of drainage facilities is desirable as drainage helps in timely agricultural operations, improving the soil physical properties and also helps in crop diversification in rice fields.

In respect of rice fields, three types of drainage systems need to be considered depending upon the particular situation.

1. A shallow surface drainage system for the removal of excess water from the soil surface. This consists of a network of open drains serving one or a group of fields and ultimately delivering the water to an outlet. This system is generally desirable for all rice fields except those on very pervious soils.

2. A shallow surface drainage system combined with deep open drains. This type of system is required essentially when the water table is to be lowered in the root zone. Lowering of the water table might be required either for leaching purposes or crops other than rice are proposed to be grown.

3. A shallow surface drainage system combined with a subsurface system to provide for groundwater table control.

Design of surface drainage systems

Surface drainage to control the water level in the rice fields is achieved by a network of open channels serving each individual or a group of fields and ultimately delivering the water to the outlet. The concept of drainage coefficient or drainage modulus is used in the design of drainage systems for agricultural lands. The drainage coefficient is defined as the depth of water to be removed in a 24 hour period from the entire drainage area. It is also expressed as the flow rate per unit area. Based upon the rainfall characteristics and soils, values of the drainage coefficient are determined such that no appreciable damage is caused to the crops grown in the area (ILRI 1973). Knowing the values of the drainage coefficient, the drainage channels are designed based on the principles of open channel hydraulics.

Subsurface drainage

Subsurface drainage refers to the removal of excess water from the crop root zone. This is achieved by means of subsurface drains located below the soil surface.

Extensive agricultural areas are under subsurface drainage in USA and the Netherlands. Detailed procedures for their design and operation can be found in Shilfgaarde (1974) and the publications of the International Institute for Land Reclamation and Improvement, the Netherlands (ILRI 1973). In the Asian context, subsurface drainage systems are being installed in the salt affected irrigated areas of the Indus basin in Pakistan and India. In respect of lowlands, subsurface drainage systems are being extensively installed in Japan (Tabuchi 1985, Nagahori 1989). Installation of subsurface systems in paddy areas have also been reported from China (Soong & Wei 1985).

Design of subsurface drainage systems

The design of a subsurface drainage system essentially consist of:
i)   determining the drainage area,
ii)   spacing and depth of the subsurface drains,
iii)   size and materials of subsurface drains,
iv)   ancillary structures,
v)   outlets,
vi)   installation procedures, and
vii)   operation of the system.

The subsurface drainage systems installed in the upland irrigated areas and those installed in lowland rice fields differ in many aspects of design and functioning. The details of the installation of subsurface drainage systems in rice fields are outlined by Tabuchi (1985) and Nagahori (1989). The different features of the subsurface drainage system as adopted in Japan are as follows.

1. Design of drain discharge: The design discharge depends upon the allowable time period of removal of surface water from the fields. It is assumed that land grading can achieve precision up to 5 cm and as such, a maximum of 5 cm of water has to be removed from the fields. With 2 to 5 cm of water and 1 to 2 days of allowable discharge period, the design discharge for subsurface drainage becomes 10 to 50 mm/day. Total volume of discharge is calculated by multiplying the design discharge by drainage area.

2. System layout: A subsurface drainage system generally consists of lateral drains, collecting drains, relief wells, outlets and ancillary structures. Typical layout of the system is shown in Figure 15.7. Lateral drains are generally of PVC, but at some locations bamboos and wood are used.

3. Spacing, depth, grade and size: The spacing of the laterals is generally about 10 m in paddy fields with heavy clay soils. Depths are in the range of 0.8 to 1 m. Grades adopted range from 1/100 to 1/600 in order to keep sufficient velocity in the pipes. Diameter of the pipes range from 50 mm to 100 mm.

4. Supplementary drains: When the average hydraulic conductivity of the soil is very low (less than 1×10^{-5} cm s^{-1}), normal subsurface drainage cannot be very effective. In such situations, supplementary drains at relatively shallow depth and close drain spacing are necessary to provide adequate drainage. Table

15.7 shows various kinds of supplementary drains under different soil conditions.

Drainage of polders
A drainage system for the polders, particularly those areas that are situated below the water level of adjoining areas is very important. The drainage system usually consists of both surface and subsurface drainage, a hydraulic conveyance system in the form of ditches, ancillary structures like culverts, weirs etc. and outlet structures like sluices or pumping stations. Design of the system is based upon crops, soils and hydrological parameters of the areas. Crop factors consist of the degree of protection needed from water table and salinity tolerance. Soil factors to be considered involve hydraulic conductivity,

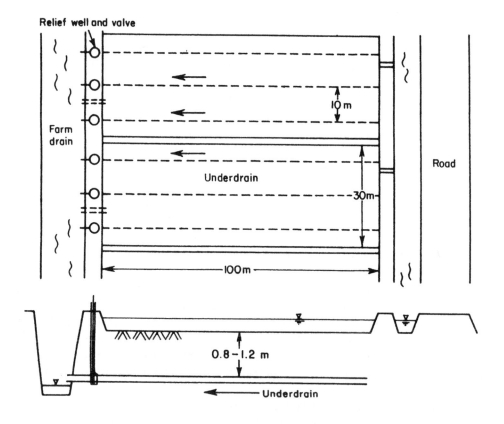

Figure 15.7. Typical layout of a subsurface drainage system in rice fields (after Tabuchi 1985).

Table 15.7. Kinds and characteristics of supplementary drains (after Nagahori 1989).

Kinds	Applicable soil	Characteristics of installation
mole drains (unfilled)	soil with good mole sustainability	easy installation by tractor
mole drains (filled with filter material)	impermeable soil with poor sustainable cracks	easy installation under weak foundation, easy rehabilitation
filter material drains	impermeable soil with poor sustainable cracks	requires trencher for installation, rice husks etc. can be utilized
subsoil breaking	hard foundation, sustainable cracks	easy operation

salt content, etc. Together with the hydrological parameters, soil factors determine the depth and spacing of the surface and subsurface drains. Suitable materials for the subsurface drains, sizes, grades, envelope materials and methods of installation are to be decided depending on the site conditions. Case studies of polders in the Netherlands as well as in other parts of the world can be found in ILRI (1982).

GROUNDWATER CONTROL

Groundwater levels affect the soil moisture conditions above the groundwater table. Shallow groundwater levels, therefore, affect crop yields and also influence trafficability of farm machinery. Control of groundwater level in lowlands should aim at balancing of the twin goals of providing moisture to the crop root zone and eliminating excess water. Groundwater table control is, therefore, very much influenced by the water quality, nature of the soil, type of crop and the range of fluctuations of the water table.

In lowlands, water table is in a dynamic state rather than in static condition. The water table rises during recharge events like rainfall or irrigation, and falls during the drainage period. The optimum ranges of fluctuation is important instead of a single optimum depth. This has to be determined for the particular site conditions, and will depend upon the factors mentioned earlier. The design and functioning of the drainage system will influence the range of fluctuations of the groundwater table. Subsurface drains are generally located at a depth of 2.5 m in arid regions, 1.5 m in humid regions and 1.0 m in the paddy fields.

The operation of the drainage system for groundwater control could be substantially different in polders and rice fields. In polders, salinity problems in soils and ground water can be expected. Groundwater control should, therefore, achieve leaching of the salts and keeping low salinity levels in the crop root zone. In rice fields, even though the drainage system is used for leaching, groundwater control is used for providing the soil moisture or the required submergence for crop growth (Nagahori 1989). For example, in case of paddy grown in acid sulphate soils, soil submergence will keep the pH values of the soil at a higher level providing better conditions for crop growth. Excessive drainage of such soils could depress the soil pH, adversely affecting the crop growth. Discussions on crop growth in relation to water table depths as well as salinity levels are given in Shilfgaarde (1974).

ISSUES FOR RESEARCH AND DEVELOPMENT

Based on a review of the existing information, the following issues could be identified in respect of water management for agriculture in lowlands:
a)   Matching water demand and supply in irrigation systems,
b)   On–farm development and layout of fields,
c)   Diversified cropping in rice fields,
d)   Land consolidation,
e)   Water quality,
f)   Drainage,
g)   Conjunctive use of surface and groundwater,
h)   Labor productivity (mechanization), and
i)   Environmental sustainability.

The issues mentioned above are location specific and development projects have to be initiated depending upon the needs of the region. In general, the crop water requirements need to be understood and irrigation supplies are to match the requirements. Field layout is to be improved to serve the needs of efficient water application and drainage. Installation of drainage systems has to be undertaken depending upon the needs and possibilities. The environmental hazards in relation to water management are to be assessed and steps to mitigate the hazards have to initiated. It may also be suggested that problems of waterlogging, salinization, surface and ground water quality as affected by agricultural pollutants, land degradation and vector–borne diseases be given appropriate research attention.

REFERENCES

Allen, R.G., T.A. Howell, W.O. Pruitt, J.A. Walter & M.E. Jensen 1991. *Lysimeters for evapotranspiration and environmental measurements.* American Society of Civil Engineers, New York.

Datta, S.K. 1981. *Principles and practices of rice production.* John Wiley, New York.

Doorenbos, J. & W. O. Pruitt 1977. *Crop water requirements.* FAO irrigation and drainage paper 24. Food and Agriculture Organization, Rome.

International Institute for Land Reclamation and Improvement (ILRI) 1973. *Drainage principles and applications.* Vol I – IV, Wageningen, the Netherlands.

International Institute for Land Reclamation and Improvement (ILRI) 1982. *Polders of the World.* Vol I – III, International symposium, Wageningen, the Netherlands.

International Rice Research Institute (IRRI) 1989. *Physical measurements in flooded rice fields.* IRRI, Manila, Philippines.

International Rice Research Institute (IRRI) 1991. *World rice statistics, 1990.* IRRI, Manila, Philippines.

Jensen, M.E., R.D. Burman & R.G. Allen 1990. *Evapotranspiration and irrigation water requirements.* American Society of Civil Engineers, New York.

Koga, K. 1992. *Introduction to paddy field engineering.* Asian Institute of Technology, Bangkok, Thailand.

Luijendijk, J., E. Schultz & W.A. Segeren 1988. *Polders.* (In) P. Novak (ed.). *Developments in hydraulic engineering–5*: 195–228. Elsevier Applied Science, London– New York.

Mather, T.H. 1984. *Environmental management for vector control in rice fields.* FAO irrigation and drainage paper 41, Food and Agriculture Organization, Rome.

Nagahori, K. 1989. Subsurface drainage. *Journal of irrigation engineering and rural planning.* Japanese Society of Irrigation, Drainage and Reclamation Engineering. 16:42–52.

Rydzewski, J.R. & S. Abdullah 1992. *Water for sustainable food and agriculture production.* International conference on water and the environment, World Meteorological Organization, Switzerland.

Shiklomanov, I.A. 1991. The World's water resources. (In) *International symposium to commemorate the 25 years of 1HD/1HP.* UNESCO Paris.

Shilfgaarde, J.V. (Editor) 1974. *Drainage for agriculture.* American Society of Agronomy, Madison.

Soong, S. & Z. Wei 1985. Subsurface drainage of lowland rice fields in China. (In) *Soil Physics and Rice*: 351–366 International Rice Research Institute, Philippines.

Tabbal, D.F, R.M. Lampayan & S.I. Bhuiyan 1992. Water efficient irrigation technique for rice. (In) V.V.N. Murty & K. Koga (eds.) *Soil and water engineering for paddy field management*: 164–159. Asian Institute of Technology, Bangkok, Thailand.

Tabuchi, T. 1985. Underdrainage of lowland rice fields. (In) *Soil physics and rice*: 147–159. International Rice Research Institute, Philippines.

Tomar, V.S. & J.C. O' Toole 1979. *Evapotranspiration from rice fields.* IRRI Research paper series, No. 34, International Rice Research Institute, Philippines.

UN 1991. *World population prospectus 1990.* New York.

WRI 1990. *World resources 1990–91.* A report by the World Resources Institute, in collaboration with the United Nations Environment Programme and the United Nations Development Programme, Oxford University Press.

Watanabe, T. 1992. Water Budget in Paddy Fields. (In) V.V.N. Murty & K. Koga (eds.),

Soil and water engineering for paddy field management: 1–11. Asian Institute of Technology, Bangkok, Thailand.

Yamaji, E. 1992. Standard evaluation of paddy field consolidation. (In) V.V.N. Murty & K. Koga (eds.) *Soil and water engineering for paddy field management*: 385–394. Asian Institute of Technology, Bangkok, Thailand.

16 INTEGRATED WATER MANAGEMENT

P. ANKUM
Delft University of Technology, the Netherlands

ORGANIZATION OF LOWLAND MANAGEMENT

Dual-managed systems

Lowland development involves a technical interference on a large-scale basis. Embankments, channels and structures have to be designed, constructed and financed. Individual farmers are normally not technically and financially capable for such a development, and a certain type of organization is required. It is often the government which plays this essential role. The involvement of the government does often not end at the construction phase of the project, but also extends into the "Operation & Maintenance" (O&M) phase.

It means that lowland development schemes are under dual-management. The government, or another entity, is charged with the operation, management and maintenance of the main infrastructure of the scheme. This part of the scheme belongs to the general interest of the population, and requires technical, financial and organizational capabilities that are beyond the capacity of the individual farmer. However, such an entity cannot be charged with the operational aspects at the lowest level, such as the operation and maintenance at farm level. This remains the task of the individual farmer, or may become the task of a group of farmers.

An example can be found in the Netherlands, where polder development started in the Middle Ages (8th – 13th century) with the protection of land against flooding. These activities were too comprehensive for the individual farmer and required an organization to solve the manpower and financial problems. At that time, a central or regional government was not well developed yet, so local communities had to play an important role. Thus, the Water Boards ("waterschap") were formed, being special corporations on polder management (Ven 1993).

The Water Board still exists in the Netherlands and is now a form of functional decentralisation within the State (Ankum 1992). The task of the individual farmer is to maintain the part of the polder drainage system that is located at or

within his farm boundaries. The technical office of the Water Board has responsibilities on water management matters, such as:
- operation and maintenance of hydraulic structures along coast and rivers,
- control of water quantities (mainly drainage, but also water supply),
- control of water quality,
- control of polder dikes and roads.

The above concept of dual-managed systems is also widely applied in irrigation schemes. The central government is responsible for the irrigation and
drainage main systems and for the flood control works. The local population is
grouped in "tertiary units", where they are responsible for the tertiary water
management systems.

It is obvious that technicians have to take an initiative role in lowland
development. However, the population have to react and participate in these
technical innovations in order to benefit from the investments made. It means
that a new organizational set-up of the lowland area have to be created, with
new tasks and responsibilities, and with new communication lines.

Planners and designers of these dual-managed lowland projects have to
clarify on the management philosophy and the Operation & Maintenance
(O&M) considerations. This is done by the preparation of an "O&M manual".
A table of contents of a typical O&M manual may comprise of:
- Introduction (e.g. pre-project condition, rationale for development);
- Description of the lowland scheme (e.g. location and mapping, boundaries
  of responsibilities ("tertiary units"), boundaries of villages, types of soils,
  hydrology and meteorology, location of canals and structures);
- Objectives of the lowland development (e.g. cropping calender, water and
  drainage requirements);
- Operation of the water management systems (e.g. types of structures, flow
  regulation, data processing, procedures, performance monitoring);
- Maintenance of the water management systems (schedule and type of
  routine works, design of special maintenance works, budgeting, force
  account and tendering, supervision);
- Organization of the O&M agency (e.g. tasks and job descriptions for the
  main system management, required staffing, human resources management);
- Infrastructure and organization of farmers-managed part of the system (e.g.
  tertiary unit development, organization of water users associations,
  extension to water users, O&M within the tertiary unit);
- Institutional setting (e.g. decision-making process, conflict handling,
  financing and cost recovery).

Experiences with lowland development

Lowland development was introduced in many Asian deltas during the 19th and
the early 20th century. It appeared that both rainfed agriculture as well as small-

scale and village–managed systems could not meet the requirements of increasing population in these deltaic areas and floodplains. Periodic mass starvation of the population was the result. Moreover, the cultivation of many cash crops required a more reliable supply of water. Irrigation and drainage systems, as well as flood control works were implemented. The operation of these large-scale systems was done successfully under dual–management by strict discipline, often under the Colonial rules.

The Second World War and the struggle for independence led in many countries to less attention to the maintenance of these systems, while also other social rules between the government and the population evolved. Thus, it was experienced that many dual–management systems were not functioning satisfactorily. General criticism concentrated on an inequitable distribution of irrigation water, especially on water shortages in the tail–ends.

The cause of these water problems in the tail–ends were initially considered to be found in the poor physical condition of canals and structures. So, "rehabilitation" programmes were carried out, which focused on restoring the conditions of the main irrigation and drainage systems as per original design. Examples are the PROSIDA programmes (initially some 800,000 ha) in Indonesia in the 1960s and 1970s, and the Ganges–Kobadak project (130,000 ha) in Bangladesh during the 1980s. These programmes were often not very successful, as for e.g., (i) initial design mistakes were made repeatedly, (ii) the population had increased considerably since the first construction, and required a more intensified irrigated agriculture, and (iii) the cropping pattern and the type of crop had often changed and required a more precise water management. It was learned that a "modernization" of the system is more appropriate than just restoring the system again in its initial condition through a "rehabilitation" programme. Furthermore, it was observed that the staff of the Operation & Maintenance agency did remain a weak link in the performance of the main systems (Ankum 1989).

Comprehensive Operation & Maintenance training programmes for the staff of the Operation & Maintenance agency were initiated since the 1970s as a next step in irrigated lowland development. These programmes were often dealing with the production of training aids and with the training of trainers. Other programmes called for "coordination and integration", but can often be considered as placebos only.

It was also felt since the 1970s that the water distribution within the farmers (tertiary) units did not follow the improvements made on the main system level. Although the infrastructure of the tertiary unit does not belong to the responsibility of the government but to the water users, it was decided that the Operation & Maintenance agency should develop the farmers system on behalf of the water users. Thus, extensive tertiary unit development programmes were initiated. The costs were borne by the government, and designs were made by consultants with limited time available for consultation of the users. Success of such a tertiary unit development programme was generally very limited.

Gradually, it was acknowledged that non–technical aspects also play a role, such as the "ignorance" of water users, poor organization of the water users, lack of discipline of the water users, absence of water users in the decision–making process, etc. Since the 1980s, training programmes for water users are undertaken in many countries in South and in South–east Asia. Major topics are: the organization of the water users association, the construction and maintenance of the infrastructure in the tertiary unit, the operation of the tertiary unit, water management at farm level (Ankum 1989).

Despite all the efforts, the performances of many large–scale irrigation schemes have often not improved in the expected manner. These activities require vast sums of money, and often vast foreign exchange. Questions arise on the financial sustainability of irrigation, and on whether water users could bear the future costs of lowland improvement and the regular operation and maintenance costs through an "irrigation service fee".

Further efforts have to be made to improve the large–scale irrigation and drainage systems in the lowlands. The question now is, whether just more rehabilitation, training and tertiary development is required, or that other ways have to be followed?

Components of lowland management

It seems obvious that a step–by–step approach should be followed in improving the performance of dual–managed systems in the lowlands. Moreover, it is important to recognize that there are more components in lowland development than only the infrastructural works. It means that all technical and non–technical components of lowland development should be improved gradually, instead of aiming directly at final solutions in a top–down approach.

Water management systems in the deltaic areas will generally comprise of six main components to make lowland development successful (Figure 16.1):
- the main water management systems, such as irrigation and drainage systems,
- the farmers–managed systems, i.e. the "tertiary" irrigation and drainage systems,
- the legal framework,
- the "highest authority",
- the agency for Operation & Maintenance (O&M) of main systems,
- the Water Users Associations for Operation & Maintenance of tertiary units.

All these components together contribute to "Good Lowland Management". When one of these components is absent or not well defined, good lowland management will not be achieved. In fact, "good" lowland management itself is not well–defined: what is good for one farmer, might not be good for another farmer. Even, what is good for all farmers, might not be good for the Operation & Maintenance agency. Thus, conflicts on the objectives of water management

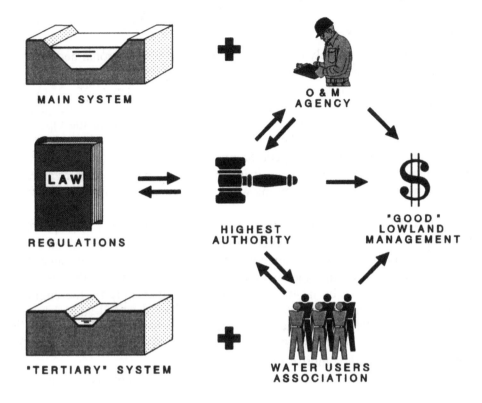

MAIN SYSTEM

REGULATIONS

'TERTIARY' SYSTEM

O & M
AGENCY

HIGHEST
AUTHORITY

'GOOD'
LOWLAND
MANAGEMENT

WATER USERS
ASSOCIATION

Figure 16.1. Components of lowland management.

might be considered normal. Basically, lowland management should follow the aspirations as set by the politicians, economists and the water users. Together, they should determine what is "good" lowland management.

Legal framework

During the end of the 18th and in the early 19th century, many large-scale water management systems were managed by engineers to the best of their knowledge. However, it was felt by them that the operation of these schemes could never be a pure technical matter, and that also socio-political decisions have to be made. Several of these decisions appeared to be long lasting, and could be written down. Thus, the North India (1873) and the Bengal Acts (1876) were promulgated in South Asia to define the roles in irrigation. Indonesia followed a bottom-up approach. Local operation and maintenance matters were covered by Operation & Maintenance (O&M) regulations concerning the Irrigation Districts, e.g. "regulations concerning the Irrigation District Pemali-Comal" (1929). The legal framework of these regulations was created by the introduction of the General Water Law (1936) and subsequently by the Provincial Water Laws (East Java in 1938, West Java in 1940, Yogyakarta in 1949 and Central Java in 1959).

It appears to be relatively easy to update or to develop National and Provincial Water Laws. They are prepared in general terms, and state that water should be controlled by the State and that delegation of power is possible. They also state that utilization of water resources shall be in the interest of the people, and may follow established priorities. It is more difficult to prepare the Operation & Maintenance (O&M) regulations for individual lowland schemes. These regulations have to outline the operation procedures and to make clear decisions on the target polder water levels, irrigation scheduling, the authorization of cropping patterns, how to react during water shortage and surplus, cost recovery, etc.

The O&M regulations should be based on the O&M manual as prepared by the designer of the system. However, the proposed operation and maintenance practices cannot be translated directly into the new regulations because they have to be placed first into the socio–political context. It means that decisions on procedures have to be taken by a wider forum than only by the engineer. Finally, the O&M regulations have to be endorsed by the local administration.

The O&M regulations of a lowland development scheme have to delineate the rights and the duties of all parties, as well as the required procedures. A typical table of contents may comprise of:
- Definitions (e.g. scheme area, responsibilities on main and tertiary systems);
- Right of control (e.g. water utilization with/without permission);
- Evacuation of drainage water (e.g. target polder water levels in summer and winter and in different polder sections, timing of pumping);
- Supply of irrigation water (e.g. manner, timing and quantity of supply, irrigation season, procedures, flow of information);
- Utilization of irrigation water (e.g. operation and maintenance in tertiary units, Water Users Associations);
- Maintenance of the water management systems;
- Financing (e.g. budgeting for Operation & Maintenance and construction, cost recovery);
- Penal provisions.

Highest authority

All lowland development schemes require a highest authority who can take not only decisions on conflicting matters, but who can also impose sanctions if necessary. It is a mistake to assign the Operation & Maintenance agency as the highest authority. It is true that the technical knowledge is available here, but the Operation & Maintenance agency should not implement decisions that have been made by them alone. Moreover, the Operation & Maintenance agency can usually not impose sanctions by means of a police involvement.

The choice for the highest authority depends to a certain extent on the ownership of the system. There are three workable options:
- the Water Users through a "Water Board",
- an Autonomous Enterprise, such as the Tennessee Valley Authority,
- the local administration through a "Water Committee".

The Water Users through a "Water Board" are usually the highest authority of the irrigation scheme, when the scheme is the property of the water users themselves. Such an organizational form is found in most of the village–managed irrigation systems, and also in the polders of the Netherlands where farmers developed the lowlands through a Water Board ("Waterschap") and without governmental involvement. The Water Boards, as applied in the Netherlands, have no fixed financial relationship with the central government, as the costs of the water management in a polder is paid by the interested parties, mainly land owners and domestic/industrial polluters. The Water Boards are supervised by the provincial governments since the early 19th century, but they have retained a high degree of independence with regard to issuing regulations concerning water control in their areas. The management of the Water Board is formed by: (i) the general assembly, (ii) the council, and (iii) the dike–reeve ("dijkgraaf") as executive. The general assembly is composed on a functional basis and represent different interest groups, mainly land owners, on an election basis. The general assembly elects the council, which does the day–to–day management of the Water Board. The dike–reeve is the chairman of the council and is employed by the assembly.

An Autonomous Enterprise, like the Tennessee Valley Authority, as the highest authority, is created when income can be generated through e.g. selling of hydropower. Thus, these schemes are financially independent from the government and the farmers. Experiences with these organizations for rural development are not very successful as they operate outside the power of the local politics.

The local administration is the highest authority in many lowland schemes in developing countries when the government considers lowland development as part of its efforts to develop the rural population. The government is the owner of these water management systems and its financial involvement is permanently required. The local administration can entrust a special "Water Committee" with the actual authority over all matters concerning water.

Such a "Water Committee", also referred to as the "Irrigation Committee" in irrigation and drainage projects, was applied in Indonesia since 1920 and is a good forum for solving problems in the irrigation and drainage schemes of the deltaic areas. Similarly, France knows the "Basin Committee" since 1964. The basin committee is a local water parliament, being the dialogue body and the decision–maker on the water management problems. The "basin agency" is the financial and technical executive body of the basin committee (Oliver 1992).

Such a Water Committee may have the following organization (Figure 16.2): (i) the Head of local administration as Chairman, (ii) the Head of the Operation

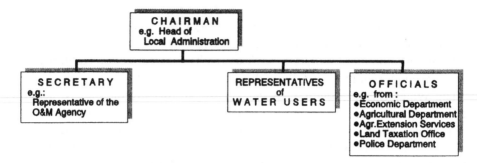

Figure 16.2. Organization chart of a water committee.

& Maintenance agency as Secretary, and the members are (iii) the represen-
tatives of the water users, and (iv) the officials from all relevant departments,
such as the Heads of Agriculture Extension Services and of the Regional Planning Office, the Police Commander. Water Committees may exist at different
levels, e.g. at the provincial, district and at the sub–district levels. The committee at the provincial level may formulates policies and provide general guidance.
The Water Committee at the district level may cover an area of one or more
irrigation and drainage schemes (25,000 – 100,000 ha) and will be in charge of
regular decision–making. At this level, policy decisions must be made operational regarding cropping pattern, planting dates, etc., and if necessary decisions
enforced. The District Water Committee may meet at least twice–a–year, but
meetings can be more frequent if special problems have to be dealt with, such
as unexpected water shortages or floods. The Water Committee at the sub-
district level may have a main task on the coordination and the implementation
of the orders received from the district level. Farmers are also expected to
actively participate in this.

Operation & maintenance agency

The Operation & Maintenance agency is the executive (technical) body between
the highest authority and the field, and is responsible for implementing the
operation and maintenance (O&M) of the main irrigation and drainage systems.
Furthermore, the Operation & Maintenance agency should provide data on water
availability, water and drainage requirements, land use, performance of irrigation
and drainage, etc., to permit proper decisions by the highest authority.

A typical Operation & Maintenance agency may have offices at different
levels (Figure 16.3):
  – Provincial level, with administrative tasks and supporting technical tasks
    on survey, design and construction, and on supervising and programming
    Operation & Maintenance;

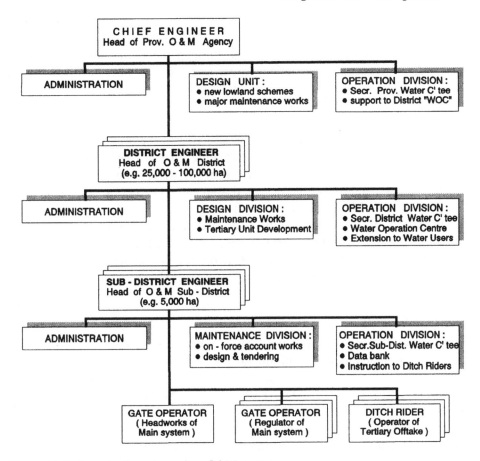

Figure 16.3. Organization chart of an O&M agency.

- District level controlling one or more irrigation and drainage schemes (25,000 – 100,000 ha), and is charged together with its sub–district offices with the actual Operation & Maintenance of the system;
- Sub–District level controlling some 5000 ha, charged with the executive tasks on Operation & Maintenance.

Basically, the O&M District office in the above set–up, is the core for the implementation of the operation decisions.

Here, a "Water Operation Centre" (WOC) has to process the data on water availability, water and drainage requirements, and should determine the required water management for the coming period. Ultimately, detailed instructions are issued to gate and pump operators. Although such a centre is not widely applied, it can be stated that its role is essential in most of the irrigation and drainage systems.

Communication with the water users on water management aspects is an important task of the O&M agency. The routine communication with the water

users concerns the day–to–day operation of the systems, such as the authorization of cropping patterns, processing water requests and information on water supply. These activities have to be done through the field staff of the O&M agency. A periodic extension to the water users should also be given by the O&M agency. Such an extension might follow the explicit requests from the water users association on certain topics and can be given by a specialized training staff. Topics may cover:
  - organization of the water users (e.g. structure of the organization, the issue of village and water boundaries, financing, external contacts);
  - tertiary unit development (e.g. layout of the water management system at field level, need for infrastructure, design and construction, assistance from Operation & Maintenance agency);
  - operation of the tertiary unit (e.g. planting scheduling, role of water master, water distribution techniques, water management at farm level);
  - maintenance of the tertiary unit (e.g. routine and incidental works, costs, maintenance techniques, organization of works).

Farmers associations

Water users have to organize themselves when the lowland development is based on dual–management, with the government in charge of the main system and the farmers in charge of the tertiary systems. A typical organization structure will comprise (Figure 16.4): (i) a board of a chairman with a secretary and a treasurer, (ii) water masters as executives for the day–to–day works in the field, and (iii) the owners and/or water users as the members.

In many countries, an organization of water users exists already within the village: the village leader and the village officials have tasks on the different functions. The operation and maintenance of the tertiary system may or may not be done by this existing village–organization. When the area of the village comprises of one or more tertiary units, it is obvious that the village leader can just appoint the board of the Water Users Associations, following the current procedures. However, the "boundary issue" plays a role when different villages

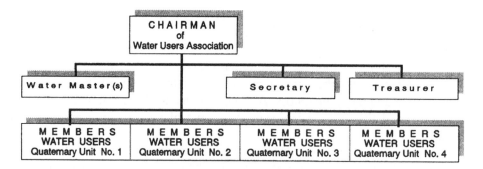

Figure 16.4. Organization chart of a water users association.

are located in one tertiary unit. The authority of the village leader then becomes no more valid.

Normally, dual-managed lowland systems are developed in a top-down approach. Technicians and economists introduce the new water management system with the objective of cultivating higher-yielding crops. The population cannot easily have an active role in such a technical innovation. The lowland development has to become sustainable, also in the meaning of cost recovery. It means that the population who are the ultimate beneficiaries, have to pay for the development direct or indirect. A continuing top-down approach will lead easily to unacceptable situations on system management, budget-spending and inequitable water distribution. It is obvious that the water users associations have to evolve gradually into partners in lowland development. A bottom-up approach should be propagated where the water users associations already play an active role in the lowland water management. Ultimately, more business-like approaches may be followed, like the introduction of water contracts, discussions with the O&M agency about system improvements, employment by the water users of third parties for tertiary unit development.

INTEGRATED POLICY PLANNING

Need for integrated policy planning

Lowland development follows a logical sequence, from the initial construction stage towards the ultimate stage of "integrated water management" (Ankum et al 1988), which aims at a sustainable development. Generally, lowlands will be developed through the following stages:
- stage 1. simple flood protection works,
- stage 2. extensive flood protection and drainage works,
- stage 3. water quantity management (irrigation and drainage),
- stage 4. integrated water management.

In its first stage, a deltaic area is flooded during high water levels of the sea and of the rivers. The people, mainly fishermen and hunters, may live on the natural levees of the rivers. Simple protection works against the water are created by constructing artificial dwelling mounds and somewhat later by construction of small dikes to protect some agricultural land.

The lowland development will change into the second stage when the population density increases. A systematic reclamation of the coastal areas is begun and polders are constructed. Flood protection of the lowlands will continue and the drainage systems will be improved as the gained areas are subsiding gradually. Also the supply of irrigation water will become a concern so as to increase the agricultural production.

In search for an increasing welfare, the third stage of the lowland development will arrive. The objectives in water management will focus on new water problems: too little water and too much pollution. So, lowland development will concentrate at the increasing water demand for agriculture, for flushing of salt, for drinking water, and for other purposes. It is felt that an enormous supply of water is required, and that such a supply is difficult to achieve. Gradually, the economic <u>need</u> for more water will be considered against non–technical actions. The result might be that many construction activities will be omitted, because they are not economically feasible. So, the activities on water management are gradually changing from building infrastructure towards the optimum manage–ment of the water resources. A start will be made in this stage of development to improve the quality of the water. A first activity will be the restoration of the oxygen balance of surface water by building sewage water treatment plants. Also, the management organizations are made more efficient, for instance by a de–centralization of the management, and by an introduction of a "Water Board" type of organization with cost recovery from the beneficiaries.

The fourth and final stage of the delta development will be based on concepts like "sustainable development" and "integrated water management" (Ankum et al 1988). These concepts are based on the consciousness that there is a compe–titive demand between the different functions of water, e.g. for agriculture, transportation, and also for recreation and for the nature itself.

The concept of integrated water management emphasizes on the strong relation between groundwater and surface water, and between water quantity and water quality (Figure 16.5).

Integrated water management, as developed in the Netherlands, places the water system on a central place, with more functions than only the transport of water. It is acknowledged that water management systems have also a function of "water–for–nature", so that the ecological objectives and functions are satis–fied as well. It means that a new set of actions, i.e. new policies have to be developed in this stage of lowland development.

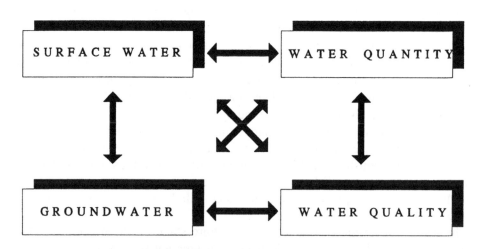

Figure 16.5. Elements of integrated water management.

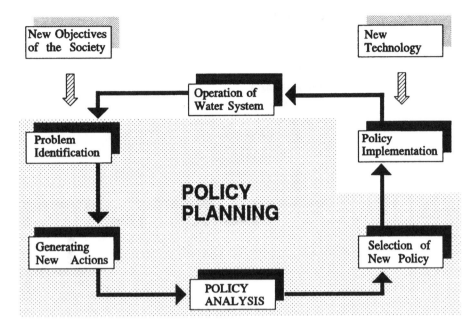

Figure 16.6. The cycle of policy planning.

Policy planning as an ongoing process

Development of new policies and the related policy planning, is an ongoing process in lowland development. The society sets constantly new objectives, for e.g. due to an increasing population pressure, an increasing desire for higher welfare and the need for sustainable development. It means that policy planning have to follow cycles (Figure 16.6). Based on the new objectives of the society, problems in lowland development are identified.

A great number of possible new actions can be generated, but not all of them will satisfy. These proposed actions have to be evaluated through a process of policy analyses before a new policy can be selected finally.

The new policy will be implemented by means of the technology and the operational rules of the water management system. However, the objectives of the society will change further and new technology may permit higher-qualified actions. It means that the valid water policies will become out of date.

Methodology to policy analysis

The analysis of a promising policy requires a series of activities, and is an iterative process (Figure 16.7). The change of a policy is effected through a coherent mix of actions. Therefore, the policy analysis starts with the generation of possible alternative actions to meet the new objectives. There are different kinds of actions possible in water management, such as:

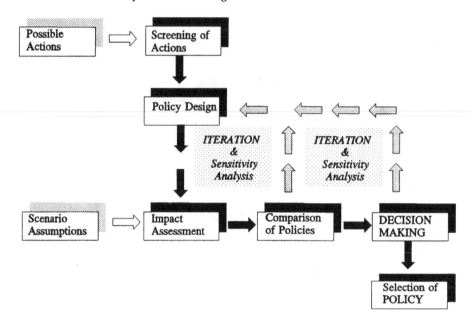

Figure 16.7. Activities in policy analysis.

- technical actions, e.g. modifying or extending the infrastructure,
- managerial actions, e.g. changing the operation rules of the infrastructure,
- pricing actions, e.g. imposing taxes,
- regulation actions, e.g. changing the laws on quotas.

Each kind of action offers many alternatives, e.g. technical actions could modify the current infrastructure in hundreds of different ways.

It is impractical to evaluate the detailed consequences of all possible actions. Therefore, a screening of the alternative actions have to be made. A few "promising" alternative actions will remain and they are grouped together into promising policies. These policies will be evaluated in terms of the full set of potential effects.

The impact of the promising policy depends on the "scenario assumptions". Each scenario assumption describes the input variables, as expected in the future, such as a changing attitude of farmers towards groundwater use, crop prices, growth rate of the industrial pollution.

Various mathematical models are required to assess the impact of a promising policy, and to determine the effects on the water management system itself, on the different users, the natural environment, and on the nation as a whole.

Once the impacts of the promising policies have been assessed, a major difficulty remains: comparing the policies and presenting them to the decision-makers. An approach is the cost–benefit analysis, which converts as many impacts as possible to monetary terms. This approach has several major disadvantages for water management policy analysis, as it loses considerable

	CASE 1	CASE 2	CASE 3	CASE 4	CASE 5
SURFACE WATER IRRIGATION	medium	medium	medium	high	high
GROUNDWATER IRRIGATION	low	low	medium	high	low
PRIORITY GROUNDWATER TO:	none	none	dr.water	dr.water	agric.
Annual Net Benefit, in million US$	750	680	850	1500	-120
Minimum Summer Level IJssellake, in m+NAP	-0.20	-0.25	-0.25	-0.45	-0.45
Drinking water from Groundwater, in %	53%	53%	77%	77%	16%
Groundwater Abstraction, in 10^6 m3/year	1707	1707	1812	1919	568
Environmental Impact of recommended Actions	NO	YES	YES	YES	YES
POLLUTION, in violation frequency: BOD:	47%	47%	49%	51%	68%
Total Phosphate:	76%	76%	76%	76%	76%
Chloride:	37%	37%	42%	53%	73%

RANKING: | Best | Medium | Worst |

Figure 16.8. Example of a "score card" in policy analysis.

information, e.g. it suppresses that an alternative has environmental problems. The approach selected for the policy analysis in water management in the Netherlands has been the "score card" (Figure 16.8). A score card is the table of impacts. Decision–makers can add their judgements about the relative importance of the different impacts, thereby weighing and trading off the impacts to select a preferred policy alternative, i.e. to make "the decision" (Goeller et al 1983).

Policy analysis of water management in the Netherlands

The policy planning of water management in the Netherlands has been strongly influenced by mathematical modelling in order to assess the effects of certain policy decisions (Luiten 1990). The first "Policy analysis of water management of the Netherlands (PAWN-1)" project was executed in 1977 – 1981 (Goeller et al 1983). The PAWN-1 project was a joint research project between the Ministry of transportation and public works, the Delft Hydraulic Laboratory and the Rand Corporation from USA. It was experienced as very fruitful to participate with foreign analysts with a fresh look on the very complex water management system of the Netherlands.

The rationale for the comprehensive PAWN-1 study was the serious drought of the year 1976, which was a 2% dry year with only 530 mm rainfall. The extensive damage to the agriculture was widely acknowledged. Farmers sprinkled only about 15% of the land under cultivation, and most of the farmers had to face large economic losses. It was realised that water shortages occur even in a normal growing season. This seems paradoxical for an annual evaporation of 500 mm/year, while the average rainfall depth is 750 mm/year and the rivers contribute 2000 mm/year as a uniform water depth. Another problem felt was on pollution, as the surface water became strongly polluted due to industrialization.

The primary objectives of the PAWN-1 project were: (i) to develop a methodology for assessing the multiple consequences of water management policies, (ii) to apply it to develop alternative water management policies and to compare their consequences, and (iii) to create a capability in the Netherlands for further analyses.

PAWN-1 considered the Netherlands (3,400,000 ha) as one large hydrological unit, located in a delta of three rivers: the Rhine, the Meuse and the Scheldt. It took the national water system as the heart of the simulation studies, i.e. the system which is managed by the Ministry. The main rivers, canals and lakes were represented in the distribution model as a network of links and nodes. The "distribution model" simulated the operation of the national water distribution system on a ten-day period, on parameters such as the river flows, lake levels, shipping depths, pollution concentrations.

Basic inputs to the distribution model of PAWN-1 were the demand and the allocation, both represented by a series of models (Figure 16.9). For example, "district models" computed the water balances and salt balances in some 80 hydrological districts of the country. Important parameters here are the ground-

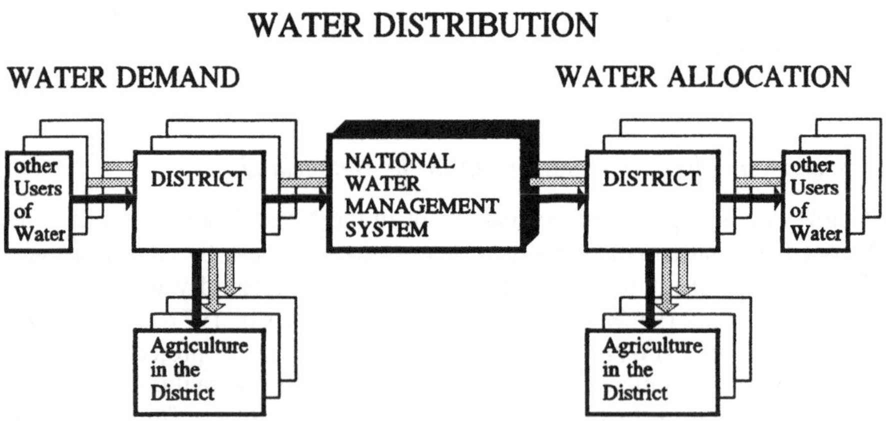

WATER DISTRIBUTION

WATER DEMAND   **WATER ALLOCATION**

Figure 16.9. The distribution model as the heart of PAWN.

water levels, damage to crops, irrigation costs, etc. But also other sectors of water users were incorporated, such as inland shipping, public water supplies, industries, power plants and ecosystems (Goeller et al 1983).

PAWN–1 calculated that the economic losses caused by damage in dry years from water shortages and salinity are far greater for agriculture than for any other sector. As a result, most of the technical and managerial actions considered in the PAWN–1 project were designed to increase the quantity and quality of surface water supplied to agriculture.

The pollutants considered in the PAWN–1 project have been limited to salt, heat, chromium, biochemical oxygen demand (BOD) and phosphate. PAWN–1 stated that the current eutrophication control strategy of phosphate reduction will be largely ineffective because the Netherlands cannot control the large amounts of phosphate brought into the country by the two major rivers, the Rhine and the Meuse. In addition, the bed sediments of eutrophic lakes are phosphate–rich. Finally, PAWN–1 concluded on water quality: "dilution is no solution to pollution", and argued for a regional approach to pollution problems (Goeller et al 1983).

The PAWN–1 study formed the basis for the 2nd National Note on Water Management of the Netherlands in 1984 (Ministry of transportation and public works 1984). The earlier 1st National Note on Water Management of 1968 did not consider the economic need of water supply for agriculture, but concentrated mainly on the question how water managers could cover the enormous water need (Ministry of transportation and public works 1968).

This 2nd Note was based on the cost – benefits relation of the water development and on the multiple use of the water. It elaborated the need for water. The result of the analyses was that half of the activities on the national water management system as proposed in the 1st Note, were omitted, because they were not economically feasible.

The policy note "Living with Water" (Ministry of transportation and public works 1985) was prepared soon after the submission of the 2nd National Note on Water Management, and elaborated the long–term vision on the water resources management in the Netherlands. It was emphasized that the past activities on water management concentrated too much on the pure "water" aspects. The policy note recommended that more attention is to be given to the "water system approach".

This development of changing objectives was widely supported by the public opinion. Problems with the agricultural sector like the manure surplus from the bio–industry, the groundwater lowering in nature reserves by sprinkler irrigation of farms, received political criticism. There is a growing political interest in the nature and in a sustainable environment, and in the functioning of the water systems on recreation. The production of drinking water from groundwater faces problems with increasing pollution of soil and groundwater. Moreover, ground–water recovery is also subject to political criticism because of damage to nature reserve areas.

The second "Policy Analysis of Water Management of the Netherlands (PAWN-2)" project was executed in 1986 – 1989 (Luiten 1990). The mathematical models of PAWN-2 incorporated not only the water quantity aspects, but also the quality of water and the quality of the channel beds. Initially, test runs with PAWN-2 concentrated on four extreme policies on water management in the Netherlands, as to define the extreme cases:

1. "nature–wet" case, focusing on the aquatic nature. Actions were related to limit eutrophication, micro–pollutants, thermal pollution, etc.;
2. "nature–dry" case, focusing on the nature reserves. Actions were related to limit the groundwater recovery, agricultural land drainage, foreign water, etc.;
3. "agricultural" case, focusing on agriculture. Actions were related to stimulate works for drainage and irrigation, etc.;
4. "other–interests" case, focusing on the other interests, such as navigation, drinking water production, flood protection.

The conclusion of the test run confirmed that the present policy is mainly based on the "agricultural" case.

The first run of the policy analyses model was based on the calibration during the test run. The first run studied nine promising actions in detail. Preliminary conclusions were: (i) it is unfeasible in many areas to use groundwater to irrigate the agriculture, (ii) it is technically not yet possible to reduce the micro–pollution by the required 90% reduction of the pollution emissions, (iii) it is cheap to improve the aquatic nature by physical and biological measures, such as fish passages at weirs, environmental–friendly banks of channels. The second run also concentrated on four different reductions of the pollution emissions, i.e. 0%, 50%, 60% and 90%.

The final runs led to major conclusions for the water management policy in the Netherlands (Luiten 1990):

- supply of water to some agricultural areas is feasible, but does not require major adaptations of the existing distribution system;
- groundwater abstractions for drinking water and agriculture have to be reduced in areas with nature reserves as to prevent further dehydration of these areas;
- the reduction of the pollution emissions to the level of "1985" gives technical problems. Nevertheless, these reductions are required to improve the quality of the water and of the channel beds.
- a combination of actions on adjacent fields, such as on pollution emissions and at the same time on physical works, gives the highest benefits.

The 3rd National Note on Water Management of the Netherlands was published in 1989 (Ministry of transportation and public works 1989) and is substantially based on the research results of the second PAWN project (PAWN-2), for both the fresh–water systems as well as for the salt–water systems. The 3rd Note outlines the "integrated water management" approach, relating groundwater with surface water, and water quantity with water quality.

The new Note elaborates extensively on the policy that "water systems have to be managed and developed so that they satisfy their ecological objectives and functions".

TOPICS ON INTEGRATED WATER MANAGEMENT

Water management for nature reserves

In the course of lowland development, the supply of water to agricultural areas may become subject to contradictious attitudes. National income depends on an optimum water management, which can be obtained at low investments in the final stage of lowland development. On the other hand, the political interest becomes increasing on the water need of nature reserve areas.

Nature reserve areas might be located next to agricultural areas and have other requirements for water management. Actions to improve the water management in these areas concentrate for e.g. in the prevention of further groundwater lowering in nature areas by reducing land drainage and groundwater irrigation in the adjacent agricultural areas. Moreover, new nature areas can be developed by minimizing the land drainage and flood control locally as to (re)create marshes and flood plains.

Water quality management

The water quality might be improved by mixing the polluted water with fresh water. However, this dilution provides only a local relief and not on a regional scale, as "dilution is no solution for pollution".

The alternative activity of building sewage treatment plants is "source-oriented", and has been successful in many countries. As the "emission control" programs processes, increasing attention will be given to the question: "what water quality should be aimed at?". Water quality management will ultimately lead to the "quality" approach, which is also inspired by the wish not to spend more money on treatment than strictly necessary (Provoost 1991). Acts on the pollution of surface waters are effective in many densely-populated lowlands. They prescribe that all surface waters should meet certain standards of basic quality. These quality standards specify the colour, smell, transparency and maximum or minimum concentrations of numerous chemicals. The standards may have different critical values for water with different functions, such as swimming water, drinking water, water for fish production.

Ecological water management

Ecology now plays a growing role in water management of well-developed lowlands (Provoost 1991). Aquatic ecology is the study of the bio-community (plants, animals) in the water systems. Ecological water management is the water management that guarantees an optimum development of the aquatic ecosystem.

Some of the most important parameters for the aquatic ecology are: (i) salt content of the water, (ii) flow velocity, (iii) depth of the channel, (iv) perennial flow, (v) acidity of the water, (vi) nutrient content of the water.

However, issues on the role of ecology in water management are related to what ecological knowledge is required, and to the knowledge on the disturbance – effect relations. It observed that researchers on aquatic ecology have knowledge that is not known by their colleagues in water management, and vice versa.

A major question is also: "what is natural in an aquatic ecosystem?" Most of the channels in lowlands have been dug by men, and many of them would gradually be filled without human interference. Their natural stage is "land" or "not existent". Also the ultimate stage of swamps is "land" with peat soils. Thus, the requested ecological stage of development must be defined by men.

The acceptance of objectives for ecological water quality means a drastic revolution, as for centuries the water managers had focused on making water serviceable to mankind, so "man–orientated". Now, also ecological or "nature–orientated" functions have to be considered. That requires mobilization of biological disciplines in water management.

Experiences in the Netherlands with ecological water management are encouraging (Provoost 1991). Initially, opposition came from water managers who were more man–oriented than nature–oriented. Consequently, the first generation water quality plans were rather reserved in defining ecological water quality objectives. These ecological objectives were often assigned only to waters located in nature reserves. There was also the initial problem of the implementation of the ecological objectives. Man–oriented objectives, such as the restoration of swimming water, can be easier explained to administrators and to the public, than bringing crayfish back into the rivers. These matters have been settled in the Netherlands.

It is no longer a question of choice between man–oriented and nature–oriented objectives, but rather of "how man–oriented and nature–oriented functions can be brought into balance again". Thus, every surface water should at least house a healthy and a viable ecosystem. The ecological water management objective applies also to fully man–made water systems. It has nothing to do with the strive for natural and original ecosystems.

Precise predictions of those bio–communities that will return, are hard to guess, and are also considered as not relevant. It is only the effort to create the preconditions for more natural bio–communities of plants and animals. The choice between man– and nature–oriented functions cannot await until a complete insight into the relations, measures and objectives have been obtained. The deterioration of the living environment is a clear signal that the excessive use of the nature by man must be curbed (Provoost 1991).

REFERENCES

Ankum, P., K. Koga & W.A. Segeren. 1988. Integrated water resources management in the Netherlands. *Proc. int. symposium on shallow sea and low land, Saga University, Saga, Japan*: 67–78.

Ankum, P. 1992. Water policy and management in the Netherlands. *Proc. int. symposium on shallow sea and low land, Saga University, Saga, Japan*: 27–36.

Ankum, P. 1989. Management of irrigation rehabilitation projects. *Proc. Asian regional symposium on the modernization and rehabilitation of irrigation and drainage schemes, Wallingford, UK*: 347–360.

Goeller, B.F. et al. 1983. *Policy analysis of water management for the Netherlands*, Rand, Santa Monica, USA.

Luiten, J.P.A. 1990. *Beleidsanalyse zoete wateren* (Policy analysis fresh water), in Dutch. Ministry of transportation and public works, the Netherlands.

Ministry of Transportation and Public Works, the Netherlands. 1968. *De waterhuishouding van Nederland* (The 1st Note on Water Management of the Netherlands) in Dutch.

Ministry of transportation and public works, the Netherlands. 1984. *De waterhuishouding van Nederland* (The 2nd Note on water management of the Netherlands), in Dutch.

Ministry of transportation and public works, the Netherlands. 1985. *Omgaan met water* (Living with water), in Dutch.

Ministry of transportation and public works, the Netherlands. 1989. *De derde nota water-huishouding* (The 3rd note on water management of the Netherlands), in Dutch.

Oliver, J.L. 1992. Incorporating environmental policies into water resources management in France. *Country experiences with water resources management World Bank technical paper No.175, Washington*: 57–60.

Provoost, K.J. 1991. Ecological Water Management and Policy Planning. Ecological Water Management in Practice. *Proc. and information No.45, TNO committee on hydrological research, Delft, the Netherlands*: 77–88.

Ven, G.P. van de (ed.). 1993. *Man–made lowlands, history of water management in the Netherlands*. ICID, the Hague, the Netherlands

17 REAL TIME CONTROL OF URBAN DRAINAGE SYSTEMS

A. J. M. NELEN
Delft University of Technology, the Netherlands

INTRODUCTION

Since the last few decades, the focus of urban drainage technology has been greatly expanded. The urban drainage problem has changed from 'simply' draining the storm water, to disposing of it in an acceptable sanitary way. In solving this problem, the modern urban drainage engineer has to consider all pathways of the water in the urban area, i.e. the sewer system, the groundwater system and surface waters. Besides improved (computerized) design methods of storm water collection, transport and treatment facilities, various methods have been developed to upgrade the performance of urban drainage systems, i.e. to reduce flooding problems and to limit the pollution outflow to receiving waters. These methods may be divided into three broad categories:

1. Source controls, i.e. measures taken to reduce the peaks and volume of surface runoff entering the sewer system. These are generally the most cost-effective methods for reduction of runoff in urban areas. Several measures can be considered to increase storm water infiltration, such as porous pavements, soak-away pits, seepage trenches, the use of cisterns to store water for e.g. garden watering, infiltration basins, etc. Although source controls may be very effective in reducing runoff volumes, the protection of groundwater quality must restrict their application to less polluted storm water originating from e.g. rooftops, backyards and residential streets.

2. 'Structural' measures taken to enlarge the system capacity. Retention and detention are the most frequently employed methods of runoff control and can be used to achieve virtually any degree of runoff control, provided that the costs of such facilities are not prohibitive. The basic principle is to provide sufficient (in-line or off-line) storage in the drainage system in order to keep the water in the system until there is sufficient capacity to lead it to the treatment plant. The advantage of off-line storage tanks is that pollutants will be removed to a certain extent by settling of suspended solids. The pollution outflow to receiving waters can be further reduced by implementing improved overflow structures,

such as the swirl concentrator, the high sided weir chamber, etc. The efficiencies of the various measures greatly depend on the hydraulic design and on local circumstances.

3. 'Non–structural' measures, i.e. by improving the planning and operation of the system. Water managers generally recognize that a proper planning and maintenance is an absolute requirement for a good systems behaviour. However, in solving urban drainage problems the common approach is still to provide sufficient system capacity (given a certain design load), rather than investigating how the available capacity can be used in a more efficient way. This is not State–of–the–Art. The limited efficiency of this approach, which is caused by the lack of flexibility of the operation of the system, is being recognized by an increasing number of urban water managers. The awareness is growing that by using 'proper' system dimensions, that meet all the design constraints, does not automatically imply that the urban drainage system performs optimally for all rains that it is exposed to. The main reasons are:

– The design method itself is by definition inaccurate, due to schematizations and assumptions that are necessary for the computations;

– The planned and actual drainage conditions will differ due to urban development, maintenance work, sewer construction, system failures, etc. As a result, some sections will have more storage and/or discharge capacity compared to other sections of the system, meaning that some parts of the system may be overloaded while elsewhere in the system some capacity is still available;

– The temporal and spatial variability of the system loading will lead to an uneven use of the available capacities. A homogeneous design storm, on the basis of which the system has been designed, will never occur as a physical event. Real storms are distributed in time and space. Although they might not reach the depth of the design storm, local storms might result in sewer overflows, while elsewhere in the system storage capacity is still available. Theoretically, in an uncontrolled system, maximum use of all available storage and transport capacity will only be achieved when the entire system is loaded with a storm greater or equal to the design storm. By definition, for every other loading some capacity will remain unused.

– The effects of the system output on the environment are variable in time and space.

Facing this situation, the conclusion can be drawn that better use of the capacities of the drainage system can be achieved by actively directing and storing the flows, on the basis of currently monitored process data (i.e. in real time). Real Time Control (RTC) provides a way to make an optimal use of all components of the urban drainage system for all storms that the system is exposed to. Generally, it can be stated that nowadays the appropriate solution to urban drainage problems is no longer only 'structural', i.e. resolved through construction of sewer systems, basins, and other auxiliaries, but also 'non–

structural', i.e. resolved through appropriate planning and operation. The potential of the latter will be discussed below.

PRINCIPLES OF CONTROL

Terminology

Real Time Control (RTC) implies continuous monitoring and controlling of the flow process. In principle, the control of a process can be schematized to a simple control loop (Fig. 17.1), comprising the following basic elements:

– A sensor to monitor the systems state and possibly variables that are used to predict the disturbances of the system (e.g. rainfall, water level, water flow). Generally, the required data depends on the configuration of the system, the specific aim of the control system and the level of control .

– A corrective flow regulator (or actuator) which is able to control the process in order to reach or to maintain a desired state of the system, which can be defined by the water levels and flow rates as a function of time and place. Regulators in a water management system could be e.g. pumps, gates, sluices, movable weirs and valves.

– A controller which controls the flow regulator, using the received data from the sensor. A large variety of controllers are known from Control Technique, such as the two point controller (on/off), the PID controller, predictive controllers, and many others. The task of the controller is to verify and process the measured data and to send instructions to the regulator. The control signals depend on the deviation of the measurement from the 'desired value' or set point.

– A (tele)communication system is needed to link the sensors, controllers and regulators in the system. It needs no argument that the reliability of the communication system is very important. In urban drainage it is most common to use (public or hired) telephone lines for this purpose.

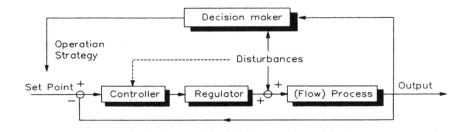

Figure 17.1 Scheme of a controlled process.

Most urban drainage systems are statically controlled, meaning that the set points (desired values) of the flow regulators (e.g. pumps, valves, weirs) are constant in time. They function independent of each other and maintain a pre-set flow related only to the water level at the regulator site. Due to the reasons mentioned in the introduction, this way of control leads, by definition, to an uneven use of the available system capacities.

To make an optimal use of the system capacities, the set points of the flow regulators have to be modified in real time on the basis of currently measured process data throughout the system. The time sequence of set points of all flow regulators (i.e. the desired systems state as a function of time) is called the operation strategy (Fig. 17.1). The decision maker can be an operator (possibly supplied with a decision support system), or an automatic control system.

The operational optimization problem is aimed at finding the optimal set points in time. This problem should not be confused with the control problem, which deals with the determination of the required adjustments of the flow regulators in time (e.g. gate settings, pump rates), to achieve minimum deviation from the (known) set points. Both problems are closely related, as in deriving an operation strategy, an important constraint is that the strategy has to be feasible. In literature, both the terms 'operation strategy' and 'control strategy' can be found, often in the same meaning. Although it is in some cases difficult to make a clear distinction between the operation and the control problem (as they may be solved simultaneously, e.g. when designing multi-variable controllers), it is generally useful to get a clear conception of the problem. For simplicity it could be stated that the formulation of the operation strategy can be dealt with by the methods known from Operations Research, whereas the behaviour of various types of controllers is a topic of research within the field of Control Technique. In this chapter only the operational problem is addressed.

Control concepts

Control systems differ in complexity. Generally, three different levels of control can be distinguished. The simplest level is called 'local control', which consists of one control–loop as illustrated in Figure 17.1, wherein the controlled variable is measured at the regulator site. This implies that regulator action is independent of other sections of the water management system. More advanced is 'unit process control', by which the control of the regulator depends on more than one measurement within the system–unit. Applying 'central or systems control' means that the actual state of the total system determines the operation of every regulator. By definition, optimal performance of the urban drainage system can only be achieved when systems control is applied.

It is noted that the data collected on the performance of the drainage system may also be used for the daily management of the system, such as further data analysis, emergency control and maintenance planning. In practice, these benefits appear in fact to be the main reason for water managers to decide for the implementation of a monitoring and remote control system.

Mode of operation

A control system can be operated in different ways, which is called the mode of operation. When a system is operated in a manual mode, it means that the regulator itself is operated directly by an operator. This type of operation is seldom found in urban drainage. When a control system is operated in a supervisory mode, the regulators are activated by local automatic controllers, but the set points of the controller (e.g. a desired water level or discharge) are specified by an operator. To perform optimal control, the operators need a full understanding of the dynamics of the system. Even for experienced operators it can be very difficult, if not impossible, to predict the result of moving a number of gates or activating a number of pumps. For this purpose, one might use a decision support system (DSS). This can be a simulation model, with which one can try possible strategies before actually executing them, or a decision support model that can be used to suggest a strategy (e.g. an expert system, an optimization model). The advantage of this type of operation is that complete operation is under human control, which makes it flexible as opposed to automatic operation. This might be an advantage, e.g. during extraordinary operational conditions. On the other hand, automatic operation might be preferred as you cannot expect operators to be alert 24 hours a day.

The system is operated in an automatic mode, when the decision on the operation strategy and the execution is fully automatic. Central automatic control requires generally more hardware compared to the other ways of operation, reliable communication lines and a computer model that can generate a suitable control strategy. In the Netherlands, a few examples of central automatic control systems can be found.

THE OPERATIONAL OPTIMIZATION PROBLEM

The technology of monitoring and remote control systems, including their hardware, are becoming well known. The key problem for the drainage engineer when applying this technology to urban drainage systems is the formulation of the operation strategy. Since the optimum setpoints vary with each flow pattern, a flexible method is required to react properly to every loading that may occur.

In a general sense, to determine an optimal operation strategy means that the 'cost' have to minimized, given a perturbation of the system, subject to the constraint that the strategy has to be feasible. Hence, to solve the operational problem, three types of information are required, namely

- a specification of the operational objectives of the system;
- a description of the physical constraints of the problem;
- information on the current systems state.

If one of these types of information is lacking, no (rational) decision on the operation strategy is possible.

Operational objectives

The general aim of operating an urban drainage system is
- to minimize flooding;
- to minimize the impact of pollution loads to receiving waters by means of minimizing combined sewer overflows (differentiated in time and place, depending on the receiving water body), while maintaining optimum flow rates to the treatment plant (depending on the conditions at the plant);
- to minimize operational costs (e.g. energy).

Ranking the operational tasks according to their priority is usually not the most difficult part of the problem. It is more complicated to indicate the relative importance of the operational tasks or to define the 'cost' of not performing these tasks. This is necessary as different objectives are considered, which may be conflicting, meaning that they cannot be fulfilled at the same time. For example, storing storm water to reduce combined sewer overflows might increase the risk of flooding. Minimizing pumping to save energy cost might increase the risk of overflows. Obviously, this problem is becoming more acute with an increasing number of operational tasks, i.e. when the operational problem is tending towards an integrated water management problem.

In principle, there are two options to make a rational decision in the set point selection: by specifying performance criteria or by formulating an objective function in terms of 'unit costs'.

Performance criteria are applied to indicate the allowable limits of the controlled output variables. In this case, solving the operational problem means that a feasible solution has to be found, where all systems state variables fall within these limits. The performance criteria determine only whether a solution is acceptable, but no criterion exists by which the optimum is defined.

Optimality can only be specified by applying 'unit cost' or 'weights', which express how a deviation of a systems state variable from its desired value is evaluated. The optimal systems state is then defined by the least cost solution. The problem is now transferred into a new problem, namely how to find proper 'unit costs' of the variables. Obviously, the specification of the actual 'costs' is difficult as most operational objectives are impossible to express in financial terms. For example, it is impossible to define the cost of x m^3 overflow or y m^3 flooding. Moreover, these 'costs' may be variable in time and space as they depend on the current systems state. How to deal with these problems is explained below.

Since every party concerned will have other priorities, it is impossible to define the best solution unambiguously. Therefore, the term 'optimality' could be misleading. Since the operation strategy is based on the evaluation of a subjectively formulated objective function, the usefulness and validity of the chosen objective function should be carefully investigated.

Physical constraints

The operation strategy has to be physically achievable. Pumping rates, flow through conduits or structures, water levels in reservoirs, etc. are limited by the capacities of the particular elements of the water management system. These are called the static or capacity constraints of the problem. Other static constraints which may be less obvious are the initial and final state of the system.

Additionally, the operation strategy has to be consistent with the physical laws of water motion, i.e. continuity and energy balances. These form the dynamic constraints of the operational problem. It is these constraints that make the problem of finding a good strategy generally difficult. Whereas the formulation of capacity constraints is rather straightforward, the hydrodynamic constraints usually imply a simplification of the governing physical laws, as approximated by the De Saint Venant Equations. In solving the operational optimization problem, flow routing is usually performed by linearized equations or a lumped storage approach, depending on the technique that is applied to solve the optimization problem.

Other constraints of the problem are related to the characteristics and behaviour of the controlled flow regulators. For example it takes a certain time before a pump reaches its desired discharge, the frequency of switching pumps is limited, a valve should not be closed too suddenly to prevent water hammer, etc. These aspects are referred to as the technical (or hardware) constraints.

System disturbances

It is evident that information on the actual systems state is required to make a decision on how to operate the system. Depending on the response time of the system, the ability of the system to correct a certain perturbation of the system, and the accuracy of the forecast, information on future disturbances may be beneficial in improving this decision. In principle, there are three options to predict the (future) behaviour of an urban drainage system:
  – Flow and water level measurements in upstream sections of the system;
  – Rainfall measurements in combination with a rainfall–runoff model, which extend the reaction time by the surface–runoff time of the catchment;
  – Rainfall forecasts, which allow to gain additional time depending on the control horizon.
It has to be considered that measurements include measurement errors. Model

calculations are inaccurate due to unknown input, unknown parameters and model simplifications. Rainfall forecasting, using radar images, is a field into which great efforts are being put, however until today, this technique includes great uncertainties. Moreover, a rainstorm develops within a few (2–3) hours, which restricts the possible time horizon of a reliable forecast. Therefore, in determining the operation strategy, it is advised to perform a sensitivity analysis to determine the extent to which the results are affected by these uncertainties.

SOLUTION TECHNIQUES

A large variety of optimization problems and related solution techniques are known from Operations Research. Possible procedures that can be applied to solve the operational problem as described above may be divided into three broad categories:

1. Heuristic methods, meaning that a solution is found by experience (gained by 'trial and error');

2. Rule based scenarios, meaning that a solution is found through comparison and evaluation of a finite number of feasible solutions. A decision tree is a tool for organizing the enumeration process;

3. Mathematical optimization techniques, meaning that the operational objectives are expressed by an objective function, which is minimized by means of mathematical calculus.

Each method has its advantages and disadvantages concerning the flexibility in formulation of objectives and constraints, computing time and needed computer resources, robustness of the control performance, etc. Obviously, the suitability of the technique depends on the specific application.

Heuristic methods

When the operational optimization problem is solved heuristically (i.e. based on experience), it is not common to formulate the objectives of the system in terms of 'costs', but certain performance criteria are obviously required to be able to evaluate the systems performance. An advantage of heuristic methods is that any kind of information can be used in the decision making, such as actual process data, intuition, rain likelihood, experience from previous events, etc. An experienced decision maker (i.e. an operator) may solve the problem effectively, by disregarding all options that are possible but not advisable. In formulating the dynamic constraints of the problem (the system description), the experienced operator in fact uses the best model available, i.e. the actual system.

A disadvantage of heuristic control is that the reasoning behind the decisions is not always clear. As a result solutions may be inconsistent and difficult to evaluate. Moreover, the experience, gained by trial–and–error, will be lost once the operator leaves his job. His successor will make (the same) mistakes all over again. A Decision Support System (DSS), which could be a flow model, an optimization model, an expert system or a combination of these models, is a helpful tool in reducing these problems. Heuristic control is extremely application oriented, since the experience gained at one catchment is not just transferable to another area.

An expert system or knowledge base system is a computer model, in which it is programmed how experience is gained and how this knowledge is applied to solve a problem. Simply stated, it is a knowledge base containing different kinds of information in which a search pattern (rules) is programmed that should lead to a good or an acceptable solution. It is not an optimization of the problem. In fact, their name implies the idea that the solutions are not found through a model of the physical system, but through a model of the expert (here: the operator).

An expert system is said to be self–learning if it is able to accumulate information. This means that the model is not only programmed to make a decision, but also to evaluate the rules that lead to this decision, analyze them in relation to the operational situation and eventually come up with more appropriate rules. They will often be used interactively with an operator, e.g. to diagnose the actual systems state, to identify 'faults' in the system, to recommend emergency procedures, etc. (Graillot 1990, Babovic 1991).

An expert system is usually considered less suited to make a decision on the operation strategy itself, since this would require a huge number of rules which are difficult to oversee and to evaluate (this problem will be discussed below). Nevertheless, attempts have been made to adapt an expert system for real time control of the sewerage network (e.g. Khelil et al. 1990). A more suited approach is to use an expert system interactively with a decision model (e.g. a mathematical optimization model) as a pre– and post–processor. In such a case, the task of the expert system is to define the actual operational problem (i.e. the objectives and constraints, which may both vary in time), and to interpret the model results.

Rule based scenarios

Rule based scenarios can be described as a hierarchy of 'if..then..else..' statements, which relate input variables to output (control) variables by means of boolean logic (true/false). Decision trees (or matrices) may be used to organize the enumeration process. In deriving the rules, the most common approach is to evaluate the system on the basis of performance criteria.

Scenarios require extensive development work. Based on a careful analyses of the water management system, the output variables have to be specified in advance for all possible states of the system. The number of operational objectives is generally limited, in order to keep overview. Obviously, it is impossible to include every single input variable and current state variable in the decision tree, meaning that they have to be divided into certain classes. As a consequence, one is never sure whether the best solution is found for each possible situation. Moreover, the set of rules has to be modified when system parameters have been changed.

A scenario is not necessarily a set of fixed rules that relate an input variable to an output variable. The rules may be formulated as a function of systems state variables, including some adjustable control parameters to manipulate the rules. The advantage of this approach is that the decisions are related to the current systems state. Besides, the rules can easily be adjusted to the actual drainage conditions by a proper attuning of the control parameters. The extent to which the rules can be modified to achieve different objectives (which may vary in time) depend on the structure of the algorithm, but generally this flexibility is limited as the rules are formulated to meet a predefined aim.

Important features of control scenarios are that they allow for fast on–line execution. Besides, most technical constraints of the problem can easily be incorporated in the decision tree. The structured enumeration process may facilitate an evaluation of rules, given the number of rules is limited. Like heuristic methods, rule based scenarios are in the first place application oriented. In fact, all present examples of central automatic control are using a rule based scenario.

Mathematical optimization

The methods described above are basically using two separate models: a decision model to formulate an operation strategy and a mathematical or conceptual model of the system to check whether the various criteria are maintained. By employing the mathematical tools of Operations Research, these problems can be solved simultaneously.

Finding a solution of the operational problem by means of a mathematical optimization model requires the formulation of an objective function in which the operational tasks are specified in terms of 'unit costs'. This objective function is minimized, subject to a set of constraints, which represent the system. Since optimality is defined for the complete system by the least cost solution of the objective function, the model will produce an operation strategy by which the objectives are met in the best possible way, within the limits of the constraints.

Main advantages of a mathematical optimization model are its flexibility and consistency in decision making. The model determines the optimal strategy, which is unique for each situation, depending only on the applied objective function. The rationale behind the solutions is clear and consistent. Obviously, these are crucial features when investigating the potential of real time control of a water management system. In other words, for a systems analysis, this approach is the most suited as it provides maximum flexibility and consistency in the decision process. Some well known optimization techniques are dynamic and (non)linear programming. (Note that the term 'programming' does not refer to computer programming but to planning in a general sense). The discussion in the next sections is confined to the way these techniques require the objective function and constraints to be formulated. For details on the mathematics behind the techniques, reference is made to the several handbooks on applications of Operations Research (e.g. Wagner 1975).

DYNAMIC PROGRAMMING

Dynamic programming (DP) is a multi–stage decision process (and not an algorithm as often assumed) based on the concept of decomposing and solving the optimization problem by a sequence of sub–problems. The key concept upon which optimal policies are obtained is "The Principle of Optimality", which originates from Bellman & Dreyfus (1962). This principle may conceptually be comprehended as follows: Given an optimal trajectory from point A to point C, then the portion from any intermediate point B to point C must be the optimal trajectory from B to C.

Applying this principle to the determination of operation strategies means that each decision at a certain time must be optimal, independent of decisions at former time steps. This leads to the following process:

– Start at the end of n control decisions, i.e. the last time step of the control horizon. Determine the optimum set points for each possible situation at this time step and store the results.

– Next, time step $n-1$ is being considered. Again determine the optimum strategy for each possible situation, but this time by taking into account the strategy as determined at $t= n$. Store the results.

– Repeat this process, until $t= 1$. As the initial state is known, it is possible to determine the optimum strategy for all time steps $1,..,n$.

Simply stated, the methodology comprises trying out all possible combinations and finding the optimum one. Since all results at each time step have to be stored, a lot of computer storage capacity is needed. Bellman called this phenomenon "the curse of dimensionality". Although the method may be very flexible concerning the formulation of the objective function and the physical constraints (theoretically there is no restriction at all), practical applications of DP to solve control problems appear to be restricted to a few state variables due to the fact that today's computers cannot handle more data. Therefore, DP is not considered to be a suitable approach for the control problem under consideration.

It is noted that, despite this dimensionality problem, DP and its many derivatives are successfully applied in the field of water resources management in solving various optimization problems, such as the determination of the optimum design or to find the best policy in solving water allocation problems. An overview of these applications can be found in Kularathna (1992).

LINEAR PROGRAMMING

A very popular solution technique is linear programming (LP), which deals with minimization (or maximization) of a linear objective function F, subject to a set of linear constraints. Due to its robustness and its capability to deal with relatively large problems, the procedure has found applications in many different fields. Moreover, basic knowledge of its main principles is sufficient for a successful use of the available standard software. A general LP problem reads

$$minimize \; F = \sum_{i=1}^{m} \sum_{j=1}^{n} c_{ij} x_{ij} \tag{17.1}$$

subject to a set of m linear constraints

$$\sum_{j=1}^{n} a_{ij} x_{ij} \leq b_i \; ; \quad for \; i = 1, 2, \dots, m \tag{17.2}$$

with all variables subject to the non-negativity constraint

$$x_{ij} \geq 0 \; ; \quad for \, all \; i \; and \; j. \tag{17.3}$$

The x_{ij} are the decision variables, the a_{ij}, b_i and c_{ij} the parameters of the model. In this case, the variables and parameters can be explained as follows:

x_{ij} systems state variable, e.g. stored volume, discharge, overflow, etc.;
c_{ij} unit cost of the particular state variable;
a_{ij} coefficient by which the particular state variable is multiplied;
b_i upper capacity constraints (and rain input).

For the past few years, various examples of applications of LP to real time control of urban drainage systems can be found in literature (e.g. Petersen 1987, Schilling & Petersen 1987, IAWPRC 1989, Neugebauer 1989), and several authors in (EAWAG 1990). It can be shown, however, that the problem under consideration cannot be handled properly by a strict linear model. In this section, the behaviour of the LP model and its limitations are discussed quite extensively to facilitate an understanding of the modifications required in formulating a more appropriate (non-linear) objective function.

A prototype example shows how the operational optimization problem can be formulated as a LP problem. Consider an urban drainage system, which is schematized as a number of reservoirs (i.e. a lumped storage approach). Each

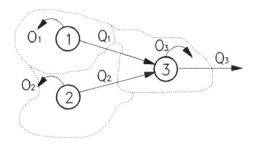

Figure 17.2. Schematization of a simplified system.

reservoir (node) has an inflow (I), which can be stored (V), discharged to a downstream node (Q) or discharged out of the system by an overflow (O). Using the symbols of Figure 17.2 the LP problem reads

$$minimize \ F = \sum_{t=1}^{T} [cv_1^t V_1(t) + cq_1^t Q_1(t) + co_1^t O_1(t) +$$
$$cv_2^t V_2(t) + cq_2^t Q_2(t) + co_2^t O_2(t) +$$
$$cv_3^t V_3(t) + cq_3^t Q_3(t) + co_3^t O_3(t)] \tag{17.4}$$

subject to a set of capacity constraints

$$0 \le V_i(t) \le V_{i,max} \ ;$$
$$0 \le Q_i(t) \le Q_{i,max} \ ; \quad for \ all \ i \ and \ t \tag{17.5}$$
$$0 \le O_i(t) \le O_{i,max} \ ;$$

and a set of dynamic constraints (the system equations)

$$I_1(t) = V_1(t+1) - V_1(t) + Q_1(t) + O_1(t) \ ;$$
$$I_2(t) = V_2(t+1) - V_2(t) + Q_2(t) + O_2(t) \ ; \quad for \ t = 1,2,...,T \tag{17.6}$$
$$I_3(t) = V_3(t+1) - V_3(t) + Q_3(t) + O_3(t) - Q_1(t-t_{13}) - Q_2(t-t_{23}) \ ;$$

where V_i is the stored volume in node i [m³]; Q_i is the discharge of node i [m³/Δt]; O_i is overflow of node i [m³/Δt]; cv_i^t is the unit cost of V_i at time t [–]; cq_i^t is the unit cost of Q_i at time t [–]; co_i^t is the unit cost of O_i at time t [–]; I_i is the inflow to node i [m³/Δt]; t is time; t_{ij} is the flow time from node i to node j [Δt]; and T is the control horizon, for which inflow is specified [Δt].

The least cost solution of Equation 17.4 consists of the optimal values of $V_i(t)$, $Q_i(t)$ and $O_i(t)$, for $t= 1,2,...,T$. If the complete inflow hydrograph $I_i(t)$ is known for all T time steps, the LP problem has to be solved only once to determine the optimal strategy for the complete event. The assumption, however, that the complete event is known in advance is not only unrealistic, but it will also lead to a huge optimization problem (and to the use of big time steps), which may be a problem regarding computer memory requirements. It will certainly restrict the application of the LP model to smaller problems.

Therefore, a better and more realistic approach is to re-formulate the LP problem each time step of the simulation. This means that an inflow hydrograph $I_i(t)$ consisting of n time steps ($n > T$) requires a succession of $n-T$ LP problems to be solved. At $t= 1$, the problem is solved for the first T time steps. The calculated values of $V_i(1)$, $Q_i(1)$ and $O_i(1)$ are the initial state for the next time step. Next, the problem is solved for $t= 2$ to $T+1$, with a new set of dynamic constraints (Equation 17.6). This procedure is repeated $n-T$ times, until the whole hydrograph is simulated. Obviously, a fast algorithm is required to keep the computational time within acceptable limits.

All applications of LP to operational problems of urban drainage systems are based on a similar concept as described above. Although these studies have demonstrated that LP may be applied in a systems analysis, it should be noted the method has some major limitations. Those concerning the linear objective function are discussed in the next section.

DRAWBACKS OF A LINEAR OBJECTIVE FUNCTION

The flexibility of LP concerning the formulation of the objective function is limited, since the function has to be linear. On the other hand, a linear function may have the advantage that, with basic knowledge of the solution routine, it is relatively easy to predict the behaviour of the model and thus to determine the best set of unit costs with respect to the desired systems performance.

To illustrate the behaviour of the LP model only one operational objective is considered, namely to minimize the total overflow volume of the three catchments. This is usually considered a main task of a combined sewer system. For the given example, the unit costs of the variables can be determined as follows. If the complete inflow hydrograph is known (consisting of T time steps) then the problem can easily be solved by applying unit costs of overflows (co_i^t) equal to 1 (or any non-negative value) and all other unit costs (cv_i^t, cq_i^t) equal to zero. In this case the problem reads

$$minimize \; F = \sum_{i=1}^{3} \sum_{t=1}^{T} O_i(t) \qquad (17.7)$$

subject to the same constraints as described by Equations 17.5 and 17.6. By definition, multiple optimal solutions exist for $V_i(t)$ and $Q_i(t)$ as no unit costs have been specified for these variables. Concerning the systems behaviour, Equation 17.7 only ensures that all storage capacity of the three reservoirs is used before an overflow may occur. Depending on the capacity constraints (Equation 17.5) several options may be possible to achieve this.

As mentioned above, the whole inflow hydrograph $I_i(t)$ can generally not be foreseen, meaning that the operational problem consists in fact of a succession of LP problems (with a limited control horizon T). This implies that we have to make a decision on how to use the storage and discharge capacity of the system elements, taking into account this uncertainty.

A desired systems behaviour which provides a suitable solution to this problem is to keep the filling degrees of the storage elements at a relatively equal level (related to the storage capacity of the particular sub–system). When using fixed unit costs of storage and transport, however, it is impossible to control the filling (and emptying) behaviour of the system in a satisfactory way. Here it is investigated whether this behaviour (or any plausible solution) can be translated into a proper set of (constant) unit costs.

To minimize overflows, the unit costs of overflows should obviously be given a value much greater than the costs of storage and transport. Besides, the unit costs of storage should be given a value greater than the unit costs of transport, in order to discharge the water out of the system. Hence, in this case the unit costs should meet the following criterion

$$co_i^t > cv_i^t > cq_i^t \; ; \quad \textit{for all } i \textit{ and } t. \qquad (17.8)$$

Furthermore, the unit costs of an overflow should slightly decrease in time to avoid multiple optimal solutions (meaning that an overflow at t and an overflow at $t+1$ would both lead to minimum cost). As inflow predictions are subject to errors, a decision to overflow should be postponed as much as possible until overflows cannot be avoided any more. Moreover, if overflow structures are uncontrolled (as in most cases), it is even physically impossible that overflows could occur as long as the current water level is below the crest level of the overflow weir. In formula, this criterion reads

$$co_i^t > co_i^{t+1} \; ; \quad \textit{for all } i \textit{ and } t. \qquad (17.9)$$

The unit costs of storage and discharge may in this case be chosen as a constant in time, or

$$cv_i^t = cv_i^{t+1} \text{ and } cq_i^t = cq_i^{t+1} ; \quad \text{for all } i \text{ and } t. \qquad (17.10)$$

In Petersen (1987) and Schilling & Petersen (1987) it was demonstrated that the LP model is quite sensitive to inaccurate inflow predictions. Furthermore, it appeared to be difficult to determine the best set of unit costs, yielding optimal performance of the system. This set was derived by trial–and–error. Although a basic knowledge of the principles of LP is sufficient to derive some criteria, such as Equations 17.8 and 17.9, it can be concluded that a sensitivity analysis will mostly be required to verify the validity of the chosen set of unit costs.

From Equation 17.4 it can be read that the optimal solution is independent of the value of the state variables at $t = 0$, or the initial systems state. In fact it does not matter whether the objective function is minimized for $t=1$ to T or $t = 0$ to T as the state variables at $t = 0$ are constants. The least cost solution of the LP model is determined by the constant unit costs of the state variables and the specified inflow during the time horizon T for which the problem is optimized. As the unit costs are constants, this implies that the best set of unit costs is depending on the inflow hydrograph that is calculated, which explains the above mentioned findings of Petersen & Schilling (1987). In the LP model, this problem obviously does not occur if the time horizon covers the whole inflow hydrograph. In that case the problem can be formulated by Equation 17.7 in which, in a strict sense, no unit costs have to be specified.

As mentioned above, the LP problem can be solved with standard software. A well known algorithm is the Simplex method. This algorithm (in its standard or a revised form) is mostly applied. To describe the dynamic constraints (Equation 17.2) of a system consisting of n nodes, A system variables and a time horizon of T time steps requires a matrix of $[((A \cdot n \cdot T)+n)(n \cdot T)]$ entries to be solved, which may restrict the method to smaller problems. It surely needs an efficient algorithm to keep the computational time within acceptable limits.

Such a model has been developed at the Institute for Operations Research at Zürich, Switzerland (Neugebauer 1989, Neugebauer et al. 1991). This model, called NOUDS (Network Optimization for Urban Drainage Systems) is based on the LP concept as described above and uses an efficient network flow algorithm to solve the problem. On a 'normal' PC, the model allows control horizons in the order of 250 time steps.

A NON–LINEAR OBJECTIVE FUNCTION

From the above it can be seen that better results are obtained when the operational problem is formulated as a non–linear programming problem, where the objective function to be minimized is dependent on the current value of the particular systems state variable (Nelen 1992). For example, to control the use of storage in the various sub–systems (nodes) it is necessary to increase the unit costs of storage with increasing filling degree, like

$$cv_i^t = \frac{V_i(t)}{V_{i,\max}} \cdot \kappa_i \qquad (17.11)$$

where κ_i is a constant denoting the maximum unit cost of V_i (which may differ for every node). Note that substitution of Equation 17.11 in Equation 17.4 yields a non–linear programming (NLP) problem.

The unit costs of overflows should be determined on the basis of the function and vulnerability of the receiving water. As overflows are to be prevented as much as possible, their unit costs (co_i^t) should obviously be given a greater value than the costs of storage (cv_i^t) and transport (cq_i^t). If necessary, the flow to the treatment plant should be modified according to the actual conditions at the plant. On the basis of such considerations, it is possible to formulate an objective function, which is valid for all operational conditions.

An effective way to solve the NLP problem is to replace it by a succession of LP problems. This means that at each time step of the simulated inflow hydrograph, the optimization problem as described by Equations 17.4, 17.5 and 17.6, is re–formulated using the results of the preceding time step. Main advantage of this approach is that it allows the use of a powerful network flow algorithm, by which the model is fast enough to simulate time series of events. Besides it provides a possibility of using variable bounds of the systems state variables (Equation 17.5), which can be used to improve the flow routing in the model. The latter is out of the scope of this paper, but the principle is based on the concept that the flows along the arcs can be calculated explicitly out of the calculated water levels of the preceding time step.

The NLP model has been incorporated in a newly developed modelling package, called LOCUS, which is an acronym of 'Local versus Optimized Control of Urban drainage Systems'. The name denotes that besides optimal controlled systems, local (or static) controlled systems can be simulated as well (i.e. the present way of operation of most urban drainage systems). The latter has been included in LOCUS to serve as a reference. As the reference model and the optimization model are based on an identical system description, the difference between the results of both models is only due to the way the system is operated and hence the effects of optimal control can be quantified by comparing these results.

CASE STUDY

We consider a simple fictitious system that is illustrated in Figure 17.2. The upstream sections of the system (nodes 1 & 2) are sized 'properly', according to the Dutch standards. The main problems are to be expected at the downstream section, at node 3, where the storage capacity is relatively small. The potential of RTC is quantified here in terms of extra storage capacity that would be required in a local controlled system to achieve the same performance as an optimal controlled system.

The systems characteristics are listed in Table 17.1. All sub-catchments are identical and the rainfall is assumed to be homogeneous. For convenience the dry weather flow (DWF) is set equal to zero. Note that in this case only the temporal variability of the system input and the inhomogeneity of the system are contemplated. The effects of the spatial distribution of the system input are in this case excluded.

The rainfall data used for the analyses are historic events that have been recorded in Lelystad (the Netherlands) in the year 1981. The rain data are transformed into inflow by applying an initial loss of 1 mm and a linear reservoir with a reservoir constant of 15 minutes. The time step used in the simulations amounts to 10 minutes. The system is simulated for 5 different cases. The storage capacity of Node 3 is increased by 1000 m³ (= 1 mm) for each successive case. All cases are simulated using three different operation strategies:

1. local control, using a fixed stage-discharge relationship:
   if $V_i \geq (0.10 \setminus 0.30 \setminus 0.50)$ $V_{i,max}$ then $Q_i = (0.33 \setminus 0.67 \setminus 1.0)$ Q_{max};
2. local control, using a fixed stage-discharge relationship:
   if $V_i \geq (0.05 \setminus 0.10 \setminus 0.15)$ $V_{i,max}$ then $Q_i = (0.33 \setminus 0.67 \setminus 1.0)$ Q_{max};
3. optimal control, using a control horizon of 1 hour (= 6 time steps). The objective function is formulated so that the use of storage is maximized at all 3 sub-systems.

Assuming the system is controlled by pumps (which is mostly the case in the Netherlands), the first strategy may be regarded as more realistic than the second one, as the maximum pump capacity is usually not yet activated at a

Table 17.1. The systems characteristics.

	node 1	node 2	node 3
Impervious Area [ha]	100	100	100
Storage Capacity [m³]	7000	8000	5000 *
Storage Capacity [mm]	7	8	5 *
Discharge Capacity [m³/h]	700	600	2100
Discharge Capacity [mm/h]	0.7	0.6	0.8

* : the capacity of node 3 is enlarged to 6000, 7000, 8000 and 9000 m³
    (= 6, 7, 8, 9 mm)

filling degree of 15%. Because only major events are simulated, it is obvious that better results are obtained by applying strategy 2 in which the maximum discharge capacity is activated at a very early stage. Therefore, for a fair comparison between the results of local and optimal control, the results of the second strategy should be used. As when optimal control is applied it is also assumed that the full discharge capacity is available from the beginning of the rain event.

At the starting point of the calculations, i.e. along the Y–axes of the graphs, the overflow volumes of nodes 1 and 2, using strategy 2, are more or less equal to the volumes as determined for the optimal controlled system. As was mentioned above, strategy 2 (= discharge as much as possible) is in this case a suitable strategy for nodes 1 and 2 to minimize overflows. In the local controlled system the overflow volume of node 3 is much greater than those of nodes 1 and 2, which is in consonant with the systems characteristics. In the optimal controlled system, however, the overflow volumes of all 3 nodes are at an equal level, despite the difference in available storage. By applying RTC, we can make use of the temporal variability of the inflow and attune the system capacities to the current loading, by which a significant reduction of the overflow volume of node 3 can be achieved, without increasing the overflow volumes of nodes 1 and 2.

Now, let us consider the effects of increasing the storage or discharge capacity at node 3. This has obviously no effect on the upstream sections of the system when local control is applied. For the optimization model, however, it means that extra capacity is added to the system that can be used to upgrade its performance. As a result, the overflow volumes of nodes 1 and 2 can be reduced, although capacity is added only at node 3.

As can be seen from Figure 17.3, the storage capacity of Node 3 has to be augmented by at least 1500 m^3 (= 1.5 mm) to reduce the overflow volume of node 3 to the level that is achieved by applying optimal control, without increasing the storage capacity.

LESSONS FROM THE CASE STUDY

Based on the results of this simple, yet illustrative, case study it can be concluded that it is worth investigating the potential of RTC before constructing extra storage in the system. For a small fictitious system with limited storage capacity at the downstream section, it is shown that the potential of RTC is comparable with increasing the storage capacity by 1.5 mm at this particular section.

In this case, the main contributing factor to this potential is the temporal distribution of the system input and the distribution of available system capacities. The effects of the spatial distribution of the sewer inflow, which will

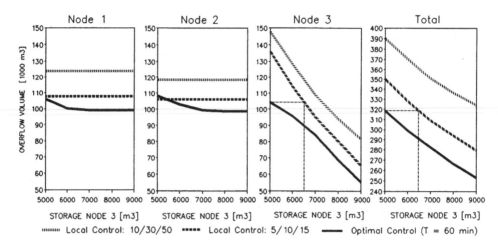

Figure 17.3. Effects of increasing the storage capacity at node 3.

have a positive effect on the potential of RTC are in this case not taken into account. (It is noted that the latter is less important as compared to the first two factors). Moreover, even when it has been decided to add some extra capacity to the system, RTC is still required to improve the performance of the entire system. Applying local control means that the systems performance is improved only locally, i.e. at the site where the capacity has been increased.

CONCLUDING REMARKS

In this chapter, the use of LOCUS in analyzing the performance of an urban drainage system that is operated in real time is illustrated. For a discussion on 'optimized control' and a more detailed description of the model reference is made to Nelen (1992), in which also the aspects that may contribute to the potential of real time control are discussed. In Nelen (1992), several case-studies are performed with the LOCUS model to get insight into the possible gains concerning the required system capacities, the importance of predicting inflows and effects of prediction errors on the operation strategy, the possibilities of controlling the system based on pollution parameters, the effects of rainfall distribution and the possibilities of reducing the peak flow to the treatment plant.

The increasing complexity of urban water management and the increasing possibilities of monitoring and remote control equipment stimulated many water managers to install computerized monitoring systems. As yet, the reasons for installing such systems are mostly data acquisition (e.g. for maintenance planning), monitoring the state of operation and alarming in case of technical trouble. Until today, the possibilities of real time control of the regulators (pumps, gates, weirs etc.), based on the actual state of the drainage system, are used only in a few cases. Although several research projects have shown the potential of RTC, it is still not widely accepted in urban drainage. Main obstacles to the use of this technique are to be found in administrative borders within the urban drainage system, funding arrangements and inflexible regulations and standards that are in favour or even prescribe static solutions to urban drainage problems. Nevertheless, it can be expected that, due to the change in mind towards environmental protection and the necessity of minimizing the costs, RTC will become a common feature in urban drainage systems in the (near) future.

REFERENCES

Babovic, V. 1991. *A control and advisory system for real-time applications.*
International Institute for Hydraulic and Environmental Engineering, Delft

Bellman R. & S. Dreyfus 1962. *Applied Dynamic Programming.* Princeton University Press, Princeton, N.J.

EAWAG 1990. *Applications of Operations Research to Real Time Control of Water Resources Systems,* (First European Junior Scientist Workshop), Luzern, Switzerland, (ed. T. Einfalt, M. Grottker, W. Schilling), ISBN 3 906484 04 1.

Graillot, D. 1990. *Application of expert system technology in drainage systems* in: (EAWAG 1990), pp. 185–197.

IAWPRC Taskgroup on RTC of UDS 1989. *Real Time Control of Urban Drainage Systems – The State of the art,* (ed. W. Schilling), Pergamon Press, ISBN 0 08 040145 7.

Khelil, A., M. Grottker and M. Semke 1990. Adaptation of an expert system for the real time control of a sewerage network: case of Bremen left side of the Weser, in: *Proc. of the 5th International Conference on Urban Storm Drainage,* Osaka, Japan.

Kularathna, M.D.U.P. 1992, *Application of dynamic programming for the analysis of complex water resources systems,* Agricultural University of Wageningen.

Nelen, A.J.M. 1992. *Optimized control of urban drainage systems.*
Delft University of Technology, ISBN 90 9005144 9

Neugebauer, K. 1989. *Steuerung von Entwässerungssystemen* (Control of waste water systems; in German), Institut für Operations Research, Zürich, Switzerland.

Neugebauer, K., W. Schilling & J. Weiss 1991. A network algorithm for the optimum operation of urban drainage systems, *Water and Science Technology, Vol. 24.,* No. 6, pp. 209–216

Petersen, S.O. 1987. *Real Time Control of Urban Drainage System.*
Technical University of Denmark, ISBN 87 89220 04 8

Schilling, W. & S. Petersen 1987. Real time operation of urban drainage systems – validity and sensitivity of optimization techniques, in: *proceedings of the symposium: Systems analysis in water quality management,* University of London.

Wagner, H.M. 1975. *Principles of Operations Research – With Applications to Managerial Decisions,* Prentice Hall, ISBN 0 13 709592 9

18 IRRIGATION AND DRAINAGE IN SAGA PLAIN, JAPAN

K. WATANABE
Saga University, Saga, Japan

INTRODUCTION

The water management aspects for agriculture in the Saga plain, situated in the Kyushu island in the southern part of Japan, are briefly outlined in this chapter. The Saga plain is a typical lowland that serves as an example where water management is achieved by a combination of irrigation and drainage aspects.

IRRIGATION IN SAGA PLAIN

Description of the area

The Saga plain, approximately 60,000 ha in extent and occupying nearly a quarter of the Chikushi Plain, is a vast land developed along the northern coastal areas of the Ariake Sea. It is an alluvial plain at the foot of the Sefuri and the Tenzan mountains, belonging to the Quaternary period. The eastern part of this plain was created mainly through deposition of sediments that had been carried down along the Chikugo and the Kase rivers. The western part is also an alluvial plain formed by deposition of sediments carried down from the hilly areas along the Rokkaku and the Shiota rivers. In addition, there are another two rivers, namely the Hayatsue and the Hattae rivers, which flow through this plain. These rivers play an important role in serving as water sources for irrigation in the Saga plain. Besides natural lands, there exist some areas that had been reclaimed from shallow sea. During the process of reclamation, gullies were left unchanged and the so-called 'creeks' were developed like a large fishing net, giving the Saga plain an unique aerial panoramic view.

Agricultural water use

Most of the rivers in the Saga plain, except the Kase river, are of small or medium size with narrow catchment areas. Usually, the amount of water from

403

these rivers is not sufficient to irrigate all paddy fields in the plain. The above mentioned creeks which are distributed throughout the plain can help store water from the rivers and subsequently supply the water to cope up with the high demand for water for irrigation purposes. During the high tides when the water levels in the rivers are relatively high, some portion of the river waters are diverted into the creeks. It is estimated that the rivers contribute about 72% of all the irrigation water required for paddy fields in the Saga plain, while waters from the creeks, groundwater wells and other sources contribute about 20%, 6% and 2%, respectively.

Distribution of creeks

The creeks in the Saga plain (Figure 18.1) occupy about 1800 ha, i.e. about 7% of the total paddy field area, which is approximately 26,000 ha. The percentage is as high as 17% in the area of Kose–machi, Saga city. The total volume of the creeks in the Saga plain is about 25,000,000 m^3, accounting for approximately 8% of the total volume of water required for paddy field irrigation each year, which is approximately 300,000,000 m^3. The mean water depth in the creeks is 1.17 m. Table 18.1 shows the distribution of the creeks in the main areas of the Saga plain. Most of the creeks are located in the Saga lowland.

The total storage capacity of the creeks equals nearly that of the Hokuzan reservoir constructed in the upstream of the Kase River. These creeks play a significant role in paddy field irrigation which can be shown by the ratio between the storage volume of the creeks and the volume of irrigation water per annum as tabulated in Table 18.1. Many creeks in the Shiroishi lowland are used as canals rather than for water storage. It is found that the creek and paddy

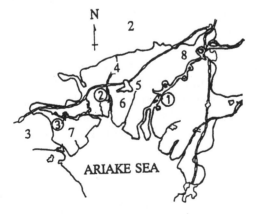

1   TENZAN MOUNTAINS
2   SEFURI MOUNTAINS
3   KISHIMA MOUNTAINS
4   SAGA PLAIN
5   SAGA CITY
6   SAGA LOWLAND
7   SHIROISHI LOWLAND
8   EAST SAGA LOWLAND
① CHIKUGO RIVER
② KASE RIVER
③ ROKKAKU RIVER

Figure 18.1. The Saga plain.

Table 18.1. Distribution of creeks in main areas of the Saga plain.

	Total area of paddy fields (A) (ha)	Total area of creeks (B) (ha)	B/A (%)	Storage of creeks (C) ($\times$ 1000 m^3)	Volume of irrigation water (D) ($\times$ 1000 m^3) per annum	C/D (%)
East Saga lowland	3666	158.6	4.3	5891	44,534	13.2
Saga lowland	15,705	1405.1	8.9	17,375	181,908	9.6
Shiroishi lowland	6625	221.5	3.3	1683	74,252	2.3
Total	25,996	1785.2	6.9	24,939	300,694	8.3

field area ratio, as well as the creek storage and irrigation water volume ratio in the Shiroishi lowland are lower when compared with the other two areas, namely the East Saga lowland and the Saga lowland (Table 18.1).

IMPROVEMENT OF AGRICULTURAL INFRASTRUCTURES IN THE SAGA PLAIN

At present, the agricultural sector in Japan is facing some problems such as excessive supply, increased imports of agricultural products, etc. To overcome these problems, some agricultural development plans are needed. The main objectives are to secure stable national food supply and to develop competitive agriculture. In order to meet these objectives, it is necessary to improve the agricultural infrastructures, especially those of the paddy fields. Among the agricultural development projects, the so–called 'land improvement project' in the Saga plain has been planned with the following objectives:

1. To improve the infrastructures necessary for rice crop management, the following plans are set:
   i) eliminating water shortage problems and securing the supply of irrigation water which will be required after the infrastructure improvement,
   ii) improving the irrigation and the drainage facilities,
   iii) improving field conditions to enable the use of agricultural machines, and
   iv) redesigning the creeks.
2. To improve the infrastructures for growing rice and other crops in rotation, the following plans are set:

i) maintaining the creek water level at about 1.0 m below the paddy fields, so that the groundwater table can be adjusted as required, and
ii) improving the drainage facilities to prevent flood damages.

IRRIGATION DESIGN IN THE SAGA PLAIN

After studying the hydrologic characteristics including rainfall distribution, effective rainfall, successive no–rain days during the irrigation periods, and river flow patterns, the year 1960 was chosen as the reference year for the irrigation design.

Irrigation

Table 18.2 shows the irrigation periods and the irrigation methods adopted in the Saga plain for growing rice, rush – a type of grass, lotus roots and forage crop.

Water requirement for rice

Results of research carried out at the site and an investigation of the groundwater levels indicated that water level lowering by setting the irrigation and the drainage canals independently (creek water level : 1 m lower than paddy field water level) would increase the coefficient of percolation by

Table 18.2. Periods and methods of irrigation in paddy fields in the Saga Plain.

TYPE OF CROP	IRRIGATION PERIODS AND METHODS	
Rice	IRRIGATION PERIOD	
	Land preparation period	June 21 to June 30
	Crop growth period	July 1 to October 30
	IRRIGATION METHOD	Ponding irrigation
Rush (a type of grass)	IRRIGATION PERIOD	
	Land preparation period	November 21 to November 30
	Crop growth period	December 1 to August 10
	IRRIGATION METHOD	Ponding irrigation
Lotus root	IRRIGATION PERIOD	April 1 to December 31
	IRRIGATION METHOD	Ponding irrigation
Forage crop	IRRIGATION METHOD	Furrow irrigation

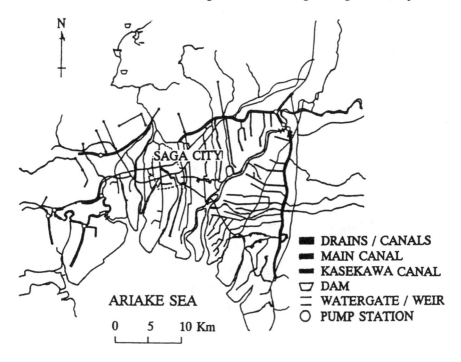

Figure 18.2. Facilities designed for irrigation drainage in the Saga plain.

approximately 20%. Therefore, the designed water requirement rate was selected at 120% of the current water requirement rate. In the case of paddy field irrigation, water from the creeks is used repeatedly by pumping. According to the information disseminated from various research activities including the on–site and lysimeter studies, as well as from the above–mentioned water requirement rate, maintaining the water level difference of 1 m between the paddy field and the creek seems to be appropriate.

By taking into consideration the rate of evapotranspiration in the paddy fields, the designed return flow rate was selected at 50% of the designed water requirement rate.

Water sources

The sources of water in the Saga plain include the storage reservoirs, the diversion rivers and the creeks. Since some parts of the Saga plain have been subjected to land subsidence, use of groundwater for irrigation is restricted. The water supply facilities in the Saga plain include pumping stations and conveyance canals. These facilities are shown in Figure 18.2.

Water gates are provided at appropriate places for water control at the river intakes as well as in the estuaries. Creeks are redesigned as shown in Figure 18.3 to suit future farming activities.

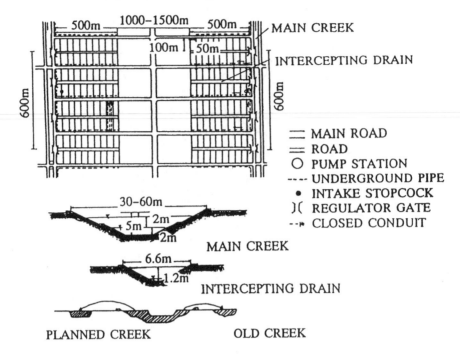

Figure 18.3. Creek redesign models.

DRAINAGE DESIGN

Overview

The Saga plain, as mentioned earlier, has several rivers draining water from the upper mountainous areas. Some of these rivers are connected to other rivers and ultimately drain into the Ariake Sea. Usually, water from the paddy fields in the downstream area is drained out to the rivers by using some facilities such as the drainage canals and the creeks. Due to tidal effect in the Chikugo river, drainage by gravity from the rivers within that area during the high tides is sometimes difficult or rather impossible, and so water remains undrained within the area. However, damages from this are of comparatively small magnitude because of the flat terrain of the area, which is so vast that the depth of the undrained water remains low.

Design precipitation

Since the region is vast, design precipitations are set separately for the Saga and the Shiroishi lowlands according to the rainfall data recorded at the Saga and the Shiroishi meteorological stations, respectively.

The design precipitation is based on the 10–year return period and 3–day continuous rainfall. The values adopted are as shown in Table 18.3.

Design of outside water level reference

As the area to be drained is vast, the outside water level references vary with the location. The water level in the Chikugo river was considered as the datum for the outside water level references. The following steps were used in the design of the outside water level references:

1. Determining the high water level at Senoshita (See Figure 18.4),
2. Selecting a suitable flood hydrograph,
3. Setting the estuary water level (water level of the Ariake Sea) equal to the mean tidal level of the ordinary neap tide, and
4. Calculating the surface water levels by flow routing at some points between the estuary and Senoshita. The peak outflow time and high tide period must be considered in the computation.

Drainage routes are as shown in Figure 18.4.

Drainage Volume

The volume of drainage is calculated so that the inundation duration can be limited within the allowable range. The following factors were taken into consideration:

1. Relationship between the internal water level and inundation volume estimated from the topography.
2. Cumulative inflow curve obtained from the inflow hydrographs that are based on the forecasted rainfall distribution.

Table 18.3. Design precipitation.

	Saga lowland (mm)	Shiroishi lowland (mm)
Rainfall/hour	68	61
Rainfall/day	204	203
Rainfall/2 days	269	276
Rainfall/3 days	321	332

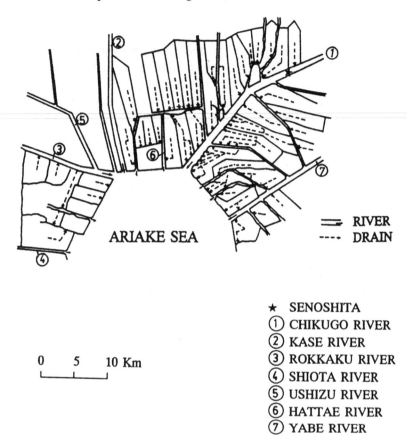

ARIAKE SEA

═══ RIVER
---- DRAIN

0   5   10 Km

★ SENOSHITA
① CHIKUGO RIVER
② KASE RIVER
③ ROKKAKU RIVER
④ SHIOTA RIVER
⑤ USHIZU RIVER
⑥ HATTAE RIVER
⑦ YABE RIVER

Figure 18.4. Drainage routes.

DRAINAGE FACILITIES

Drainage canal

The size of the drainage canal is determined so as to allow the flow of the peak drainage obtained from the forecasted precipitation, which is around 5 to 10 mm/hr at uniform flow, and 10 to 20 mm/hr at the end of the drainage canal.

Pump

The capacity of a pump is often determined according to the inundation level and the inundation duration rather than on the peak drainage. It is usually one-half to one-fourth of the peak drainage.

Drainage gate

The major factors for the drainage capacity are the cross–sectional area of the flow and the sill. The sill is determined according to the lowest reference altitude, i.e. bottom of the main drainage canals, and their relations with the outside water level in order that the drainage function be constantly maintained. The cross–sectional area of the drainage canal is roughly estimated based on the past records and further corrections are incorporated by drainage analysis.

FUTURE DRAINAGE DESIGN

Communities are formed at relatively high places in the Saga plain and it is said that the temporary inundation of the paddy fields does not cause too much problems for rice production. However, with the need for growing some other crops other than rice, multi–purpose fields are required. With the progress of the modern society and the economy, agricultural communities are changing day by day with the expansion of the residential areas, increase of the impermeable areas like roads, and introduction of industries. Under these circumstances, it is important to establish a flexible drainage design that can accommodate the future changes.

REFERENCES

Watanabe, K. 1979. Farming fields and drainage in areas around the Ariake Sea. *Kyushu branch of the Japanese Society of Irrigation, Drainage and Reclamation Engineering.*
Watanabe, K. 1980. Drainage design for agricultural water use in the downstream areas of the Chikugo River. *Japanese Institute of Irrigation, Drainage and Reclamation Engineering.*

19 LAND SUBSIDENCE PROBLEMS IN JAPAN

I. TOHNO

The National Institute for Environmental Studies, Japan

INTRODUCTION

Lowering of ground surface, called subsidence, results from the heavy withdrawal of groundwater, oil, natural gas, geothermal fluids, extraction of coal and mineral resources, etc. Subsidence is also caused by consolidation of sediments (as in reclaimed ground), oxidation and shrinkage of organic deposits, development of sinkholes and caverns in limestone deposits and various other phenomena. Subsidence has been occurring in various parts of the world, notably in Japan, Mexico and U.S.A. Many new areas of subsidence are developing as a result of accelerated extraction of natural resources, in particular groundwater, to meet the increasing needs of the people and for industrial development.

Major subsidences due to withdrawal of groundwater are relatively large in highly developed areas, e.g. Shanghai, China; Taipei basin, Taiwan; Tokyo and Osaka, Japan; Mexico city, Mexico; Bangkok, Thailand; Venice and Po delta, Italy; San Joaquin and Santa Clara valleys, U.S.A. etc. In fact development and subsidence seem to be closely interrelated. Types of damages that may result from subsidence are (i) increased soil wetness leading to loss in agricultural and farm production; (ii) reduced safety of dikes and protection works; (iii) decrease in quality of water and management in regions of subsidence; and (iv) ecological damages due to changes in hydrological regime, sediment transport, etc. Prediction of subsidence allows appropriate planning and management of engineering activities and performance of drainage systems. The history of land subsidence phenomenon since the early part of this century and the present state of land subsidence in many areas in Japan are briefly discussed in this chapter.

During the 1930s, the observed subsidence rates were rapid in Tokyo and Osaka areas, at more than 10 cm/year. During the Second World War, subsidence rates were very small due to a reduction in the withdrawal of groundwater from the wells. But around 1950, rapid subsidence was again observed due to the industries being rebuilt and rapid industrial development in both Tokyo and Osaka areas. Since 1955, land subsidence has become a nation-

413

wide phenomenon in Japan. However, since 1979, the general tendency of land subsidence in major areas across the nation has slowed down considerably as a result of various countermeasures and regulations enforced by the central and the local governments (Tohno 1992).

The effect of annual changes of groundwater level on land subsidence in the Muikamachi Basin is being reviewed as a case study (Tohno & Iwata 1988). Simulation of land subsidence at the Muikamachi Basin was carried out by using the results of the repeated and the normal consolidation tests. The calculated values are found to be in good agreement with the observed values of subsidence.

HISTORY OF LAND SUBSIDENCE IN JAPAN

Land subsidence phenomenon has a long history in Japan. Since the early part of this century, particularly since the 1910s, many deep wells were drilled with the development of modern drilling machines that pumped out groundwater excessively from the wells. Thus, the ground water table was getting lowered continuously (Figure19.1). After the Kanto Earthquake in 1923, excessive subsidences of some of the bench marks were observed by level surveying in the Tokyo lowland in the Kanto Plain. The maximum subsidence was 63 cm from 1923 to 1926, while it was 56 cm between 1926 to 1930 (Imamura 1931).

Tokyo lowland comprises of the Recent Quaternary layers, the so called "Chuseki–So". These layers formed in the Recent Pleistocene time (about 25,000 years B.P.), named the Nanagohchi and the Yuraku–cho members, are composed mainly of very soft cohesive soils.

In the 1930s, rapid subsidence, at a rate of more than 10 cm/year was observed in the Osaka and the Tokyo lowlands. See Figures 19.2 and 19.3. The Osaka lowland in the Osaka plain is also one of the typical land subsidence

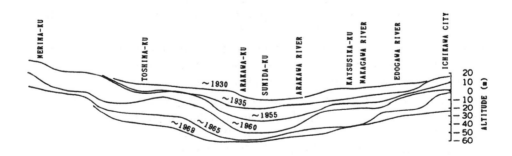

Figure 19.1. Change of groundwater table in Tokyo area between 1930 and 1969 (Shindo 1972).

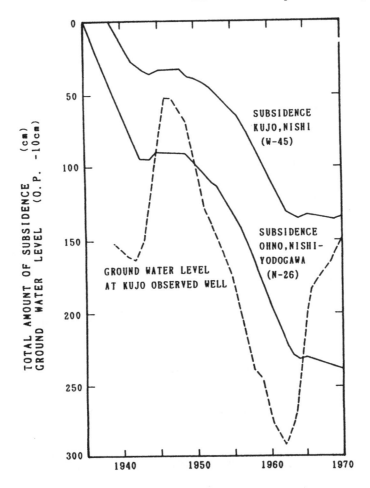

Figure 19.2. Groundwater level and subsidence of principal bench marks in Osaka city (Data from the Environment Agency of Japan).

areas, where subsidence had accelerated since around 1930 due to rapid industrial development. Figure 19.2 shows subsidence of principal bench marks and groundwater levels in the Osaka lowland. The groundwater level was getting lowered continuously and the observed subsidence rate was more than 10 cm a year until 1942. It was pointed out that the land subsidence was caused by the extraction of groundwater (Hirono & Wadachi 1939, Wadachi 1940, Wadachi & Hirono 1939).

When the Second World War broke out, the pumping out of groundwater from the wells decreased. It can be seen from Figure 19.2 that the groundwater level had risen rapidly between 1942 and 1946. Around 1945, there was no significant change in the levels of the ground surface in the Osaka and the Tokyo lowlands (Figures 19.2 and 19.3).

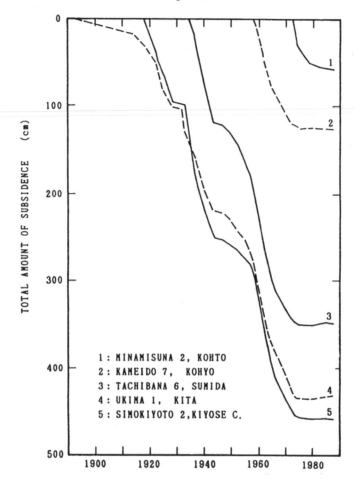

Figure 19.3. Subsidence of principal bench marks in Tokyo area (Data from the Tokyo Government).

Since around 1950, the discharge of pumped out groundwater again started to increase excessively with the rebuilding of the industries and industrial development. Groundwater tables were lowered substantially and rapid land subsidences were observed in these lowlands once again (Figures 19.1, 19.2 and 19.3). The groundwater level dropped by about 2 meters a year between 1949 and 1960 in Osaka lowland (Figure 19.2).

From 1955 onwards, land subsidence has become a nationwide phenomenon in Japan, observed not only in the Osaka and the Tokyo lowlands, but also in the areas adjacent to these lowlands, and other areas including the Haranomachi, Niigata, Nobi and the Saga Plains.

Table 19.1 shows main land subsidence areas in Japan. In the Niigata Plain, remarkable subsidence of about 54 cm was observed in the year 1960, which

Table 19.1. The extent and magnitude of maximum subsidence in major land subsidence areas in Japan, based on level surveying (Data from the Environment Agency of Japan).

Prefecture	Area	Recognized Subsidence Zone (Km²)	Zero–Meter Zone (Km²)	Maximum Subsidence a year		Maximum amount of Total Subsidence	Starting year of leveling
				Subsidence (cm)	Year	Subsidence (cm)	
Hokkaido	Ishikari Plain	250		10.1	1978	45	1975
Aomori	Aomori Plain	65	3	6.8	1973	52	1958
Aomori	Hachinohe Plain	10		3.4	1984	28	1975
Akita	Kisakata–Konoura	10		13.8	1970	57	1968
Yamagata	Yamagata Basin	60		13.7	1975	33	1974
Miyagi	Sendai Plain	290		9.6	1975	81	1974
Fukushima	Haranomachi Plain	40		10.3	1975	163	1955
Ibaraki	Kanto Plain	350		9.5	1978	90	1974
Tochigi	Kanto Plain	240		6.7	1990	46	1977
Gunma	Kanto Plain	20		2.5	1984	28	1975
Saitama	Kanto Plain	1650		27.2	1974	170	1961
Chiba	Kujukuri Plain	855	18	12.2	1971	74	1968
Chiba	Kanto Plain	2010	15	24.9	1970	211	1962
Tokyo	Kanto Plain	955	124	23.9	1968	453	1918
Kanagawa	Kanto Plain	230	6	26.3	1982	139	1931
Kanagawa	Shonan Plain	130		7.3	1984	35	1967
Niigata	Niigata Plain	805	207	53.7	1960	275	1957
Niigata	Nagaoka Basin	70		3.7	1993	20	1975

Cont....

Table 19.1 (Cont.)

Prefecture	Area	Recognized Subsidence Zone (Km²)	Zero-Meter Zone (Km²)	Maximum Subsidence a year		Maximum amount of Total Subsidence	
				Subsidence (cm)	Year	Subsidence (cm)	Starting year of leveling
Niigata	Takada Plain	240		10.1	1984	31	1968
Niigata	Muikamachi Basin	60		9.2	1984	43	1975
Yamanashi	Kofu Basin	80		1.6	1978	13	1974
Nagano	Suwa Basin	20		6.7	1986	38	1977
Aichi	Okazaki Plain	65	57	4.2	1978	33	1975
Aichi	Nobi Plain	735	286	23.5	1973	149	1963
Gifu	Nobi Plain	170	61	4.6	1974	27	1963
Mie	Nobi Plain	120	55	21.3	1961	160	1961
Osaka	Osaka Plain	635	55	25.0	1961	290	1935
Hyogo	Osaka Plain	100	16	20.8	1970	283	1932
Tottori	Tottori Plain	10		4.5	1981	33	1974
Saga	Saga Plain	320	207	13.0	1973	109	1957
Kumamoto	Kumamoto Plain	35	9	4.3	1978	28	1974

Table 19.2. Maximum subsidence in major land subsidence areas of Japan based on level surveying (Data from the Environment Agency of Japan).

PREFECTURE	AREA	Maximum Subsidence (mm)																
		1975	1976	1977	1978	1979	1980	1981	1982	1983	1984	1985	1986	1987	1988	1989	1990	1991
Hokkaido	Ishikari Plain		57	33	101	86	44	32	59	52	84	67	48		12	15		38
Aomori	Aomori Plain	37	30	20	17	17	22	10	15	69*	24	29	24	20	28	8	29	11
Aomori	Hachinohe Plain		25	22	33	13	7	10	27	19	34	34	13	12	41	16	7	5
Akita	Kisakata–Konoura	71	106	42	53		7	7	38	23	23	18						
Miyagi	Sendai Plain	96	71	55	189*	56	54	37	43	33	66	51	42	42	37	26	25	23
Yamagata	Yamagata Basin	137	25	66	32	29	9	16	13	17	16	22	21	14	5	20	12	14
Fukushima	Haranomachi City		103	45	29	8	19	12	30	15	17	11	8	6	2	4	7	6 3
Ibaraki	Kanto Plain	86	71	72	95	81	54	49	48	44	56	46	41	40	45	40	43	43
Tochigi	Kanto Plain			27	41	31	29	31	18	27	55	30	27	43	33	32	67	36
Gunma	Kanto Plain		14	12	18	21	19	16	13	20	25	13	20	19	22	20	23	23
Chiba	Kujukuri Plain	54	91	44	45	36	56	54	43	51	30	32	31	93*	42	30	31	30
Chiba	Kanto Plain	77	74	59	95	86	53	45	35	34	53	29	32	45*	36	26	23	31
Saitama	Kanto Plain	147	141	98	125	96	79	67	53	52	60	56	47	48	54	46	44	42
Tokyo	Kanto Plain	54	56	43	56	22	18	25	11	13	32	18	13	15	11	10	14	8
Kanagawa	Kanto Plain	111	89	61	48	48	41	107	263	168	64	52	33	32	28	50	29	51
Kanagawa	Shonan Plain	30	22	32	33	33	20	16	21	22	73	23	20	19	23	15	52	34

Cont.....

* Subsidence caused mainly due to earthquake

Table 19.2 (*Cont.*)

PREFECTURE	AREA	Maximum Subsidence (mm)																
		1975	1976	1977	1978	1979	1980	1981	1982	1983	1984	1985	1986	1987	1988	1989	1990	1991
Niigata	Niigata Plain	49	25	12	30	15	33	17	35	11	23	38	26	17	30	21	21	23
Niigata	Takada Plain	16	20	42	25	4	20	37	23	22	101	66	54	14	16	3	12	9
Niigata	Muikamachi Basin		44	67	22	14	30	44	18	16	92	37	35	14	11	21	9	52
Yamanashi	Kofu Plain		14	6	16	16	7	10	15	12	10	8	9	13	12	8	14	8
Nagano	Suwa Plain				41	41	36	33	53	54	47	25	67	40	39	29	25	24
Aichi	Okazaki Plain		33	23	42	23	30	35	27	27	24	11	15	17	14	17	22	11
Aichi	Nobi Plain	99	84	73	46	35	22	32	24	26	23	15	10	18	6	22	20	14
Gifu	Nobi Plain	20	32	15	22	30	12	16	14	22	25	23	9	22	12	25	21	20
Mie	Nobi Plain	93	75	53	48	45	31	22	33	22	27	15	9	17	6	13	10	13
Osaka	Osaka Plain	101	93	85	63	23	17	20	25	18	25	25	13	19	14	15	14	13
Hyogo	Osaka Plain	108	75	26	24	34	57	14	28	15	12		16	17	8	28	13	17
Tottori	Tottori Plain		38	39	36	37	38	45	29	26	24	19	18	13	13	16	13	12
Saga	Saga Plain	93	69	59	103	49	35	46	54	38	59	67	39	37	56	42	45	38
Kumamoto	Kumamoto Plain	36	38	30	43	28	22	23	20	24	29	18	15	12	18	12	31	14

* Subsidence caused mainly due to earthquake

is the maximum value ever observed anywhere (Table 19.1). A subsidence rate of more than 20 cm a year was observed in 8 areas between 1960 and 1974.

Two laws, i.e. "Industrial Water Law" established in 1956, and "Law Concerning Regulation of Pumping Out of Groundwater for Use in Buildings" established in 1962, have been enforced in order to reduce the groundwater utilization in the rapid land subsidence areas. At present, 17 areas in 10 prefectures are designated under the former law, and 4 areas in 4 prefectures under the latter. Besides these laws, control of groundwater pumping is also being exercised under the ordinances and regulations of local autonomies i.e. 20 prefectures and 229 municipalities.

Since 1979, the tendency of land subsidence in major areas across the nation has reduced considerably (Tables 19.2 and 19.3), as a result of the various countermeasures invoked by the governments, users of groundwater, etc. Since 1985, the pumping out of groundwater from the wells in many subsidence prone areas, and as a consequence, the number of areas of severe subsidence (with more than 2 cm subsidence a year), have been decreasing (Table 19.3).

Past and present land subsidence areas in Japan are shown in Figure 19.4. At present, major land subsidences occur in 61 areas in 36 prefectures and this total area amounts to about 3% of the entire land area of the country. Recently, severe land subsidences were observed mainly in areas subject to seasonal alternation of the groundwater levels and in areas undergoing rapid urbanization. In Tokyo area, subsidence rate has been observed to be less than

Table 19.3. Total extent of land in Japan with more than 2 cm subsidence a year based on level surveying (Data from the Environment Agency of Japan).

Year	Area (km^2)
1978	1946
1979	624
1980	467
1981	689
1982	616
1983	594
1984	814
1985	499
1986	396
1987	500
1988	617
1989	285
1990	360
1991	467

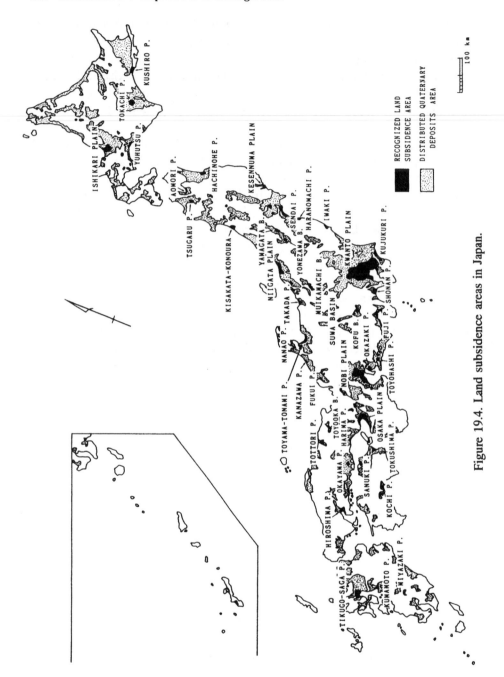

Figure 19.4. Land subsidence areas in Japan.

15 mm a year since 1986 (Table 19.2). But the total amount of subsidence accumulated since 1918 has reached about 453 cm and the total area of the "Zero–Meter Zone" at present is about 124 km^2, as shown in Table 19.1. It should be noted that the ground level of the "Zero–Meter Zone" is lower than the sea level at full tide. In the Nobi Plain,the total area of the "Zero–Meter Zone" at present is about 400 km^2.

LAND SUBSIDENCE PROBLEMS OF SOME MAJOR SUBSIDENCE AREAS IN JAPAN

Saga Plain

The Saga Plain, which is situated in the south of the Saga Prefecture has a coastline to north of the Ariake Sea. It is one of the typical land subsidence

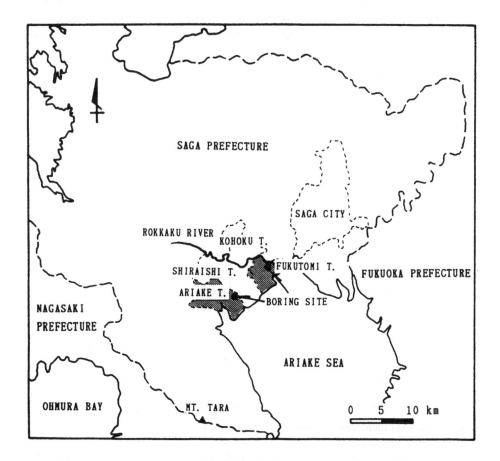

Figure 19.5. Location map of the Saga Prefecture.

areas where significant subsidence has occurred continuously and adequate countermeasures are needed urgently for the present. Figure 19.5 shows the location of the Saga Prefecture.

In this Plain, land areas whose elevations are less than 5 meters above sea level are extensively distributed, and the total area amounts to about 400 km^2. The tidal range of the Ariake Sea is about 6 m. River and coastal dikes are constructed to prevent flooding of those land areas whose elevations are lower than the mean high water level.

Saga Plain is composed mainly of deltaic plains, and is known for its large reclaimed areas. Paddy fields (paddies) are found extensively, drainage from which occurs in the topset of delta along the coast of the Ariake Sea. Very few rivers exist that can supply surface water for growing rice in the reclaimed areas. Thus, irrigation in these areas has to depend mainly on the groundwater supply.

Figure 19.6 shows the typical geological profiles of this Plain and the boring sites are shown in Figure 19.5. The top layers consist of very soft cohesive layers and are named Ariake Clay Formation, which are composed mainly of marine clay and silt Holocene deposits. K–Ah bed is contained in the Ariake

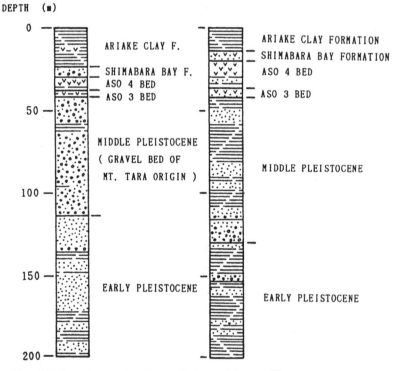

Ariake Higashi elementary school.   Fukutomi town office.

Figure 19.6. Geological columnar sections at Ariake Higashi elementary school site and the Fukutomi town office site in Shiraishi district.

Clay Formation, and is 2 cm thick and 13.05 m deep at Ariake. But, the K–Ah
bed was not found at Fukutomi. The K–Ah bed indicates wide–spread tephra
and well known key bed. This tephra has come from Mt. Kikkai, that arose 73
ka (kilo–age). The next is the Shimabara Bay Formation, latest Pleistocene
deposits, composed mainly of marine deltaic sands. The deposits of this
Formation are not marine, but of fluvial origin at Ariake (Tohno et al. 1990).
The next is late Pleistocene Group. This Group contains much intercalated
pumice bearing volcanic ash. This volcanic ash has come mainly from Mt. Aso,
i.e. Aso–4 bed and Aso–3 bed which arose about 70 and 110 ka, respectively.
Both beds are again indications of wide–spread tephra. The middle and early
Pleistocene Groups are composed of sand and silt with gravel. At Ariake, the
middle Pleistocene is composed mainly of gravel, which came to Mt. Tara, as
mud and/or debris flows. High quality groundwater had been pumped out
excessively from many wells in these sand and gravel beds from depths of
about 50 to 200 meters.

The thickness of the Ariake Clay Formation is variable in this Plain.
Equivalent value lines of the thickness of this Formation are shown in Figure
19.7. It also shows waste–filled river valleys, namely, east–side one for old
river Chikugo and west–side one for old river Rokkaku (Oshima 1988). At
Shiraishi and Ariake Towns in Shiraishi District especially, this Formation is
more than 20 meters thick (Figures 19.6 and 19.7).

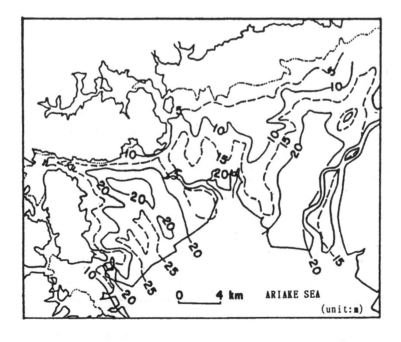

Figure 19.7. Thickness of the Ariake Clay Formation (Oshima 1988).

Land subsidence was noticed in this Plain in 1958. In 1960, the ground had cracked for about 5 km forming depressions, mainly in the paddies. In 1970, the Shiraishi District was severely damaged by land subsidence. About 3000 houses were destroyed completely or partially. About 20 km^2 paddies were cracked and settled differentially. About 40 km length of railway line was shifted, buckled and cracked. In the year 1973, subsidence based on level surveying was a maximum of 13.0 cm at the most severely affected places in this Plain (Table 19.1). Since 1979, the rate of subsidence has been observed to be about 4 to 7 cm/year (Table 19.2).

In 1985, the Ministerial Conference of the Japanese Government on "General Plan of Countermeasures for Land Subsidence of the Chikugo–Saga Plain" was held. As a result of remarkable land subsidences in the already extremely low ground levels of this plain, "Zero–Meter Zones" were formed on a large scale. The total area of the "Zero–Meter Zone" at present is about 207 km^2 (Table 19.1).

Figure 19.8 shows an example of the changes in the ground and the groundwater levels at Shiraishi Town. The data were obtained from an observation well. The groundwater level decreased remarkably in summer due to excessive pumping out of the groundwater primarily for agricultural purposes. As a consequence, rapid subsidence of the ground occurred. It was also observed that though the groundwater level ascended during winter to spring, the ground scarcely rebounded. Consequently, the subsidence accumulated to a large amount.

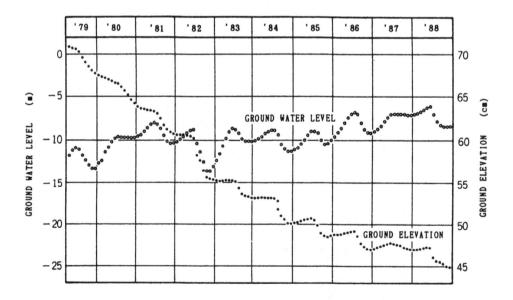

Figure 19.8. Ground elevation and groundwater level at Shiroishi town (Data from the Saga Prefecture authorities).

Figure 19.9. Lift up of the Fukutomi town office building due to land subsidence.

Figure 19.9 shows the Fukutomi Town Office building as on October 1989. This building, which was built in 1967, was constructed on piles supported on the upper sand bed of the middle Pleistocene. In 1989, seven risers had to be added to the front steps of this building as the ground had subsided about 2 meters.

Osaka Plain

The subsidence rate in the Osaka Plain had accelerated since around 1930 due to rapid industrial development. This Plain has a coastline to the east of the Osaka Bay, where the Yodo river discharges into the sea. It has a very large area of about 1600 km², and is composed mainly of flood plains, deltaic plains and man–made grounds.

The groundwater level was getting lowered continuously and the observed subsidence rate was more than 10 cm a year until 1942 (Figure 19.2). It was found that the total amount of subsidence, based on level surveying was more than 70 cm in and around the mouth of the Yodo River between 1936 to 1940 (Figure 19.10). But the groundwater level increased rapidly between 1942 to 1946 during the Second World War due to a decrease in the groundwater consumption. During that period no significant changes in the levels of the ground surface were observed (Figure 19.2).

In 1947, the groundwater level was again lowered, due to excessive pumping out of the groundwater. The groundwater level dropped by about 2 meters a year and the ground surface has been continuously subsiding, since 1949. Rapid subsidence of more than 10 cm/year was observed between 1956 and

1962. Maximum subsidence, based on level surveying, was 25.0 cm in the year 1961 at the most severely affected places of the Osaka City, Osaka Plain (Table 19.1).

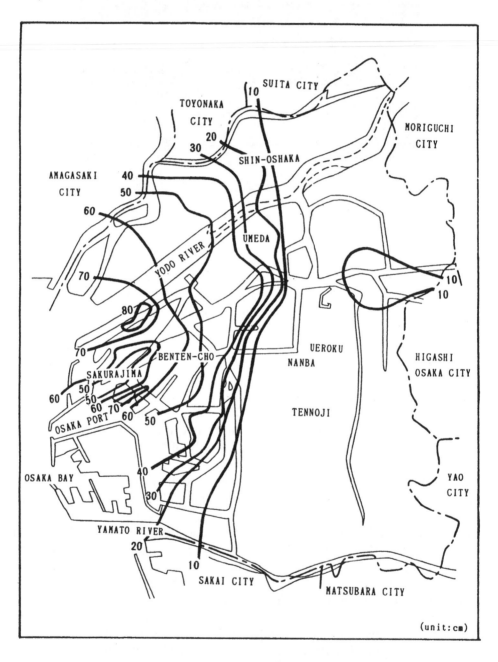

Figure 19.10. Contour lines of total amount of subsidence in Osaka City between 1936 and 1940 based on level surveying (Osaka Committee 1972).

In 1959, an ordinance for the prevention of the land subsidence was enforced for controlling the groundwater discharge in Osaka City. In 1962, two more laws, restricting the groundwater discharge were enforced in order to reduce the rapid land subsidence. Since then the groundwater discharge has decreased remarkably and the groundwater table started to increase again, and the tendency of the subsidence has decreased.

Since 1981, maximum subsidence has been observed to be less than 30 mm/year in this Plain (Table 19.2). The total amount of subsidence is 290 cm at the most severely affected places and the total extent of recognized subsidence areas at present is about 735 km² (Table 19.1).

Three fairly strong typhoons had hit Osaka. These were the Muroto Typhoon in September 1934, the Jean Typhoon in September 1950 and the Second Muroto Typhoon in September 1961. The flood tides due to these typhoons caused extensive and severe damages to life and property in the lowland areas and to ships, as shown in Table 19.4.

Figure 19.11 shows the submerged areas in Osaka City due to the Jean Typhoon of September 1950. As a result of the investigations after this Typhoon, it was found that the total submerged area was about 61.2 km² in Osaka City (Osaka Committee 1972), which agreed fairly well with the total area (in extent) of those areas having more than 10 cm of land subsidence, based on level surveying between 1936 to 1940 (Figures 19.10 and 19.11).

Nobi Plain

Nobi Plain is also one of the typical land subsidence areas which includes the Nagoya City, where the settlement rate had accelerated since around 1960 due to rapid industrial development.

Table 19.4. Damages caused by Typhoons in Osaka City (Osaka Committee 1972).

Date	Name of Typhoon	Maximum Tide Level (OP.+m)	Damages to		
			Persons	Houses	Ships
Sept. 21, 1934	Muroto Typhoon	5.10	17,898	142,845	2739
Sept. 03, 1950	Jean Typhoon	3.85	18,794	123,209	899
Sept. 16, 1961	Second Muroto Typhoon	4.12	2165	85,811	464

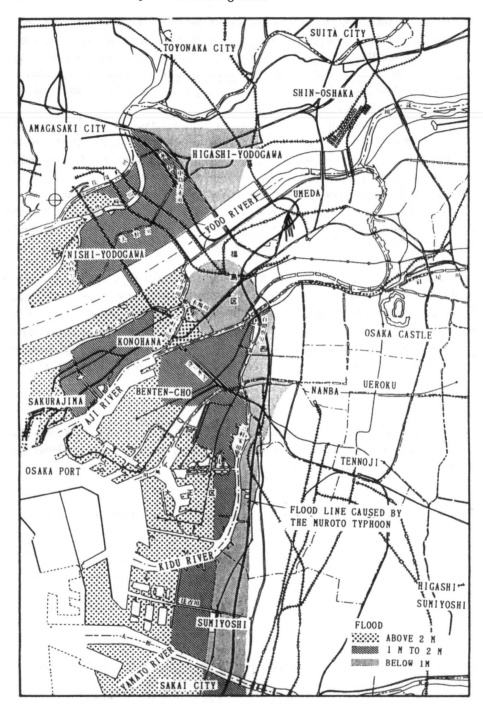

Figure 19.11. Areas submerged in the Osaka City due to the Jean Typhoon of September 3, 1950 (Osaka Committee 1972).

In the Nobi Plain, the Quaternary System has a very large exposure of about 1300 km². This Plain has formed a westward dipping sedimentary Quaternary system. This Plain has a coast with the Ise Bay, where the Ibi, Nagara, and Kiso rivers find their passage into the sea, and is composed mainly of flood plains, deltaic plains and man–made grounds. Latest Quaternary layers, which are composed mainly of very soft cohesive soils, are about 50 m thick in and around the mouth of these rivers.

In this Plain, the land subsidence phenomenon was recognized when the Isewan typhoon hit and caused severe damages by way of flood tide disaster in September, 1959. As a result of this typhoon, 5122 persons lost their lives; 15,384 persons were wounded; 127,625 houses were destroyed completely or partly, and 3846 houses were washed away (Ueshita & Daito 1984). After this typhoon, extensive areas of this plain were submerged for a long time (Figure 19.12). As a result of level surveying after this typhoon, it was found that the total area below the mean sea level was about 186 km² at that time.

Around 1970, more than 10 cm of subsidence a year was observed over a large area based on level surveying. In 1973, it was found that the maximum subsidence was 23.5 cm at the most severely affected places in Aichi Prefecture (Table 19.1 and Figure 19.13). Since 1974, the groundwater table has been rising due to regulations on groundwater use, and as a consequence, the

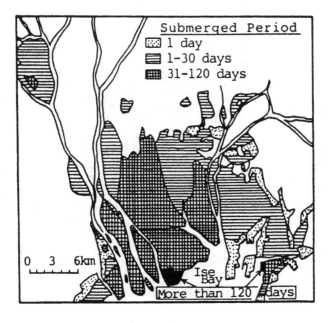

Figure 19.12. Submerged areas of the Nobi Plain due to the Isewan Typhoon of September 26, 1959 (Ueshita & Daito 1984).

tendency of the land subsidence phenomenon has calmed down generally (Table 19.2).

In 1985, the Ministerial Conference of the Japanese Government on "General Plan of Countermeasures for Land Subsidence of the Nobi Plain" was held. Desirable groundwater levels to prevent land subsidence were investigated by one–dimensional finite element models which were formulated at several places in the subsidence area (Ueshita & Sato 1981). As a consequence of the regulations and strict control enforced to keep the groundwater levels within the desired levels obtained from the results of the model study, land subsidence in the Nobi Plain has been greatly reduced.

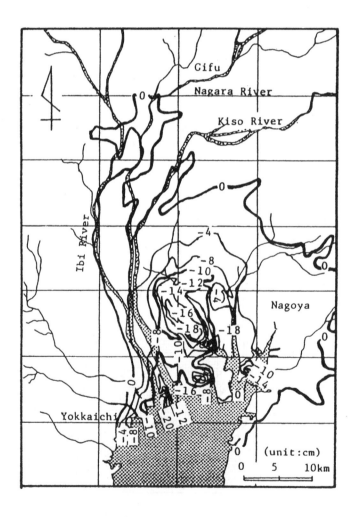

Figure 19.13. Subsidence of the Nobi Plain between November 1972 and November 1973 (Ueshita & Daito 1984).

In this Plain, the total amount of subsidence is about 160 cm at the most severely affected places. The total extent of recognized subsidence areas is more than about 1000 km² at present and about 80 % of this area is found to be underlain by Quaternary system. As a result of the remarkable land subsidence, Zero–Meter Zones were formed on a large scale in this plain. The total area of the Zone is about 400 km² at present (Table 19.1).

LAND SUBSIDENCE CAUSED BY THE ANNUAL CHANGE OF GROUNDWATER LEVEL IN THE MUIKAMACHI BASIN, NIIGATA PREFECTURE

Areas along the Japan Sea experience heavy snowfall during winter. The maximum depth of snowfall in the main areas is more than 3 meters. In order to maintain the daily life of the residents smoothly during the winter season, various methods have been adopted to get rid of the snow. In 1968, a new method for snow melting was adopted in Nagaoka City which is becoming rapidly popular. In this method the snow is melt by sprinkling groundwater from the pipes set up along the roads (Figure 19.14).

In Muikamachi Basin, Niigata Prefecture (Figure 19.4), which is one of those areas with the heaviest snowfall in Japan, the groundwater level is lowered rapidly in winter due to excessive pumping out of groundwater to melt snow

Figure 19.14. Set–up for snow melting along the roads, shown adjacent to the Hokushin elementary school in Muikamachi, Niigata Prefecture.

[AMOUNT OF GROUND WATER DISCHARGE :   2.3x10⁷ m³]

Figure 19.15. Consumption of pumped out groundwater from wells between December 1985 and March 1986 in Muikamachi, Niigata Prefecture (Data from the Niigata Prefecture authorities).

on roads, parking spaces etc. Temperature of groundwater is about 13°C throughout the winter in the basin. At Muikamachi Town, about 1900 wells pumped out nearly 18,700,000 m³ of groundwater for snow melting between December 1985 to March 1986 as estimated by the Niigata Prefecture Government (Figure 19.15).

Figure 19.16 shows groundwater levels and subsidence data obtained from an observation well at Muikamachi urban area. The average groundwater level is decreased by about 7 m in winter and the ground surface subsides about 3 cm a year. Serious land subsidence is thus caused by the repeated loading due to the changes in the groundwater levels.

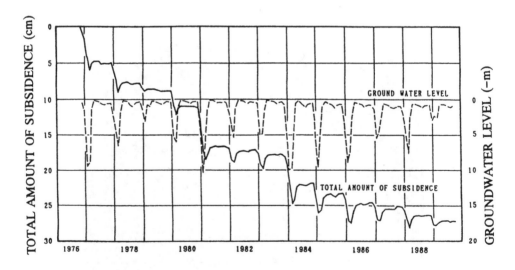

Figure 19.16. Measured land subsidence at site of old Muikamachi junior high school in Muikamachi Town, Niigata Prefecture (Data from the Niigata Prefecture authorities).

In order to investigate the mechanism of land subsidence caused by the annual changes of groundwater levels, undisturbed samples of the Quaternary deposits were obtained by means of thin walled tube and core samplers at three different sites in urban area between 1982 and 1984. Laboratory, repeated and standard consolidation tests were performed on the saturated undisturbed samples of cohesive soil. One of the boring logs and typical soil test results are shown in Figure 19.17. The upper clay layer is very soft, having an N-value of nearly zero as observed from Standard Penetration Tests. This layer is of Holocene age, extending from the ground surface to a depth of about 10 meters. The lower clay layer, with humus, is of latest Pleistocene age extending from about 18 to 28 meters depth. Both layers were found to be somewhat over-consolidated by past repeated loading.

Figure 19.18 shows the results of the repeated and the standard consolidation tests. This sample was taken from the Late Pleistocene cohesive deposits at the Hokusin elementary school site. Repeated loadings of 12.8 kgf/cm^2 (1260 kPa) and 6.4 kgf/cm^2 (628 kPa), and standard loading of 12.8 kgf/cm^2 were used in these tests. Compressive strain of the repeated loading after 28 cycles was found to be more than that of the static loading. Based on past findings, it is believed that the settlement due to repeated loading (unloading and reloading) is more severe than normal loading in the consolidation tests on saturated samples of cohesive soil.

Simulation of land subsidence at Muikamachi was carried out using the results of the repeated and the normal consolidation tests on samples taken from underground clay layers at the site. This simulation was done using one-dimensional consolidation analysis and finite element method (Tohno et al. 1989).

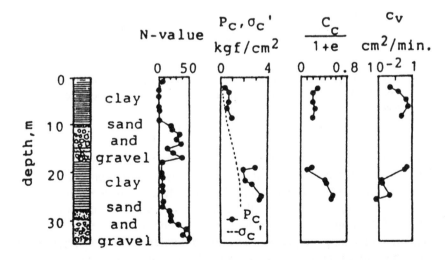

Figure 19.17. Boring log and results of soil testing at site of old junior high school in Muikamachi Town, Niigata Prefecture (Tohno et al. 1989).

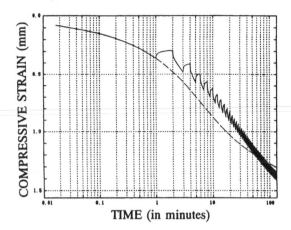

Figure 19.18. Results of the repeated and the standard consolidation tests on samples of cohesive soil obtained from about 38.7 m depth at the Hokushin elementary school site in Miukamachi town, Niigata Prefecture (Tohno & Iwata 1988).

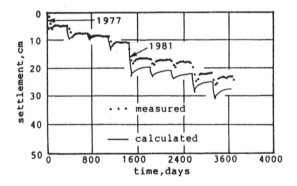

Figure 19.19. Results of the land subsidence simulation compared with the measured subsidence values in Muikamachi Town, Niigata Prefecture (Tohno et al. 1989).

Calculations of consolidation settlements were done as follows. Both clay layers were divided into eight elements. Distribution of water pressure in each element was assumed to be a linear one. The groundwater surface was assumed to be at the same level as the top of the upper clay layer. Groundwater head at the bottom of the upper clay layer, and at both ends of the lower clay layer, were considered to change equally. The groundwater level was regarded to decrease by the maximum value for one month in winter and was assumed to be at the normal value for the rest eleven months (Figure 19.16).

The results of the analysis and the measured settlements between 1977 to 1985 are shown in Figure 19.19. The calculated results fit very well with the measurements in the first four years between 1977 to 1980. A slight deviation is evident in the winter of 1981, but the relative settlements from 1982 to 1985 did not differ significantly. Therefore, it could be said that good agreement was obtained between the numerical calculations and the experimental results.

REFERENCES

Hirono, T. & K. Wadachi 1939. *On the land subsidence in Western Osaka (1).* Report of the Research Institute for Disaster Prevention, Osaka, Japan (in Japanese).

Imamura, A. 1931. Observed mass movements before and during the Tokyo earthquake of March 21, 1928. *Jisin* 3(3) (in Japanese).

Osaka Committee 1972. *Report of countermeasures for land subsidence in Osaka.* The Osaka Committee on Countermeasures of land subsidence, 559p.

Oshima, T. 1988. Geological structure of Saga Plain. *Proc. intl. symp. on shallow sea and low land, Saga.* pp.139–143.

Shindo, S. 1972. The groundwater in South Kanto. *Tsuchi–to–Kiso, JSSMFE.* 20(5):25–36 (in Japanese).

Tohno, I. & S. Iwata 1988. Present state of land subsidence phenomenon at Muikamachi, Minami–Uonuma, Niigata Prefecture. *Journal of Groundwater Hydrology.* 30(2):95–101 (in Japanese).

Tohno, I., S. Iwata & Y. Shamoto 1989. Land subsidence caused by repeated loading. *Proc. of the 12th intl. conf. on SMFE.,* Vol.3, pp.1819–1822.

Tohno, I. et al. 1990. Middle to Late Quaternary stratigraphy and paleoenvironments in the inner lowland of Ariake Sea, Saga Prefecture. *The 20th Japan National Conference on Quaternary Research.* pp.104–105 (in Japanese).

Tohno, I. 1992. Present state of land subsidence phenomenon and subsidence caused by repeated change of groundwater level in Japan. *Proc. of the ILT seminar on problems of lowland development, Saga university, Saga, Japan.* pp.303–308.

Ueshita, K. & T. Sato 1981. Study on subsidence of the Nobi Plain. *Proc. of the 10th Intl. conf. on SMFE.,* Vol.2, pp.387–390.

Ueshita, K. & K. Daito 1984. Land subsidence in Japan: A case study of the Nobi Plain. *Intl. symp. on geotechnical aspects of mass and material transportation, Bangkok, Thailand.* pp. 439– 460.

Wadachi, K. 1940. *Land subsidence in Western Osaka (2).* Report of the Research Institute for Disaster Prevention, Osaka, Japan (in Japanese).

Wadachi, K. & T. Hirono 1939. *Land subsidence in Western Osaka (3).* Report of the Research Institute for Disaster Prevention, Osaka, Japan (in Japanese).

20 SUBSIDENCE DUE TO NATURAL GAS AND OIL WITHDRAWAL

T. ESAKI[*], K. SHIKATA[*] and K. AOKI[**]
[*] *Kyushu University, Fukuoka, Japan*
[**] *National Research Institute for Pollution and Resources, Tsukuba, Japan*

INTRODUCTION

The land subsidence phenomena which results from over–development of groundwater, oil and natural gas resources is spreading in many areas of the world. Especially in lowland areas, if subsidence continues, it compels us to construct both a drainage system and a dyke system to protect the area from flooding.

In ancient times, it is known that only shallow deposits of oil–derived asphalt "pitch" were mined and used for road–making, boat construction and as a building cement. Shallow seepage of oil and gas were used mainly for medical purposes and in religious ceremonies, and was called "liquid fire" in ancient Egypt. In August, 1859, the first true oil–well was drilled by a percussion rig in Pennsylvania, United States. Subsequently, oil was found in Iran in 1908, in Iraq in 1927, in Venezuela in 1930 and in many countries in turn (Tiratsoo 1976). Since then, the petroleum industry has been flourishing. But simultaneously the prosperity has caused serious damages in oil and natural gas fields.

The typical habitat of oil and gas is the sedimentary basin. The withdrawal of oil or gas from this geological feature caused the land subsidence phenomena over the oil or natural gas reservoir. Damages due to land subsidence have been spreading in many oil and natural gas fields. The most important kinds of damages that might occur due to land subsidence in lowland areas are: less production in farming and agriculture due to an increase of soil wetness; decrease of safety as a result of lowering dikes and the erosion of dunes; ecological damage due to other hydrological changes in the currents and associated sand transport; decrease of water quality in the subsidence region (Pottgens & Frits 1991).

For example, in the Netherlands, 60% of which lies below high tide level, gas extraction has caused a very slow and smooth, but extensive subsidence. In Venezuela, the greater part of the eastern shore of Lake Maracaibo extends over an area of low ground and swamp. Compaction is presently the principal production mechanism for oil recovery in heavy oil reservoirs. Therefore, in order to protect this area from flooding, it is necessary to construct a drainage and dyke system (Nunez & Escojido 1976). In Inglewood Oil Field, Los Angeles, California, the ground surface subsidence and horizontal movements were caused by fluid removal. Though there still appears to be no consensus of opinion as to whether or not the failure was caused predominantly by the oil field operations, the earth dam which was located in the subsidence area failed because of the outbreak of a V–notch through the embankment on Dec.14, 1963. The flood damaged a large portion of the city and killed 5 people. The cost of damage caused by it was estimated at about 15 million dollars at that time (Lee 1976). A number of other such case histories are available in literature (e.g. Hepple 1966, Erickson 1976, Murakami & Takahashi 1976).

DISTRIBUTION OF GAS & OIL FIELDS

The typical habitat of oil and gas is the sedimentary basin, of which about 600 are known (Figure 20.1), 160 of which produce commercial quantities of hydrocarbons. Because the basin floors continue to subside under gravitational or tectonic forces, sedimentary basins are locations especially favorable for petroleum generation. They are likely to contain a substantial volume of source rocks. In considering the history and evolution of these basins, it is necessary to summarize some of the ideas of modern plate tectonics theory (Figure 20.2). These suggest that the Earth's surface is covered by a mosaic of rigid crustal blocks or plates, which are separated by linear active bands in which volcanic

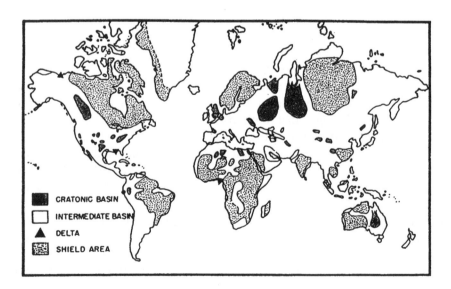

Figure 20.1. Cratonic and intermediate types of world sedimentary basins (Klemme 1975).

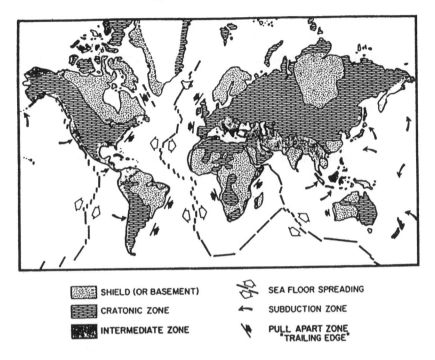

SHIELD (OR BASEMENT)	SEA FLOOR SPREADING
CRATONIC ZONE	SUBDUCTION ZONE
INTERMEDIATE ZONE	PULL APART ZONE "TRAILING EDGE"

Figure 20.2. Diagram illustrating the world's basin zones and different types of continental margins (Klemme 1975).

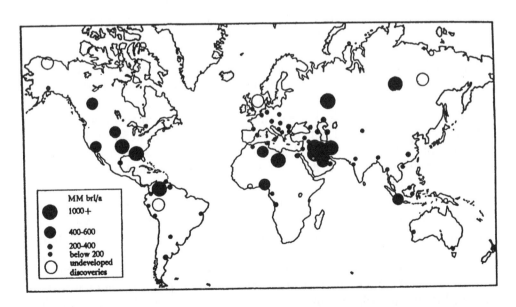

MM brl/a
1000+
400-600
200-400
below 200
undeveloped discoveries

Figure 20.3. Map illustrating relative importance of world oil producing areas (Tiratsoo 1976).

and seismic activities are concentrated. The plates float on the uppermost liquid layer of the Earth's generally solid mantle, and are in slow and intermittent motion. Therefore, these sedimentary basins are downwarpings in the earth's crust which are filled with sediments. As shown in Figures 20.1 and 20.2, basins seem to be divisible into two main groups: basins underlain by continental type crust which have been formed by forces acting in cratonic or near–cratonic areas, and basins underlain by intermediate–type crust which is along or near the continental margins which have been produced by lateral forces connected with plate movements. Most of the main oil and gas fields in the world, which are called "giants", are located in these sedimentary basins (Figure 20.3) (Tiratsoo 1976, 1979).

THE NATURE OF OIL FIELDS

Petroleum is found at the surface in the form of seepage of hydrocarbon gases or liquid crude oil. In the subsurface, accumulations of gaseous and liquid hydrocarbons, usually associated with saltwater, occur filling the pore spaces of reservoir rocks, and are confined within some forms of geologic trap (Figures 20.4, 20.5, 20.6). The most typical form of subsurface accumulation is the anticlinal trap which consists of an arch of porous rock, bounded above, below and laterally by impervious beds (Figure 20.4). Within such a trap, there are usually three layers of rock distinguished by the fluids they contain: gas–saturated rock near the apex of the anticline; a layer of rock which contains oil, and the lowest layer of saltwater saturated rock. The layers of rock in such an accumulation which are filled with hydrocarbons, in fact, also contain a considerable proportion of saltwater. This water forms a film around the individual rock particles and fills the intercapillary passages of the rock. The interstitial water is retained by adhesive and surface tension forces, and is therefore normally fixed. The interstitial water may occupy up to 50% of the pore volume of a reservoir. In general, oil field water contains a much higher concentration of dissolved salts than average seawater (up to 30%). Associated gas is defined as the free natural gas in contact with crude oil in reservoirs in the form of a gas cap, so that the production of the gas must be related to the production of the oil.

The term reservoir is used to describe the parts of a geologic trap which are filled with hydrocarbons and are under a single pressure system. An accumulation can form a single reservoir or a number of reservoirs which may be separated in either the horizontal or vertical planes (Figure 20.5). The process by which hydrocarbons leave the fine–grained clays and shales in which the hydrocarbons are believed to have been formed and enter the coarse–grained reservoir rocks in which they accumulate is termed primary migration. The subsequent movement, concentration and separation of gas and oil within the reservoir rocks in some form of geologic trap is secondary migration. The

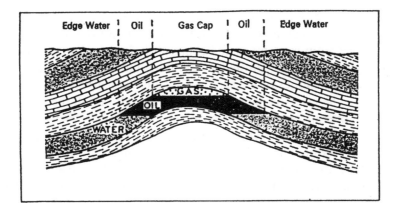

Figure 20.4. Idealized cross–section of an anticlinal trap showing gas, oil and water layers (Tiratsoo 1976).

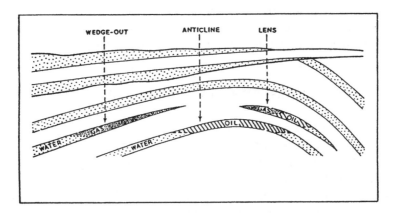

Figure 20.5. Cross–section showing several types of traps for hydrocarbons (Tiratsoo 1976).

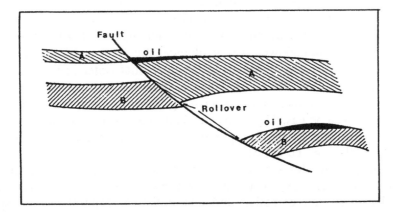

Figure 20.6. Typical growth fault and rollover traps (Tiratsoo 1976).

most important types of reservoir rocks are sandstones and carbonate rocks. One survey showed that about 59% of the production from a number of major oil fields came from sandstones and about 40% from carbonates. Other rocks occasionally also provide petroleum reservoirs, but these are relatively rare.

The essential properties of a reservoir rock are its relatively high porosity and permeability. The effective porosity which is the ratio of all the interconnected pores to the overall external volume of an unit of rock, depends essentially upon the way the particles which constitute the rock have been deposited and consolidated.   While the volume of fluid contained in an unit of rock is determined by its porosity, the permeability of rock is the measure of its capacity to transmit its fluid content. The rock pores typically form a network of interconnected pores surrounding the individual rock particles, and many of these passages are of capillary or even sub-capillary proportions. The great majority of sedimentary rocks have been laid down in an aqueous environment, so that their pore passages are likely to be water-wet. Accordingly, it is impossible for droplets of oil to move along the narrow, tortuous passages except under sufficiently great pressure to overcome the considerable surface tension resistance they would encounter (Tiratsoo 1976).

THE NATURE OF NON-ASSOCIATED NATURAL GAS

Natural gas is very commonly found in close association with liquid petroleum. In fact, natural gas is considered to be the gaseous phase of crude oil. However, there are some natural gases which do not appear to be associated with crude oil. Therefore, natural gases must be differentiated into associated and non-associated varieties. This report is intended for the non-associated gas, which is water-soluble natural gas.

This type of natural gas is different from "associated gas" produced within the typical geological system in oil fields, that is, anticline traps (Figure 20.4). These natural gas fields usually extend over plains resulting from sediment basins (Figure 20.1), which contain enough organic substance to produce natural gas and are endowed with geological structures restricting gas seepage. The natural gas usually dissolves in the interstitial water in the aquifers. It is also known that the density of iodine is relatively high in the gas dissolved water. Therefore, in most non-associated gas fields, iodine is also one of major products as well as gas. This substance is needed for the oxidation proofing of synthetic resin and for medical purposes and is used in the refining process of special metals, e.g. Ti, Zr, Si.

Natural gas provides a clean and very flexible source of heat. It burns with a clear flame when mixed with the correct proportion of air and its combustion byproducts are non-corrosive and non-polluting. Natural gas is especially valuable as a utility gas in domestic heating and cooking applications; its instant availability and absence of storage requirements are particularly attractive. Natural gas is widely used in industry for space heating, product drying and steam production with either firetube or watertube boilers (Tiratsoo 1979).

MECHANISM OF SUBSIDENCE

In a series of compacting sediments, there are fine–grained (clayey) aquitards and coarse–grained (sandy) aquifers. Land subsidence is caused by the withdrawal of fluids (e.g. groundwater, oil, gas) from weakly consolidated aquifers. The withdrawal causes a drop in the hydraulic pressure of the confined aquifers. The total stress on an aquifer is supported by the grain–to–grain skeleton of the aquifer and water pressure (Figure 20.7). Therefore, the increase in the effective pressure due to groundwater withdrawal, expressing the net load on the matrix and defined as the overburden pressure minus the water pressure, causes compaction of sand and clay stratum (Figure 20.8). However, the pressure in the permeable sand drops more rapidly than that in the almost impermeable clay. The aquifer may remain saturated, but the decrease in head in the coarse–grained aquifers creates a hydraulic gradient from the clayey aquitards to the aquifers. Therefore, the vertical escape of water and consequent decrease in pore pressure is slow and time–dependent. The compaction within clay aquitards rather than sandy aquifers is responsible for most of the land subsidence. The magnitude of compaction in each stratum depends on the coefficient of volume compressibility. In the case that the fluid pressure is greatly reduced due to large groundwater withdrawal, the grain–to–grain contact may be increased beyond the aquifer's ability to spring back elastically, and hence permanent land subsidence will occur (Kamata et al. 1976, Nunez & Escojido 1976).

COUNTERMEASURES

These days, the fluid resources management for sedimentary basins is recognized as a kind of system management. This system consists of five subsystems: monitoring, controlling, simulation, designing and reinjecting.

Monitoring: The measuring system to monitor both the subsidence phenomena with levelling surveys, and the variation of groundwater levels using a set of measuring wells in order to allow early prediction of the phenomena.

Controlling: Setting up a committee to supervise the water resource management.

Simulation: With suitable prediction methods, it would be possible to evaluate the phenomena of groundwater basins under a variety of initial and boundary conditions. Details of it are described in the following section.

Designing: The organization which form a definite policy for groundwater

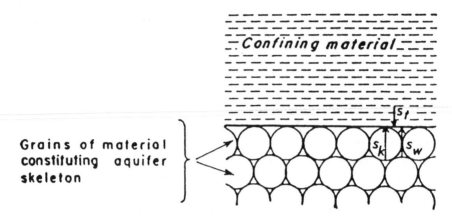

Figure 20.7. Microscopic view of forces acting at artesian aquifer interface (Rahn 1986).

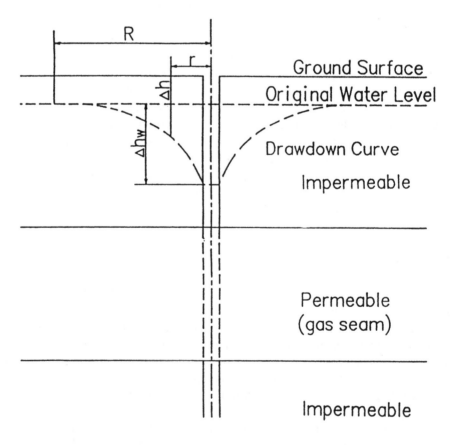

Figure 20.8. Schematic diagram showing variation in drawdown curve due to fluid withdrawal.

management, based on the monitoring results and the simulation.

Reinjecting: Repressuring of the reservoir rocks by controlled injection does not only stop subsidence but also results in rebounding or recovery of bench mark elevations. The injection of a large volume of water requires careful treatment of the water because suspended solids in the water plug the pore channels of the reservoir rock restricting injection and causing high injection pressures (Fujisaki 1989).

PREDICTION METHOD

Land subsidence is due to aquitard and aquifer compaction, which results from fluid withdrawal. In usual soil, the compaction of the aquitard is greater, because of its higher coefficient of volume compressibility than aquifers. This compaction mechanism can be seen in oil and natural gas fields, as well as groundwater basins. Therefore, simulation methods for groundwater basins can be also applied to other fluid withdrawal in a similar way. Based on the geohydrological research, the simulation model can be divided broadly into some models, in each of which there are two types. One of them is a two–step model which consists of a model used to calculate the hydraulic head drawdown and a consolidation model, in which the calculated head values are used as boundary conditions. This type of model neglects any coupling effects between the grain skeleton and the fluid. Another is a one–step model, which is termed a fully coupled model. Using a fully coupled consolidation model, it is possible to obtain satisfactory correlation between the subsidence and the piezometric decline along the whole section considered and not only at discrete points. Whether a one- or two–step model is used, setting nodes in the aquitard as well as in aquifer, water head in clayey stratum can also be computed. Therefore, it is possible to obtain the difference in groundwater head between each time steps, which is equivalent to the change in storage ability of clayey beds, that is, the squeezing water mass.

a) Horizontal two–dimensional model: In a system in which stratums above and below aquifers are quite low in permeability, the amount of land subsidence can be assumed to be equivalent to the elastic deformation in aquifers.

Example: In Miyazaki, Japan, subsidence due to the pumping of groundwater containing natural gas has been observed (Nishida et al. 1981, Esaki et al. 1991). The aquifers are deep enough, the aquitards are thick enough to prevent the gas seepage through them, and the characteristic in stratum deformation is elastic. In this case, the simulation method can be adapted with a two–step model: two–dimensional flow analysis and a certain method to evaluate the ground deformation due to subsurface compaction. Considering the elasticity in stratum deformation and the depth of the compacting stratums, it is necessary to evaluate the land subsidence as a result of the three dimensional propagation of aquifer deformation. Then, the influence function method was used to estimate the incidental land subsidence, which is characterized by an influence factor within a limited angle (Esaki et al. 1987). The practical approach to this example is shown in the following section.

b) Vertical two–dimensional model: The simulation model is constructed along a certain geological cross–section of the land subsidence area. It is assumed that flow is two–dimensional in the vertical plane (x–z plane). Dividing the geological profile into elements, it will be possible to set nodes in the aquitards. The groundwater head decline in each element which is caused by pumpage is obtained by solving the water balance equation. The total deformation is obtained by integration of influences of each element.

Example: Subsidence due to withdrawal of gas and water over Venetian area has been investigated (Figure 20.9). Venice sits on an island in the Venetian lagoon, at the northwest end of the Adriatic Sea. The cities of Marghera and Mestre occupy the mainland. At the industrial center of Porto Marghera, 7 km from Venice, there are large groundwater withdrawals from about 40 wells that tap five aquifers within the depth of 290 meters. There is one major well in Tronchetto of the Venetian islands. The history of the development of these wells parallels the occurence of measured subsidence. Gambolati & Freeze (1973) have chosen a two–step procedure to analyze the subsidence in the complex aquifer–aquitard system that exists there. First, the regional hydraulic head drawdown was calculated in a two–dimensional vertical cross–section in radial coordinates, using an idealized 10–layer representation of the geology. The calculated head values in the aquifers were then used as time dependent boundary conditions in the one–dimensional vertical consolidation model solved with a finite difference technique and applied to a more refined representation of each aquitard. Lewis & Schrefler (1978) have used a fully coupled consolidation model, of which the finite element model is given in Figure 20.10, showing clearly the idealized geological stratigraphy. On account of the spatial density of the active wells, it was more realistic to consider the source points distributed over a certain area.

Then, it was possible to obtain satisfactory correlation between the subsidence and the piezometric decline along the whole section considered and not only at discrete points.

c) Quasi three–dimensional model: This scheme consists of the horizontal (x,y) and vertical (x,z) two–dimensional simulation model. Under the condition of clear contrast in permeability between aquifers and adjacent aquitards, the flow can be assumed to be horizontal in the aquifers and vertical in the aquitards.

Example: At Ravenna, Italy, the land subsidence has been caused by both groundwater withdrawals from the Quaternary multi–aquifer system and gas production from a number of deep pre–Quaternary reservoirs. The underlying Quaternary deposits consist mostly of sandy and silty–clayey layers. The thickness of the Quaternary soils ranges between 1500 and 3000 m. Gambolati

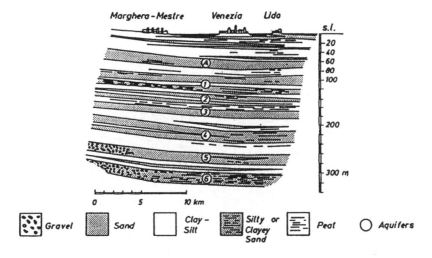

Figure 20.9. Geological stratigraphy of Venetian area with the tapped aquifers (Lewis & Schrefler 1978).

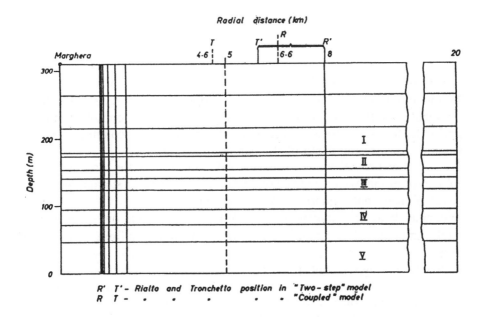

Figure 20.10. Finite element discretization of 2–D vertical cross–section of Venetian area (Lewis & Schrefler 1978).

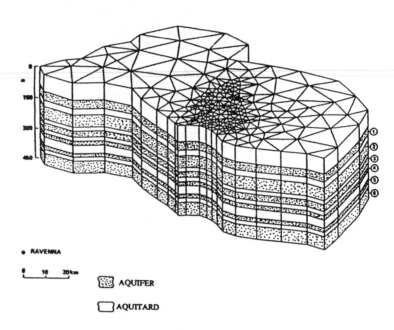

Figure 20.11. View of the regional quasi 3–D flow model of the aquifer system underlying the southeastern Po river basin (Gambolati et al. 1991).

et al. (1991) simulated this phenomena with the aid of two separate models. The model consisted of a quasi 3–D hydrologic model of subsurface flow on a regional scale (Figure 20.11) followed by a 1–D vertical consolidation model applied to the site where an accurate lithostratigraphic column of soil was available down to the depth of the pumped interval.

d)Three–dimensional model: It is very difficult to solve the three dimensional problem of groundwater hydrology, because the computer storage required for a three dimensional numerical solution is quite large.

Example: Bear & Corapcioglu (1981) demonstrated the one–step model by simple example of a single well pumping from an infinite homogeneous confined aquifer. The regional land subsidence model was obtained by employing Terzaghi's concept of effective stress, and assuming only vertical soil compressibility. First, a three–dimensional equation of water mass conservation was developed for compressible fluid and solid matrix. This equation was then integrated over the aquifer's thickness, taking into account conditions on the top and bottom surfaces bounding the aquifer. The result is a flow equation in terms of averaged piezometric head. By relating changes in head to land subsidence, a single equation was then obtained for land subsidence.

PRACTICAL APPLICATION

Subsidence over Miyazaki gas field, Japan

In Japan, natural gas is coming into wide use as energy for power plants, public use, and so on. The development of domestic natural gas fields has been accelerated, yet most of natural gas demand in Japan is now served by other countries. The gas field examined in this section is one of the largest in Japan, located in the northern area of the Miyazaki plain in Kyushu island as shown in Figure 20.12. In about 16 years time, about 50 wells were developed at intervals of roughly 500 m in an area of 40 km^2 . The geologic formation of the aquifer system is classified into five layers, which includes two sandy gas beds, of Pliocene age, and intermediate between muddy beds. The sandy ones contain water–soluble natural gas and concentrated iodine. These layers are 180–310 m, 400–490 m in thickness, respectively. The daily volume of pumpage water increased from 6600 kl in 1975 to 7700 kl in 1989. The subsequent decline in water level resulted in a corresponding increase in effective stress on the sandy layer sequence in the aquifer system. The compaction of the permeable layers affected the above layers in succession. As a result, land subsidence occurred at the surface.

Due to these circumstances, it is necessary to predict the ground movement and to consider how to minimize the movement from the viewpoint of environmental control. In this study, a new environmental system which can predict and check surface subsidence is proposed. This analytical method consists of two stages:

a) Water flow analysis by the finite element technique gives the distribution of water level in a gas seam;

b) The gas seam deformation is given by the above calculated change of water level head. Next, by using influence function method, the surface subsidence due to the gas seam deformation at arbitrary points can be obtained. This new analytical system is applied to the Miyazaki gas field.

Groundwater flow – subsidence simulation system

Many earlier studies on groundwater flow and land subsidence due to extraction of water have been done at different withdrawal sites (e.g. Nishida et al. 1981). However, land subsidence will occur as a result of three–dimensional propagation of the deformation of aquifer within a limit angle, as shown in Figure 20.13. In this section, groundwater flow, aquifer compaction, and land subsidence are systematically simulated using a useful simulation system, which is formed by combining the following three stages:

a) Simulation of groundwater flow and the water–level decline due to water pumpage from the aquifer system are given using a finite element computer program. The model uses a square grid with a certain horizontal spacing. This

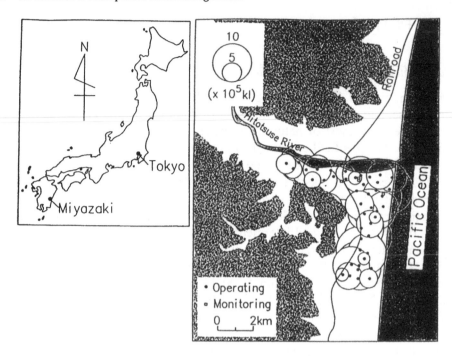

Figure 20.12. Distribution of operating and monitoring gas wells, and total volume of pumpage water from 1974 to 1988.

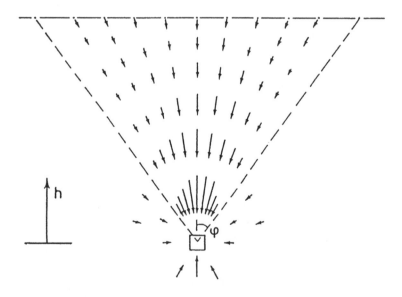

Figure 20.13. Movement of points within a circular cone of the overlying rock mass caused by the deformation of an element (Kratzsch 1983).

size of spacing is dependent on the nature of the problem.

b) The decline in water level is directly related to the increase in effective stress, while the total stress on the solid phase of the ground remains constant. So, the gas seam compaction is caused by this increase of effective pressure. While most analytical methods for predicting subsidence are based on the well–known Terzaghi's consolidation theory which is suitable for cases of relatively shallow and/or unconsolidated strata such as soft clay, the proposed method is based on the assumption that elastic behavior dominates for the relatively deep and/or compact strata. In other words, the calculation of the gas seam deformation is based on generalized Hooke's law;

$$d = (\frac{m}{E})(1-2v)\gamma_g \Delta H \tag{20.1}$$

where d is deformation of gas seam, m is thickness of gas seam, E is Young's modulus of gas seam, v is Poisson's ratio of gas seam, γ_g is unit weight of water containing natural gas, and ΔH is the amount of water level decline in the aquifer system. This is for a single seam. If there are multiple gas seams, we have to sum up the effects of each gas seam deformation.

c) Surface subsidence is not proportional to the decline in water level in the aquifer system, while the gas seam deformation is. This is because the influence of gas seam deformation at a certain depth propagates and spreads three–dimensionally to surface. Surface subsidence attributed to gas seam compaction is obtained by the influence function method, which is a three dimensional method that can be used to obtain displacement at a large number of points and distribution of subsidence, both accurately and efficiently (Esaki et al. 1987).

$$s = e \cdot a \cdot d \cdot z \tag{20.2}$$

where s is land subsidence at a point, e is an influence factor, a is coefficient of subsidence, d is amount of seam compaction, and z is a time factor. When analyzing a gas field in which many wells have been developed one by one, time factors should also be considered. Surface subsidence increases over long periods of time, while the change in water level is small except for the initial steep loss of head. It is shown that surface subsidence is related to both the change in water level and the time since development. The time factor should be used appropriately.

Simulation

In Japan, since 1932, groundwater containing natural gas has been pumped out in different places, such as Hokkaido, Chiba, Niigata, Miyazaki, and Okinawa. Following the industrial advancement, the amount of pumpage water and extracted natural gas have increased rapidly. This has resulted in a type of mining damage, i.e. land subsidence. In Niigata, especially, subsidence has exceeded 50 cm per year, and in Chiba the total subsidence since 1961 exceeds 100 cm. However, the subsidence phenomena have decreased steadily since 1970 when laws and ordinances restricting the groundwater usage were established (Yamamoto & Kobayashi 1984).

The area studied here is located in the northern portion of the Miyazaki plain in Kyushu, southern Japan. This area is an alluvial lowland, approximately 20 km long and 2 km wide facing the Pacific ocean in the east and divided by the river Hitotsuse into northern and southern regions. The typical geological sequence underlying this area is shown in Table 20.1. This geological formation has packed groundwater containing water–soluble natural gas that allows no natural recharge to the aquifer system. The locations of 47 operating wells are shown in Figures 20.14 and 20.15. Each well is located at intervals of roughly 500 m, at depths of 400–1300 m. In 1989, the daily volume of pumpage water and extracted natural gas amounted to 7700 kl and 11,300 Nm^3, respectively. The subsequent decline in water level resulted in a corresponding increase in effective stress on the sandy layer sequence in the aquifer system. The compaction of the permeable layers affected the above layers three dimensionally. As a result, land subsidence occurred in the vicinity of well field as shown in Figures 20.14 and 20.15.

First, the distribution of water level decline was evaluated using a quasi–three dimensional computer program, under the following conditions:

a) initial condition: water level at all nodes is the surface.

b) boundary condition: water level at boundary nodes is the surface.

Input data for water head decline at each of the wells were modified according to Thiem's equation:

Table 20.1. The typical geological sequence underlying the study area in the Miyazaki plain.

Depth (m)	Bed		Permeability	
surface – 300	Takanabe	muddy	impermeable	
300 – 600	Sadohara	sandy	permeable	gas seam
600 – 1100	Niinazume	muddy	impermeable	
1100 – 1500	Uryuno	sandy	permeable	gas seam
1500 –	Ikime	muddy	impermeable	

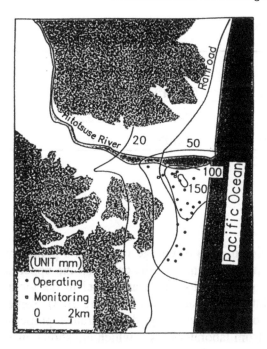

Figure 20.14. Measured contour map of subsidence from 1974 to 1978.

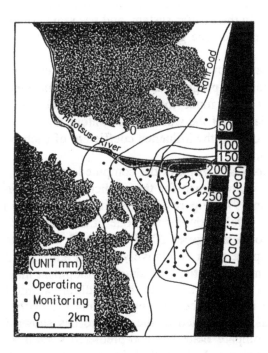

Figure 20.15. Measured contour map of subsidence from 1974 to 1988.

Table 20.2. Hydraulic properties.

Seam	m	E	v	K
	(m)	(kg/cm^2)		(cm/sec)
Sadohara	240	2×10^4	0.25	4×10^{-5}
Uryuno	480	4×10^4	0.25	4×10^{-5}

$$\Delta h_w = \frac{Q}{2\pi \cdot K \cdot m} \ln(\frac{R}{r}) \tag{20.3}$$

where Δh_w is local drawdown, Q is discharge rate, K is a permeability index, r is radius, and R is the influence limit. In the case that the upper seam of the aquifer system is of low permeability, the water level in the vicinity of well will decline locally and intensively (as shown in Figure 20.8). Some input data are modified according to the interval of grids. Hydraulic properties used as input data were deduced from laboratory tests, which are shown in Table 20.2.

Interactively, a contour map of water level, which was obtained by the post–processing program, was proportional to the contribution of gas seam deformations. This was done because water level withdrawal will lead to the increase of effective stress in the aquifer system, and gas seam compaction will be caused by the stress increase. The gas seam deformation was estimated by Equation 20.1.

Next, land subsidence caused by the seam compaction was obtained using an influence function method. The program used is performed in three steps:

a) a pre–processor for data input together with a digitizer, accurately reproduces the irregular layout of the water level contour map;

b) a main–processor calculates the influence coefficient of each point on the surface, at 25 m intervals;

c) a post–processor graphically displays the computed results.

Figure 20.16 shows the contour map for the predicted land subsidence four years after development in 1978. Figure 20.17 is after 14 years, in 1988. The shape of the contour lines is fairly similar with one of the field data in Figures 20.14 and 20.15.   As shown in Figures 20.16 and 20.17, subsidence is large where the volume of pumpage water is relatively large (Figure 20.12). Even if the pumpage volume is similar, the time lag since development may affect the subsidence phenomena. In order to cope with this condition, a time factor was applied, according to the period since development and the decline in water level. This procedure was also applied to the two profiles shown in Figures 20.18 and 20.19, which compare field data and predicted subsidence in 1988, corresponding to cross sections A–A' and B–B', respectively. This system provides the simulation of land subsidence in the aquifer system appropriately.

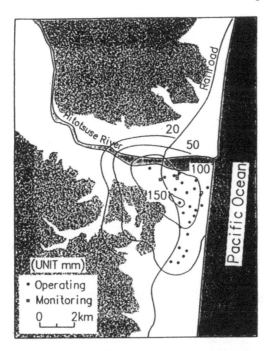

Figure 20.16. Calculated contour map of subsidence from 1974 to 1978.

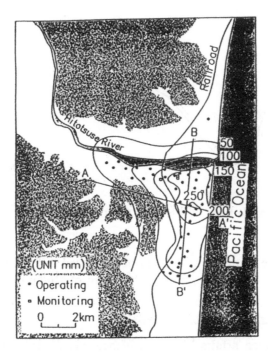

Figure 20.17. Calculated contour map of subsidence from 1974 to 1988.

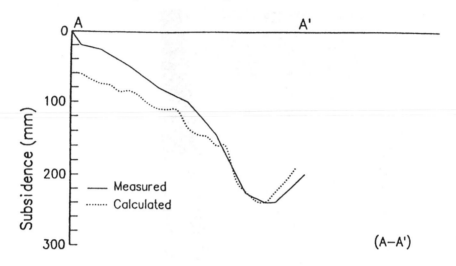

Figure 20.18. Subsidence profile from 1974 to 1988 (section A–A').

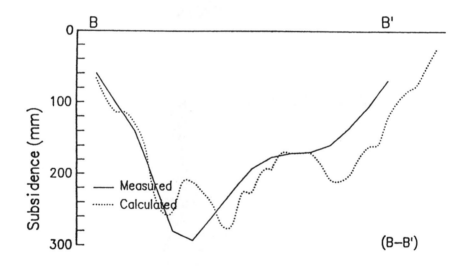

Figure 20.19. Subsidence profile from 1974 to 1988 (section B–B').

The differences in shape of two profiles, can be attributed to local variance in the geological structure, e.g. small faults or the depth of alluvium.

Subsidence monitoring and environmental control

In order to protect surface affairs from subsidence and to develop valuable domestic resources in agreement with local inhabitants, the Technical Committee for natural gas development has been established at Kyushu Branch of MITI in 1974, just before development of the Miyazaki field. As per the recommendations from the committee, level surveying (126 measuring points over a total length of 53 km), measuring of ground water level at the monitoring wells (four wells: one 1300 m and three 60 m), and prediction of subsidence have been carried out in collaboration with the authorities and the mining company every year. The progress of ground water withdrawal and subsidence have been checked and the simulation system also has been modified several times following the field data. During the 14 years since the beginning of operation, subsidence has been restricted to within 30 mm per year and withdrawal from each well has also been limited to within 200 m from sea level. In the next decade, 20 new wells are being planned in the northern field. The committee will continue the efforts of environmental control by the use of monitoring system.

SUMMARY

The approaches discussed above reveal the fact that some cases of subsidence are related to fluid withdrawals in oil and gas reservoirs. The same controlling policy, geological characteristics and prediction methods discussed about oil and gas field subsidence can also be applied to fresh water aquifers. The above management system has to make it a rule to be up–to–date and always has to provide a practical plan to control the regional environment.

ACKNOWLEDGMENT

The authors are much indebted to the staffs of the Ministry of International Trade and Industry in Japan and Ise Chemical Industries Co. Ltd. for providing the valuable data and to Associate Professor M. Nishigaki, Okayama University for offering us PC–GWAPG computer program.

REFERENCES

Bear, J. & M. Y. Corapcioglu 1981. Mathematical model for regional land subsidence due to pumping: 1.Integrated aquifer subsidence equations based on vertical displacement only. *Water Resour. Res.* 17(4): 937–946.

Erickson, R. C. 1976. Subsidence control and urban oil production – a case history Beverly Hills oil field. *Proc. 2nd int. symp. land subsidence*: 285–297. Anaheim, USA.

Esaki, T., N. Kameda, T. Nishida, T. Kimura & K. Shikata 1987. Subsidence analysis for complex irregular tabular extraction by use of personal computer. *Proc. 7th Japan symp. on rock mechanics*: 413–418.

Esaki, T., K. Shikata, K. Aoki & T. Kimura 1991. Surface subsidence in natural gas fields. *Proc. 4th int. symp. land subsidence*: 109–118. Houston, USA.

Fujisaki 1989. Land subsidence – analysis of land subsidence. *J. Groundwater Hydrology* 31(1): 39–44.

Gambolati, G. & R. Allan Freeze 1973. Mathematical simulation of the subsidence of Venice. *Water Resour. Res.* 9(3): 721–732.

Gambolati, G., G. Ricceri, W. Bertoni, G. Brighenti & E. Vuillermin. 1991. Numerical analysis of land subsidence at Ravenna due to water withdrawal and gas removal. *Proc. 4th int. symp. land subsidence*: 119–128. Houston, USA.

Hepple, P. 1966. *Natural Gas*: The Elsevier publishing company, Amsterdam, London, New York.

Kamata, K., K. Harada, & H. Nirei 1976. Analysis of land subsidence by the vertical two dimendional multi–aquifer model. *Proc. 2nd int. symp. land subsidence*: 201–210. Anaheim, USA.

Klemme, H. 1975. Giant fields related to their geologic setting. *Bull. Canad. Pet. Geol.*: 30–66.

Kratzsch, H. 1983. *Mining subsidence engineering.* Springer–Verlag.

Lee, K. L. 1976. Calculated horizontal movements at Baldwin Hills. *Proc. 2nd int. symp. land subsidence*: 299–308. Anaheim, USA.

Lewis, R. W. & B. Schrefler 1978. A fully coupled consolidation model of the subsidence of Venice. *Water Resour. Res.* 14(2): 223–230.

Murakami, M. & Y. Takahashi 1976. Land subsidence research and regional water resource planning of the Nanao Basin. *Proc. 2nd int. symp. land subsidence*: 211–221. Anaheim, USA.

Nishida, T., T. Esaki, K. Aoki & N. Kameda 1981. On the surface subsidence in natural gas fields. *Proc. int. symp. on weak rock*: 701–705.

Nunez, O. & D. Escojido 1976. Subsidence in the Bolivar Coast. *Proc. 2nd int. symp. land subsidence*: 257–266. Anaheim, USA.

Pottgens, J.J.E & F.J.J. Frits 1991. Land subsidence due to gas extraction in the northern part of the Netherlands. *Proc. 4th int. symp. land subsidence*: 99–108. Houston, USA.

Rahn, P.H. 1986. *Engineering geology – an environmental approach.* Elsevier, 303–319, New York, USA.

Tiratsoo, E.N. 1976. *Oilfields of the World:* Second edition. Houston, Texas.

Tiratsoo, E.N. 1979. *Natural gas:* Third edition. Houston, Texas.

Yamamoto, S. & A. Kobayashi 1984. Groundwater resources in Japan with special reference to its use and conservation. *Proc. 7th Japan symp. on rock mechanics*: 381–389.

21 MONITORING GROUND SUBSIDENCE BY AERIAL PHOTOGRAPHS

T. TANAKA and N. FUKUDA
Fukken Co. Ltd. (Consulting Engineers), Hiroshima, Japan

INTRODUCTION

Ground subsidence occurs over a wide area owing to the reclamation on soft foundation, or the extraction of groundwater under soft ground, or the collapse of voids due to mining. A conventional method of assessing such subsidence is by means of ground surveying or instrumented measurement. This method is, however, basically accomplished by point by point surveying and hence insufficient to assess a wide–scope ground subsidence. The present research aims at developing an application method of digital plotting on the basis of aerial photographs, which is a new surveying technique, to counter such a wide–scope subsidence and present the results of site studies based on the method of topographic interpretation together with verification of surveying precision. Furthermore, the applicability and advantages of the proposed method are discussed in this chapter.

The modern technology of surveying and mapping is expected to undergo drastic changes accompanied by the use of computer and softwares as shown in Figure 21.1 (National Land Planning Adjustment Bureau 1986, partial modification by Geographical Survey Institute 1991). As a matter of fact, the change is from analog to digital, and from ground surveying to cosmic surveying, and from basic data collection to intelligent data processing and the changes are toward the high–speed, automatic, without manpower, high–precision and high value–added quality surveying.

On the other hand, as far as the monitoring of subsidence behavior of wide area is concerned, the conventional subsidence analysis by point by point surveying or instrumented measurement is far less speedy and efficient than expected. The present research in this chapter proposes a method of combination of the analytical plotter system, a new surveying technique that makes use of aerial photographs, and the Global Positioning System (hereinafter GPS) using artificial satellites for monitoring of the wide–scope ground subsidence. This system is composed of the following subsystems. Refer Figure 21.2 (Sugimori et al. 1993) for general outline of the system.

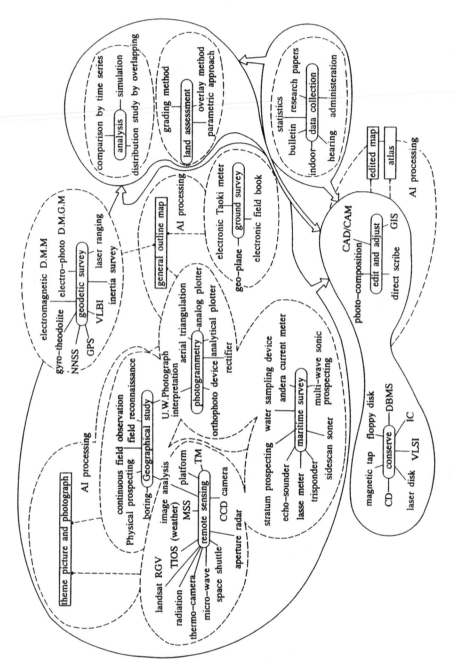

Figure 21.1. Technological systems and their inter-relation in the field of surveying and mapping (future).

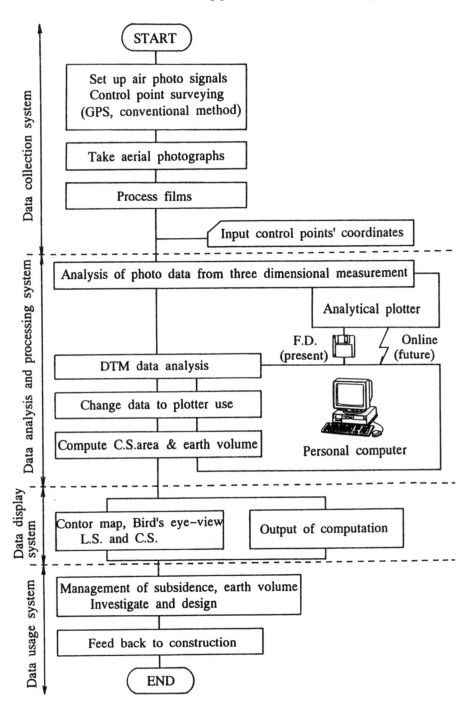

Figure 21.2. General outline of the system.

(1) Data collection system: This system relies on the aerial control point surveying to determine the positions of aerial photograph signals by means of GPS or conventional ground surveying.

(2) Data analysis and processing system: This system relies on an analytical plotter to obtain topographic data from three dimensional measurements and to compute earth–volume.

(3) Data display system: This system is mainly to display the outputs of the processed data such as contour map, bird's eye–view, longitudinal and cross sections.

(4) Data usage system: This system is mainly to use three dimensional topographic data for design, construction management and estimation of differential settlements, etc.

In the present chapter, in order to fully comprehend the proposed system, studies are conducted on analysis and verification of surveying precision and furthermore a case history of application of the system to the construction management of reclamation on soft foundation is introduced. And the condition for application of the system to monitoring of ground subsidence over a wide area are discussed.

STUDY ON ACCURACY OF TOPOGRAPHIC ANALYSIS BY AERIAL PHOTOGRAPHS

The test procedure

The various factors bear an influence on the precision of the aerial photogrammetry. These factors include the flight altitude, the function of aerial camera, the precision and location of the target, the presence or absence and type of air photo signal, number of successive models and method of adjustment computation. In this chapter, out of these factors the flight altitude and the adjustment computation together with the number and location condition of control points are brought under scrutiny by making comparison of the precisions obtained under varied conditions. The flow chart of the test procedure is as shown in Figure 21.3 (Tanaka et al. 1992a,b) and the contents of the test are as follows.

(1) Comparative studies are made as to the analytical results of aerial photographs for three types of photo scale by flight (hereinafter photo scale): 1/3000, 1/4000 and 1/5000 under the varied flight altitudes.

(2) Comparative studies are made as to the analytical results of aerial photographs under the varied conditions of location and number of control points in the aerial triangulation using analytical plotter system that is shown in Figure 21.4.

(3) The computation for aerial triangulation is conducted in the cases of the presence and absence of the additional parameters (24 error models) for the

distortion correction including lens distortion correction in the bundle adjustment program with self–calibration (hereinafter additional parameters) and the effectiveness of the computation is verified.

The test site is a reclaimed ground in Hiroshima City (T.P.+4 m~+7 m) and the scope of area under investigation is 1.8 km² (600 m × 3000 m) as shown in Figure 21.5. The standard elevation is taken as the datum T.P.±0 m and the flight altitudes are 450 m, 600 m and 750 m. Strip photographing is carried out to make stereoscopy of each model (a set of two photographs) by overlapping 60% with an adjacent photograph as shown in Figure 21.6. Table 21.1 shows the flight parameters used in aerial photographing. Out of the index maps with their respective photo scales, the index map (photo scale 1/4000) is as shown in Figure 21.7. The number of established control points is 105 and out of these the air photo signal (square type ⊠ of 30 cm × 30 cm) are laid at 60 points. At the locations of 45 points, the white traffic marks on the road are used to represent air photo signals.

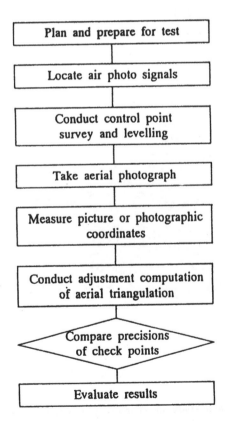

Figure 21.3. Flow chart of test procedure.

Figure 21.4. Analytical plotter system.

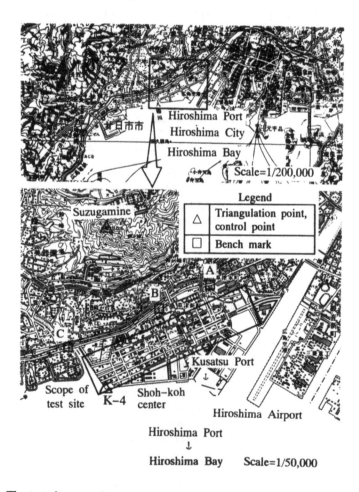

Figure 21.5. The test site to study on accuracy of topographic analysis by aerial photographs in Hiroshima City.

Table 21.1. Flight parameters used in aerial photographing in Hiroshima City.

Flight parameters	Photo scale by flight		
	1/5000	1/4000	1/3000
Model	Strip, 8–model	Strip, 9–model	Strip, 13–model
Flight altitude	750 m	600 m	450 m

Notes 1) Camera: WILD RC–10, 23 cm 23 cm, f=150 mm, Photographed by Cessna PT–206.

2) Film: High definition aero film (Panchromatic).

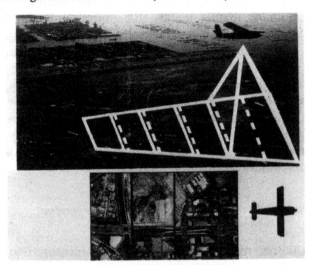

Figure 21.6. Method of taking aerial photograph with 60% overlapping to make stereoscopy.

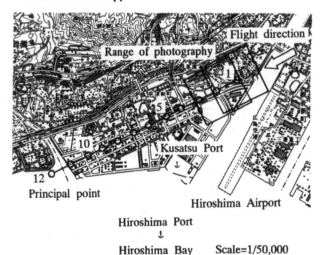

Figure 21.7. Index map (photo scale 1/4000 by flight).

Ground control point surveying

For the control point survey, the single–point single–direction method that has little impact on the precision of the predetermined point location is employed. The observations are made by means of the closed traversing and radial triangulation taking the control point K–4 and the third order triangulation toward Suzugamine. The control conditions are as shown in Figure 21.5. As to the traverse route, the horizontal position is determined by the average computation method of precise horizontal network. The levels are checked using three first order bench marks: A, B and C. As it was confirmed that nothing is wrong with the altitudes of three points, 105 control points are determined by the direct levelling method in three courses from A to K–4 using A as a basic standard point. With respect to the above determined coordinates (X,Y,H) of ground control points, the earth curvature correction is done and they are changed to three dimensional orthogonal coordinates. The survey precisions are in conformity with those of the third class control point survey and the third class levelling (three courses) according to the Public Survey Works Specification. The precisions attained in the present test for the control point survey (horizontal position) and the levelling (three courses) are within the allowable limiting values as given in Table 21.2.

Analysis of aerial photographs by analytical plotter system

Method of analysis
The first step in the method of analysis is to measure the photographic coordinates of 105 control points for one course and 10 pass points for one model using the analytical plotter system. The measuring precision of the

Table 21.2. Precisions of the control point survey and the levelling.

Coordinates	Division	Precision	Allowable error
Horizontal position	Standard deviation of unit weight	2.56"	15" (Third class control point survey)
Altitude	Section (1) A~K4 (one trip 3.5 km)	Closing error 9 mm	18 mm (Third class levelling)
	Section (2) A~K4 (one trip 3.6 km)	Closing error 5 mm	18 mm (Third class levelling)
	Section (3) A~K4 (one trip 3.3 km)	Closing error 7 mm	18 mm (Third class levelling)

Notes
1) The instruments used are Total station GTS–3, Data collector FC–3 and Auto level AT–M3.
2) The allowable limit of the third class levelling's discrepancy is $10\sqrt{s}$ mm, where s is the distance (unit : km) traversed in one trip.

analytical plotter is ±2 μm which is 3 to 5 times that of analog plotter system as against the least measuring unit of 1 μm for the case of observed photographic coordinates. Here, using a well defined and clear photograph is quite an important point to obtain reliable observation results. And again the observation is done by using the floating mark with a diameter of 20 μm together with the observation magnification of 16 times. Thus, the analytical results of aerial triangulation computed by the bundle adjustment program with self–calibration are compared with the results of the conventional control point survey. The precision comparison and the adjustment computation of aerial triangulation are conducted with regard to the following items.

(1) The actual measured data of 105 ground control points are taken as true values.

(2) Out of 105 control points, the 5–points, 6–points, 7–points, 8–points, 9–points, 11–points, 14–points, 17–points and 20–points courses are used for every photo scale as control points for the adjustment computation of aerial triangulation and the remaining control points are considered as check points.

(3) As to the number of control points of 6–points, 7–points, 8–points and 9–points courses, two kinds of computation are conducted changing the positions of control points. As to the positioning of these control points, two points are laid at both ends of the model and the basic rule is to position them at uniform interval throughout the course.

(4) In each computation, the cases of presence and absence of the additional parameters are taken into consideration.

(5) With regard to the check points, the maximum value, the minimum value, standard deviation and relative precision of the residual between the coordinate actually measured on the ground (true one) and the coordinate obtained from adjustment computation are computed and the precisions are compared. The initial values of the adjustment computation are as shown in Table 21.3. Here the principal distance of the camera and the principal point of autocollimation (PPA) are assumed as the specified values of the camera.

Results of the analysis
With respect to aerial triangulation, the values of the residual (horizontal direction ΔX, ΔY), the altitude (ΔH), the standard deviation (σ) and the absolute precision (dividing the standard deviation by flight altitude: (σ/z)) of the check points in case the adjustment computation is done by using 8 control points as shown in Table 21.4. From this table the impact on the result of the computation in both cases of presence and absence of the additional parameters cannot be observed. And the residual and direction of the point's horizontal position are expressed by vectors and the residual values of height are shown with $\bigcirc$ notation for positive values and $\square$ notation for negative values and their maps are as shown in Figures 21.8 and 21.9 respectively. And the relationship between the control point number and the standard deviation with respect to horizontal position and the same relationship with respect to altitude are shown

Table 21.3. Initial values of adjustment computation.

Adjustment computation	Initial values
Camera principal distance	$f = 153.12$ mm
Principal point of autocollimation (PPA)	$X = 0.008$ mm $Y = -0.004$ mm
Standard deviation of PPA and principal distance	0.010 mm
Standard deviation of photographic coordinates	0.005 mm
Standard values for end of iteration	0.001 mm 10 times
Standard deviation of control point	1.0 mm

Table 21.4. The residual, standard deviation and relative precision of the check point in case of 8 control points.

Results in 1/3000	Residual component			
	ΔX (mm)	ΔY (mm)	ΔS (mm)	ΔH (mm)
Maximum	98 (115)	69 (81)	104 (129)	−164 (−197)
Minimum	−1 (0)	0 (0)	2 (5)	1 (0)
Standard deviation(σ)	33 (41)	27 (31)	43 (51)	60 (62)
Relative precision (σ/z)	1/13,636 (1/10,975)	1/16,666 (1/14,516)	1/10,456 (1/8823)	1/7500 (1/7258)

Results in 1/4000	Residual component			
	ΔX (mm)	ΔY (mm)	ΔS (mm)	ΔH (mm)
Maximum	144 (−130)	139 (116)	158 (156)	−236 (−211)
Minimum	±1 (±1)	−1 (0)	6 (5)	±1 (2)
Standard deviation(σ)	41 (45)	43 (37)	59 (58)	75 (75)
Relative precision (σ/z)	1/14,634 (1/13,333)	1/13,953 (1/16,216)	1/10,169 (1/10,344)	1/8000 (1/8000)

Results in 1/5000	Residual component			
	ΔX (mm)	ΔY (mm)	ΔS (mm)	ΔH (mm)
Maximum	148 (118)	142 (120)	164 (120)	−211 (−187)
Minimum	0 (±2)	±1 (0)	4 (2)	1 (±2)
Standard deviation(σ)	43 (35)	45 (37)	62 (51)	70 (74)
Relative precision (σ/z)	1/17,441 (1/21,428)	1/16,666 (1/20,270)	1/12,096 (1/14,705)	1/10,714 (1/10,135)

Notes 1) z : flight altitude,   2) ( ): without additional parameters
       3) $\Delta S = \sqrt{\Delta X^2 + \Delta Y^2}$

in Figures 21.10 and 21.11, respectively, for both cases of with and without the additional parameters.

The number of standard control points (*N*) in the strip adjustment method is given by the following Equation 21.1 as specified by the Public Survey Works Specification, Ministry of Construction.

$$N = M / 2 + 2 \qquad\qquad (21.1)$$

where *M* is the number of models.

In the present test, the values of *N* are *N*=6 for 8–model, *N*=7 for 9–model and *N*=9 for 13–model respectively. Viewed from these standard numbers of control points, the appropriate number of the control points for each scale is 8

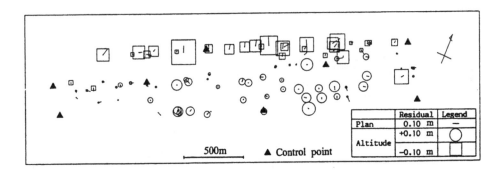

Figure 21.8. Map showing residual values, photo scale 1/3000 by flight; 13–model, 8 control points without the additional parameters.

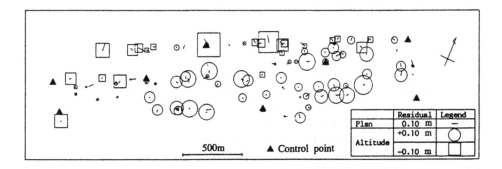

Figure 21.9. Map showing residual values, photo scale 1/5000 by flight; 8–model, 8 control points without the additional parameters.

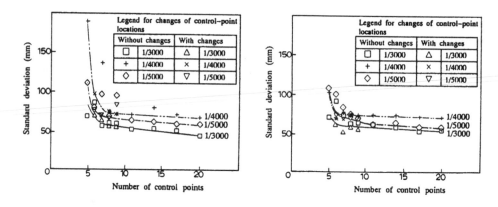

Figure 21.10. Relationship between standard deviation and number of control points with respect to horizontal position.

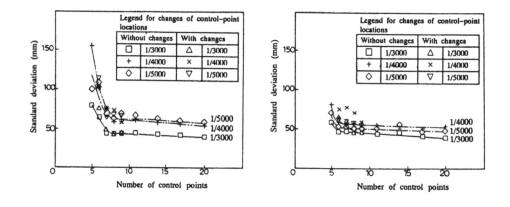

Figure 21.11. Relationship between standard deviation and number of control points with respect to altitude.

in case the additional parameters are considered. But the number of control points should be more than 8 if the additional parameters are to work effectively. Furthermore, in case the additional parameters are not taken into consideration, the precision changes with the change of arrangement of control points, even if the number of control points used for the adjustment computations remains the same. In this case it is considered important to secure two or three points more than the number of control points specified in the Public Survey Works Specification.

As is common to all the photo scales, the tendency is that the absolute value of residual becomes small converging to the precision of control point survey when accompanied by the increased number of control points. And even if the control points are laid more than required, the precision increased smoothly just like a first order function, but a rapid increase cannot be expected.

In general, it is considered that the precision of photographic information increases with the increase of the photo scale. In this test, however in some cases, the precision is reversed as the photo scale changes from 1/5000 to 1/4000. The final precision is considered to be affected by the good or bad photo–coordinate measuring precision which is in turn governed by the final makeup condition of the photographs (clarity of image), etc. Consequently, since the photo scale hardly bring an effect on precision, some precautions should be taken to secure the photographs as clear and defined as possible.

In the present test the maximum number of model is 13 with the strip photographing. It is hard to say that the additional parameters in the adjustment computation are effective enough. It was confirmed that more than 7~8 control points are required in the adjustment computation to secure a steady precision using the additional parameters in the cases of strip photographing with the photo scales 1/5000 (8–model), 1/4000 (9–model) and 1/3000 (13–model).

In the case of the photo scale: 1/3000 the standard deviation for the horizontal position (horizontal direction) is 40 mm and for the height (vertical direction) is 60 mm. This means that this value corresponds to the error of 0.009% to 0.013% for the flight altitude of 450 m. The limit of the residual of the control points is specified to be within 0.02% of the flight height above the ground. Consequently, the result of the adjustment computation in the present test is, it is noted, of a high precision. Moreover, it is presumed that at least 8 control points are required to secure the above mentioned precision.

The bird's eye–view of the east area of the present test site is shown in Figure 21.12 for example, as it is made up by editing the available digital data of topographic information.

CASE STUDY OF PROPOSED MONITORING SYSTEM

Objective of system application

Nagoya Port Island, a man–made island undertaken as a part of port and harbor development works, is reclaimed from the sea by using the dredged soil: such as clayey soil, sandy soil, etc., left over from civil construction works. In managing the fill volume and ground height, the direct measurement of ground height of super–soft land mass is quite difficult. And, also the setting up of control points is hardly possible on such a man–made island. Hence the GPS and the digital mapping system that makes use of aerial photographs are applied to construction management of the island (Sugimori et al. 1993).

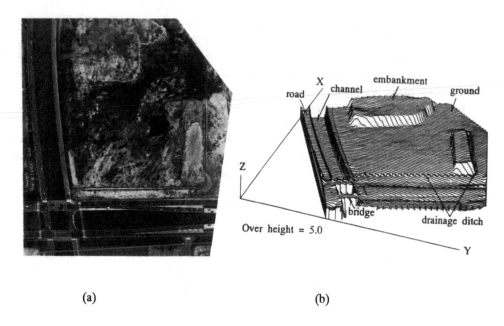

(a)   (b)

Figure 21.12. Bird's eye–view of the east area of the test site. (a) Aerial photograph;
(b) Bird's eye–view.

Method of study

In the present study, the strip–2–model near–vertical photographs are taken from a Cessna PT–206 airplane loaded with Wild RC–20 (wide–angle lens camera, 23 cm × 23 cm size) flying at an altitude of 1200 m taking T.P. ±0.0 m as the standard level of the horizontal plane as shown in Figure 21.13. Here the conditions such as simplicity of works, ease of utilization, speed, economic viability, required precision and entry into the study area are taken into consideration. Aerial photographs of 1/8000 photo scale are obtained.

As a required condition in the use of GPS control point surveying, three points outside the port island are taken to check two existing control points A and B on the sea wall. The GPS observation is conducted in such a way that seven new points (Figure 21.14) on the outer–perimeter revetment of the port island are fixed by three dimensional analysis of the results from three observation sessions (Figure 21.15) by static method using two control points A and B. Seven new points of air photo signal are surveyed to fix by GPS static method as shown in Figure 21.16.

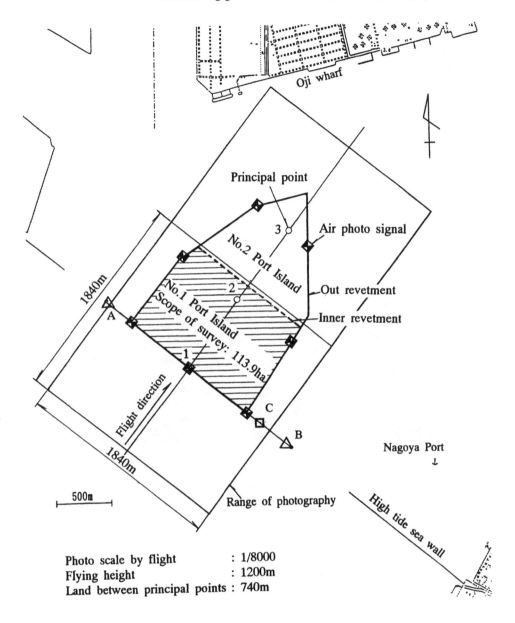

Photo scale by flight   : 1/8000
Flying height   : 1200m
Land between principal points : 740m

Figure 21.13. Aerial photographing and location of control points of Nagoya No.1 Port
           Island.

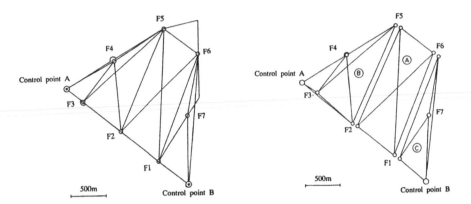

Figure 21.14. Network of control points.   Figure 21.15. Number of GPS observation
session.

Conditions of observation
1. Three sessions using static method.
2. Observation on more than 5 satellites.
3. Duration of observation : more than 90 min.
4. Satellite position : more than 15 degree above horizontal.
5. Data receiving interval : 30 sec.

Figure 21.16. GPS control point survey for fixing the locations of air photo signal points.

Results of system application

As a result of system application, the topographic map of Nagoya No.1 Port
Island with contour interval 20 cm of reclaimed ground and its bird's eye-view
derived from the analysis of mesh data (20 m mesh, 2600 points) are shown in
Figures 21.17 and 21.18 respectively. From the results of the studies the
following points should be taken into consideration in the system application:
(1) securing the subsidence pattern of a reclaimed land, (2) calculation of
quantity of earth work, (3) obtaining deformation characteristics of a structure.

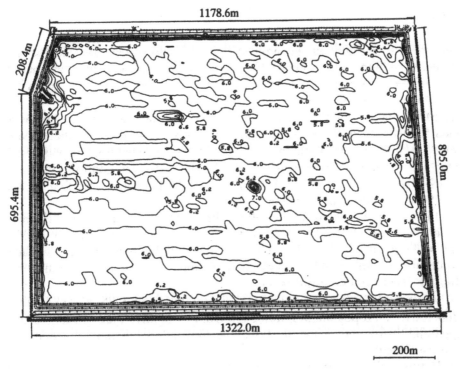

Figure 21.17. 20 cm contour interval map derived from 20 m mesh topographic data.

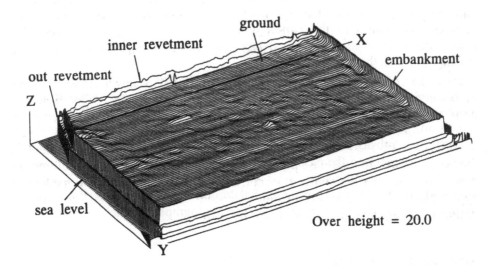

Figure 21.18. Bird's eye–view derived from 20 m mesh topographic data.

CONCLUSION

In the present research on the proposed monitoring system for wide–scope subsidence region, the precision, an important issue in the aerial photogrammetry, is assessed under several varied conditions and the following tangible results are obtained.

(1) More than 8 control points are required for the additional parameters needed for lens distortion correction to work effectively at every photo scale (1/5000, 1/4000 and 1/3000). In case of without consideration of additional parameters, a steady stable precision can be obtained when control points are arranged with 2 to 3 points more than the standard number specified in the Public Survey Works Specification.

(2) In case of the photo scale 1/3000, the standard deviation in horizontal direction is 4 cm and that in vertical direction is 6 cm. Hence, when 1/3000 photo scale is used, the topographic or plane–table management is made possible for the topographic changes of reclaimed land or natural foundation within the 10 cm range of precision.

(3) By using the analytical plotter system combined with GPS a large number of digital data on topographic information can be obtained much faster, easier, simpler and likely more economical than the conventional method of actual on–site measurement.

(4) Various relevant expressions such as the bird's eye–view, the topographic plan, etc. can be made up by using the digital data on topographic information obtained from the aerial photographs.

(5) It is to be considered that this analytical method of reading aerial photograph is, provided that the precision requirement is within that of the method, applicable to the earth–volume management and the topographic management of a large reclaimed land area where the predicted topographic change is more than 10 cm.

Here the foundation subsidence recently recorded is in the order of 5 cm (maximum) per year for Kanto plane and Shiroishi plane. As viewed from this point and taking into consideration of the system's test results (10 cm range of precision), photographing frequency of 2 to 3 years is considered necessary for the effective subsidence measurement. Again if the check points determined by the above model and the pass points are used as the control points required for controlling the photographs, it is considered that the analytical method is applicable to the observation of digital terrain model (DTM) and securing of digital mapping data after making up the photographic model.

ACKNOWLEDGEMENT

The authors would like to express their heart–felt thanks to Professor Norihiko Miura of Saga University for his advice and recommendation to submit this report, and also to Dr. Nobuo Tajiri and Mr. Aung Swe of Fukken Co., Ltd. for their collaboration in the preparation of this report.

REFERENCES

Geographical Survey Institute, Ministry of Construction. 1991. *Image survey: Earth watching:* 18–19. (in Japanese)
National Land Planning Adjustment Bureau. 1986. *Geographical information system.* (in Japanese)
Sugimori, K., T. Tanaka, A. Waga & T. Ezaki. 1993. *Application of digital mapping system to offshore man–made island's reclamation management, Annual Conf. JSCE, Div. 6:* 466–467. (in Japanese)
Tanaka, T. & H. Tsunezumi. 1992a. *Study on the treatment and application of aerial photographs to instrumented management, APA No.52–14,* Japan Society of Survey Investigation and Technology: 97–104. (in Japanese)
Tanaka, T., H. Tsunezumi, M. Yamamoto & N. Fukuda. 1992b. *Study on the application of aerial photographs to securing topographic changes of land, proceedings of the ILT seminar on problems of lowland development, ILT'92 on POLD,* Saga: 287–294.

SUBJECT INDEX

T - #0056 - 071024 - C0 - 246/174/28 [30] - CB - 9789054106036 - Gloss Lamination